高等学校教材

Engineering Basic Mechanics I Statics 工程基础力学Ⅰ 静力学

○ 主　编　郭　晶　王超营
○ 副主编　吴国辉　齐　辉　杨在林

中国教育出版传媒集团
高等教育出版社·北京

Abstract

Engineering Basic Mechanics I Statics is divided into 14 chapters, including fundamentals of statics and force analysis of objects; coplanar concurrent force system and couple system; coplanar general force system; space force system; friction; fundamental concepts of bar deformation; axial tension and compression; shear and torsion; geometric properties of cross sections; plane bending; state of stress and theories of strength; combined deformation; energy method; stability of columns.

In each chapter, the principles are applied first to simple, then to more complicated situations. The fundamental conception, formula and method are explained in detail. This book can provide the student with a clear and thorough presentation of the theory and applications of engineering mechanics.

This book can be used as a textbook for the basic course of mechanics in mechanical, civil engineering, shipping, mechanics, aviation, water conservancy and other related majors. It can also be used as a reference for engineering technicians of related areas.

图书在版编目(CIP)数据

工程基础力学. Ⅰ,静力学=Engineering Basic Mechanics Ⅰ Statics:英文/郭晶,王超营主编;吴国辉,齐辉,杨在林副主编. --北京:高等教育出版社,2023.11

ISBN 978-7-04-060496-2

Ⅰ. ①工… Ⅱ. ①郭… ②王… ③吴… ④齐… ⑤杨… Ⅲ. ①工程力学-高等学校-教材-英文②静力学-高等学校-教材-英文 Ⅳ. ①TB12②O312

中国国家版本馆 CIP 数据核字(2023)第 088965 号

策划编辑 安 莉　　责任编辑 安 莉　　封面设计 裴一丹　　版式设计 杨 树
责任绘图 于 博　　责任校对 张 然　　责任印制 存 怡

出版发行 高等教育出版社
社　　址 北京市西城区德外大街 4 号
邮政编码 100120
印　　刷 保定市中画美凯印刷有限公司
开　　本 787mm×1092mm 1/16
印　　张 21
字　　数 520 千字
购书热线 010-58581118
咨询电话 400-810-0598
网　　址 http://www.hep.edu.cn
　　　　 http://www.hep.com.cn
网上订购 http://www.hepmall.com.cn
　　　　 http://www.hepmall.com
　　　　 http://www.hepmall.cn
版　　次 2023 年11月第1版
印　　次 2023 年11月第1次印刷
定　　价 44.60元

物 料 号 60496-00

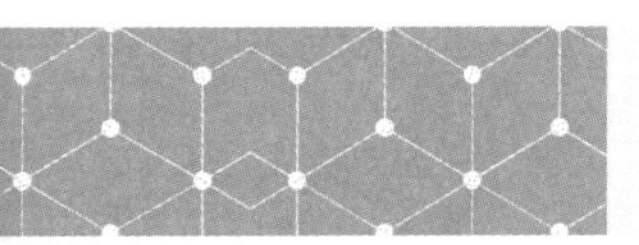

新形态教材网使用说明

gineering
sic Mechanics Ⅰ
atics

程基础力学Ⅰ 静力学

1 计算机访问https://abooks.hep.com.cn/60496，或手机扫描下方二维码，访问新形态教材网小程序。

2 注册并登录，进入"个人中心"，点击"绑定防伪码"。

3 输入教材封底的防伪码（20位密码，刮开涂层可见），或通过新形态教材网小程序扫描封底防伪码，完成课程绑定。

4 在"个人中心"→"我的图书"中选择本书，开始学习。

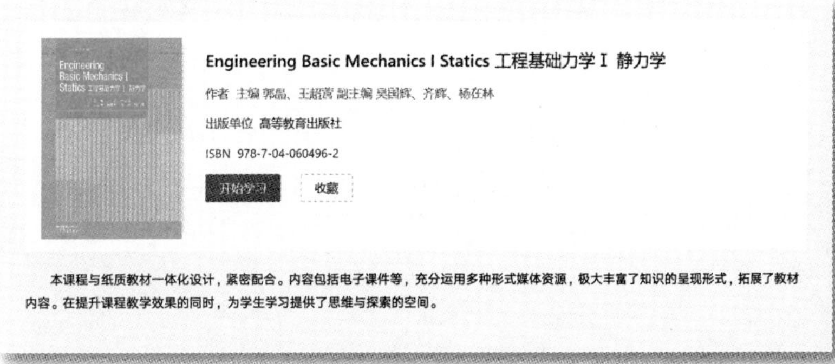

绑定成功后，课程使用有效期为一年。受硬件限制，部分内容无法在手机端显示，请按提示通过计算机访问学习。

如有使用问题，请发邮件至abook@hep.com.cn。

https://abooks.hep.com.cn/60496

Preface

Engineering basic mechanics is an important professional foundation course in higher education institutions. With the internationalization of higher education in China, a number of Sino-foreign cooperative schools have emerged, and some institutions have carried out bilingual teaching of the course. Since there are relatively few English textbooks for engineering basic mechanics, some universities directly adopt Chinese textbooks, which cannot meet the needs of bilingual teaching. Some other institutions directly use multiple English textbooks, resulting in the increasing of the burden of students and the lack of relevance of the knowledge system, and the order and focus of the content is very different from the domestic teaching process. Referring to a large number of excellent textbooks at home and abroad, this textbook is prepared in accordance with the characteristics of the domestic curriculum and the process of teaching the knowledge system, and on the basis of meeting the requirements of engineering basic mechanics in higher education.

This set of textbook is divided into two parts: Ⅰ Statics and Ⅱ Dynamics. This book is Ⅰ Statics part, including 14 chapters. Chapter 1 ~ 5 are about rigid body statics, and mainly focus on basic concepts of statics and the equilibrium problems of different force systems. Chapter 6 ~ 14 are about deformable body statics. Chapter 6 introduces the fundamental concepts of bar deformation. Chapter 7 ~ 10 are mainly focus on four basic forms of bar deformation. Chapter 11 is the state of stress and theories of strength. Chapter 12 ~ 13 are mainly focus on the strength and displacement problems of combined deformation, respectively. The stability of columns is introduced in Chapter 14.

This set of textbook combines the characteristics of excellent foreign and domestic textbooks and has some of the following features and characteristics:

1. Concise language. Strive for concise language on the basis of satisfying professional knowledge, and try to simplify the impact of language on the learning process and reduce the burden on students.

2. Reasonable knowledge system. Based on foreign classical textbooks, the equilibrium force analysis and material deformation are compiled together as the Statics part, and the kinematics and dynamics are compiled together as the Dynamics part. The knowledge structure and emphasis of the two parts are consistent with the domestic mechanics courses.

3. Focus on application. Absorb the advantages of foreign textbooks to simplify some of the complex mathematical derivation process, and strengthen the engineering background and practical application analysis process. Each chapter is equipped with a large number of practical engineering examples and exercises to improve the ability about how to solve practical problems.

The authors of this textbook include Guo Jing (preface and chapter 1 ~ 5), Wang Chaoying (chapter 6~11), Wu Guohui (chapter 12), Qi Hui (chapter 13) and Yang Zailin (chapter 14).

In the process of preparing this set of textbooks, we consult various published literature and teaching materials at home and abroad (see references for details), and the experience and achievements of educational reform in others colleges and universities. We would like to express our sincere gratitude to the above-mentioned colleagues, teachers and staff.

We wish to thank Higher Education Press, and Harbin Engineering University who funded the publication of this book.

Due to the limited level of the editors and the shortage of time, there are still some shortcomings in the book, and we sincerely invite teachers, students and readers to criticize and correct any inappropriate points.

Editor

2022. 12

Symbols

A	area
A_{jy}	bearing surface area
E	modulus of elasticity
$\boldsymbol{F}$	force
$\boldsymbol{F}'$	reaction force
$(\boldsymbol{F}, \boldsymbol{F}')$	couple
$\boldsymbol{F}_{d}$	dynamic friction
$\boldsymbol{F}_{N}$	axial force
$\boldsymbol{F}_{R}$	resultant force
$\boldsymbol{F}'_{R}$	principal vector
$\boldsymbol{F}_{s}$	static friction
$\boldsymbol{F}_{x}, \boldsymbol{F}_{y}, \boldsymbol{F}_{z}$	component force in the x, y, and z axes
F_{jy}	bearing pressure
F_{cr}	critical compressive force of column
$\boldsymbol{F}_{Q}$	shear force
f_{s}	coefficient of static friction
f_{d}	coefficient of dynamic friction
G	shear modulus of elasticity
I	moment of inertia
I_{x}, I_{y}, I_{z}	moments of inertia with respect to x, y, and z axes
I_{x_1}, I_{y_1}	moments of inertia with respect to x_1 and y_1 axes (rotated axes)
I_{xy}	product of inertia with respect to xy axes
$I_{x_1y_1}$	product of inertia with respect to x_1y_1 axes(rotated axes)
I_{p}	polar moment of inertia
$\boldsymbol{i}, \boldsymbol{j}, \boldsymbol{k}$	unit vectors of x, y, and z axes
i_{x}, i_{y}, i_{z}	radius of gyration with respect to x, y, and z axes
l, L	length, distance
M	bending moment, couple moment
$\boldsymbol{M}$	couple moment vector
M_{n}	torque
M_{O}, $\boldsymbol{M}_{O}$	principal moment for the simplified center O

M_x, M_y, M_z	bending moment with respect to x, y, and z axes
m	mass
$M_O(\boldsymbol{F})$	moment of the force $\boldsymbol{F}$ about point O
$\boldsymbol{M}_O(\boldsymbol{F})$	moment vector of the force $\boldsymbol{F}$ about point O
M_s	rolling resistance couple moment
$M_x(\boldsymbol{F})$, $M_y(\boldsymbol{F})$, $M_z(\boldsymbol{F})$	moment of the force $\boldsymbol{F}$ about x, y, and z axes
n	factor of safety; number; rotational speed
n_w	safety factor for stability
O	origin of coordinates
$\boldsymbol{P}(\boldsymbol{G},\boldsymbol{W})$	gravity
P	power
p	total stress
p_m	average stress
q	intensity of distributed load (force per unit distance)
$R(r)$	radius
$\boldsymbol{r}$	vector diameter
S_x, S_y, S_z	first moment of area with respect to x, y, and z axes
s	distance along a curve
U	strain energy
u	line displacement, strain energy density
V	volume
v	line displacement, deflection of a beam
v', v''	dv/dx, d^2v/dx^2
W	work
W_n	section modulus of torsion
W_x, W_y, W_z	section modulus of bending with respect to x, y, and z axes
x, y, z	rectangular axes (origin at point O)
x_C, y_C, z_C	rectangular axes (origin at centroid C)
$\bar{x}$, $\bar{y}$, $\bar{z}$	coordinates of centroid
ω	angular velocity
α	angle, nondimensional ratio
α_0	angle to a principal plane
α_1	angle to a plane of maximum shear stress
β	angle
γ	angle, shear strain
γ_{xy}, γ_{yz}, γ_{zx}	shear strains in xy, yz, and zx planes
δ	rolling resistance coefficient, percentage elongation, generalized displacement
Δl	absolute deformation of the bar in the axial direction

ε	normal strain
ε'	transverse normal strain
ε_{m}	average normal strain
ε_x, ε_y, ε_z	normal strains in x, y, and z directions
ε_1, ε_2, ε_3	principal normal strains
θ	angle, angle of rotation of beam cross section, angle of twist per unit length
Θ	volumetric strain
λ	slenderness ratio
μ	Poisson ratio
ρ	radius, radius of curvature, distance mass density (mass per unit volume)
σ	normal stress
σ_{b}	ultimate strength
σ_{e}	elastic limit
σ_{jy}	bearing stress
σ_{cr}	critical stress
σ_{m}	average principal stress
σ_{p}	proportional limit
σ_{s}	yield stress
σ_x, σ_y, σ_z	normal stresses on planes perpendicular to x, y, and z axes
σ_{xd}	equivalent stress
σ_α	normal stress on an inclined plane
σ_1, σ_2, σ_3	principal normal stresses
$[\sigma]$	allowable stress
$[\sigma_{\mathrm{c}}]$, $[\sigma^-]$	compressive allowable stress
$[\sigma_{\mathrm{jy}}]$	allowable bearing stress
$[\sigma_{\mathrm{t}}]$, $[\sigma^+]$	tensile allowable stress
$\sigma_{0.2}$	nominal yield stress
τ	shear stress
τ_{xy}, τ_{yz}, τ_{zx}	shear stresses on planes perpendicular to the x, y, and z axes and acting parallel to the y, z, and x axes
τ_α	shear stress on an inclined plane
$[\tau]$	allowable stress in shear
φ	angle, angle of twist of a bar in torsion
φ_{m}	friction angle
Z	percentage reduction in area

Contents

Chapter 1 Fundamentals of Statics and Force Analysis of Objects

Teaching Scheme of Chapter 1

1.1 Basic concepts of statics

Statics is the science of studying the equilibrium of objects. To study this problem, some basic concepts are first introduced below.

1. The concept of rigid body

The objects studied in statics are mainly rigid bodies. A rigid body is an object that does not deform under the action of a force, i.e., the distance between any two points inside the rigid body remains constant.

2. The concept of equilibrium

Equilibrium means that the object is at rest or moving in a uniform velocity linear motion with respect to the inertial reference system. It is a special state of mechanical motion of an object.

3. The concept of force

An important concept in statics is the concept of force. Force is the mutual mechanical action between objects.

The effect of a force on an object depends on three elements of the force:

① the magnitude of the force;

② the direction of the force;

③ the point of action of the force. The magnitude of the force reflects the strength of the mechanical interaction between objects. In the international system of units, the unit of force is the newton (N); in the engineering system of units, the unit of force is the kilogram force(kgf)[①].

Based on the above, a vector can be used to represent the three elements of a force. As shown in Fig.1-1, the force vector (force vector) is a directed line segment; the mode of the vector indicates the magnitude of the force; the direction of the arrow indicates the direction of the force; the beginning or end of the vector indicates the point of action of the force, and the line coinciding with the segment represents the line of action of the force. In this book, the vector is represented by bold letter, such as $\boldsymbol{F}$; the magnitude of the vector is represented by ordinary letter, such as F.

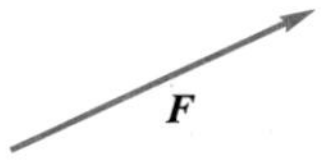

$\boldsymbol{F}$

Fig.1-1

4. Concept of force system

A group of forces acting on an object is called a force system. If a system of forces acts on a rigid body without changing its state of motion, the system of forces is called an equilibrium system of forces. If two force systems act on the same object and their effects are the same, the two force systems are called equivalent force systems. If a force is equivalent to a force system, this force is said to be the resultant force of this force system, and each force in this force system can be called a component of this resultant force.

---

① The following relationships exist between the two systems of units:
1 kgf = 9.80 N

1.2 Fundamental principles of statics

Principle 1-1 The parallelogram law of force: Two forces acting on the same point on the object can be combined into a resultant force, and the point of action of the resultant force is also at that point. The magnitude and direction of the resultant force is determined by the diagonal of the parallelogram with the two forces as sides, as shown in Fig.1-2a. This property of the force is called the parallelogram law of force and is written as a vector expression:

$$\boldsymbol{F}_R = \boldsymbol{F}_1 + \boldsymbol{F}_2$$

That is, the resultant force vector $\boldsymbol{F}_R$ is equal to the vector sum of the two partial force vectors $\boldsymbol{F}_1$ and $\boldsymbol{F}_2$.

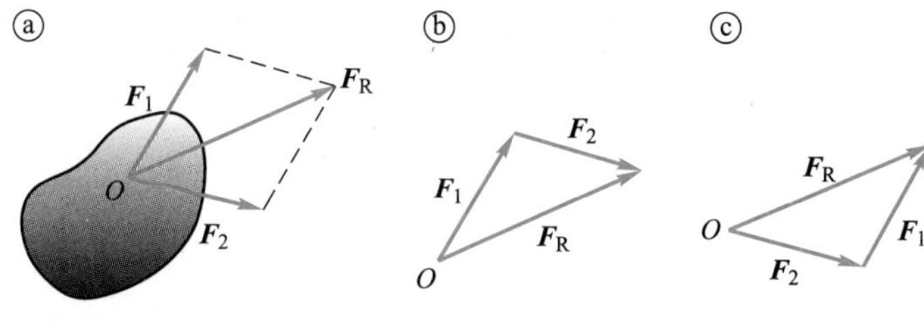

Fig.1-2

The law of vector summation can also be attributed to the triangle law of force. As shown in Fig.1-2b, $\boldsymbol{F}_1$ is made first, and then $\boldsymbol{F}_2$ is made, finally to connect the beginning of $\boldsymbol{F}_1$ with the end of $\boldsymbol{F}_2$ to get the combined force vector $\boldsymbol{F}_R$. This kind of graphing rule for the combined force vector is called the triangle rule of force. The triangular diagram of the force represents only the individual force vectors and not their positions of action. As shown in Fig.1-2c, the combined force vector $\boldsymbol{F}_R$ can also be obtained by first making $\boldsymbol{F}_2$ and then $\boldsymbol{F}_1$. This means that the combined force vector is independent of the order in which the two component force vectors are plotted.

Principle 1-2 Sufficient condition for two forces equilibrium on a rigid body: the two forces are equal in magnitude, opposite in direction and act in the same line (common line).

This principle only applies to rigid bodies.

Principle 1-3 Principle of plus or minus an equilibrium force system: The addition or subtraction of a set of equilibrium force systems to the force system acting on a rigid body does not change the effect of the original force system on the rigid body.

Inference 1 The transmissibility principle of force: The force acting on a point on the rigid body can be moved to any point within the rigid body along the line of action and it does not change the effect of the force on the rigid body.

As shown in Fig.1-3, the force $\boldsymbol{F}$ is acting on the rigid body at A point. Based on the principle of plus or minus an equilibrium force system, an equilibrium force system ($\boldsymbol{F}_1$ and $\boldsymbol{F}_2$) is added on the rigid body. Moreover, $|\boldsymbol{F}| = |\boldsymbol{F}_1| = |\boldsymbol{F}_2|$, it means that $\boldsymbol{F}$ and $\boldsymbol{F}_1$ can form another equilibrium force system, so this equilibrium force system can be removed. As a result, the force $\boldsymbol{F}_2$ is acting on the rigid body at B point. Therefore, the effect of a force on a rigid body is independent of the position of the point of action of the force on its line of action, i.e., the force can slip arbitrarily within the rigid body along its line of action without changing its effect, and this property of the force is called force transmissibility. Therefore, for a rigid body the force is the slip vector. Therefore, for the force acting on a rigid body, the three elements of the force should be the magnitude, the line of action, and the direction.

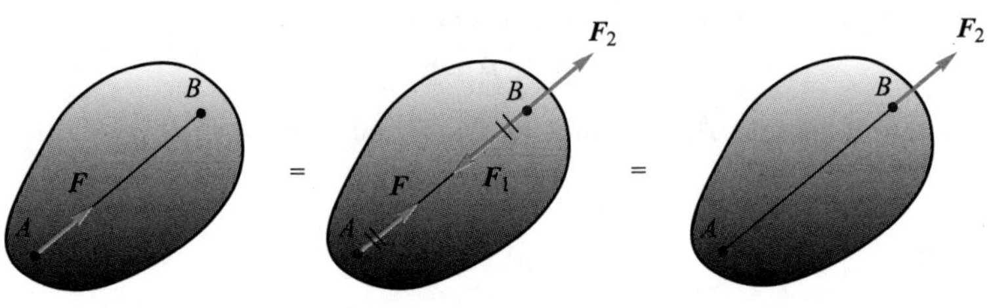

Fig.1-3

Inference 2 Three-force equilibrium confluence theorem: A rigid body is in equilibrium under the action of three forces. If the lines of action of two of the forces intersect at some point, the line of action of the third force passes through the point, and the three forces are coplanar.

Proof See Fig.1-4, suppose a rigid body is subjected to three nonparallel and coplanar forces $\boldsymbol{F}_1$, $\boldsymbol{F}_2$ and $\boldsymbol{F}_3$. Two of them, $\boldsymbol{F}_1$ and $\boldsymbol{F}_2$ must intersect at some point O and have a resultant $\boldsymbol{F}_{12}$ passing through point O. Based on two-force equilibrium principle, $\boldsymbol{F}_3$ must also pass through point O.

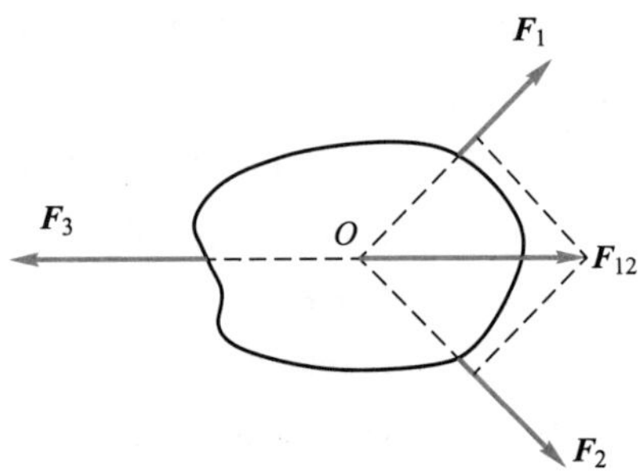

Fig.1-4

Principle 1-4 The law of action and reaction forces: The action and reaction forces of two objects interacting with each other are equal in magnitude and opposite in direction, along the same line, acting on two objects respectively. As shown in Fig.1-5, the two forces can be written as

$$\boldsymbol{F} = -\boldsymbol{F}'$$

Principle 1-5 Solidification principle: The deformed body is in equilibrium under the action of a force system, and if this deformed body is stiffened to a rigid body, the equilibrium state remains unchanged.

As shown in Fig.1−6, if F_{T1} and F_{T2} make the flexible cable in equilibrium state (the flexible cable is in small deformation), according to the Solidification principle, the flexible can be considered as a rigid body.

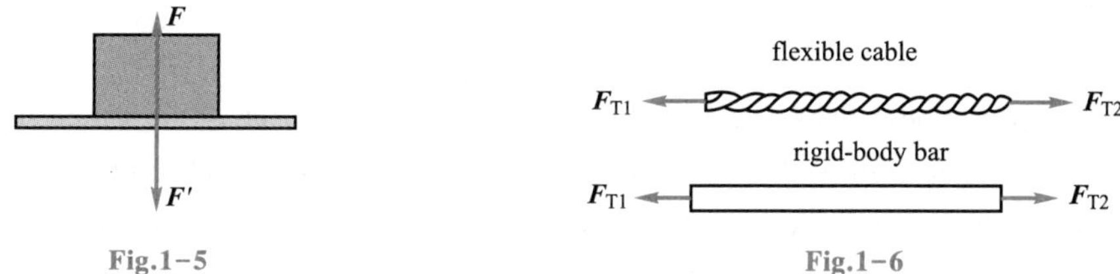

Fig.1−5

Fig.1−6

For deformed body after chapter 6, the deformation is continuous. It is theoretically impossible to apply the solidification principle. However, when the rate of loading is very slow, any moment can also be considered statically equilibrium and the solidification principle can be applied. Even if the deformed body is stiffened to a rigid body, the application of principles such as plus or minus an equilibrium force system, moving force along the line of action and the translation theorem of forces are strictly limited.

1.3 Constraint and reactive forces

Force is a mutual mechanical interaction between objects, when analyzing the individual forces acting on an object, it is necessary to understand the form of interaction and connection between the object and other objects around it. Objects are divided into two categories according to whether they directly contact with other objects: one is objects whose displacement is not limited in any way in space and can move arbitrarily in space. Such an object is called a free body. For example, a bullet in flight, a spaceship underway. Another is that some displacements of the object are limited by the surrounding objects. For example, a pendant lamp suspended by a hanging rope, such an object is called a non-free body.

Any other object that limits the displacement of an object is called a constraint of that object.

The action of the constraint on the object is essentially the action of the force. The force acting on the object by the constraint is called the constraint reaction force or constraint force, also referred to as the reactive force. The point of action of the restraint reaction force is the point of contact between the constraint and the object. The direction of the constraint reaction force must be opposite to the direction of the constraint impeding the displacement of the object.

In addition to the constraint reaction force, the gravity, pushing force and other active forces can change the state of motion of the object, such forces are called the applied force. The applied force is different from the reaction force in that its magnitude and direction are generally pre-determined and independent of each other. The magnitude of the constraint reaction force, on the other hand, is usually unknown and depends on the magnitude and direction of the applied force, which needs to be determined from the equilibrium conditions or dynamical equations of the object.

Several common typical constraints and how to determine their constraint reaction force are described below.

1. Flexible cable

Constraints made up of ropes, belts, chains, relatively soft wire ropes, etc., are called flexible cable. The flexible cable is characterized by softness and deformability, non-stretchability, weightless, and inability to resist bending and pressure, which limits the displacement of objects along the elongation direction of the flexible cable. The constraint reaction force of the flexible cable acts on the connection point with the object, and the direction leaves the object along the center line of the flexible cable and is usually tension. $\boldsymbol{F}_{\mathrm{T}}$ is usually used to represent it, as shown in Fig.1-7.

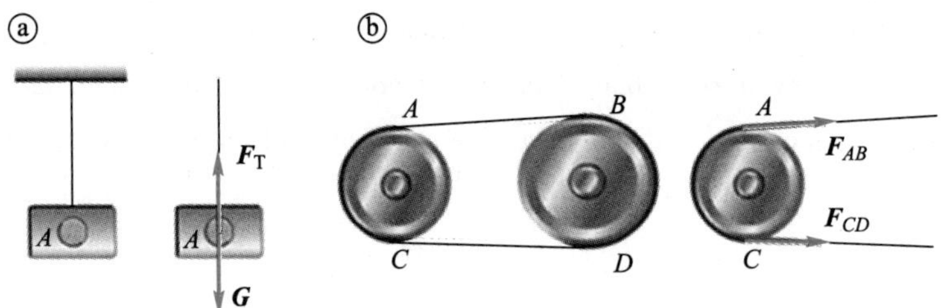

Fig.1-7

2. Smooth contact surface

The direction of the constraint reaction force for a smooth contact surface is vertical to the contact surface and points toward the object and is usually compressive stress, which is also called the normal reaction force. See Fig.1-8.

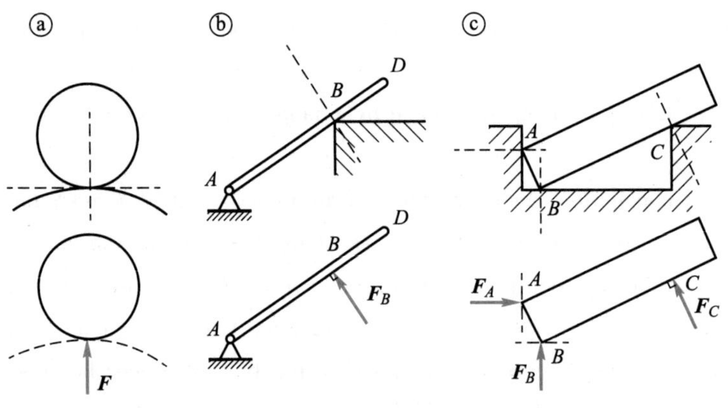

Fig.1-8

3. Smooth pin support

The constraint formed by two objects with holes connected through a pin and ignoring the friction between the pin and the holes of the two objects is called a smooth pin or hinge, as shown in Fig.1-9.

This type of constraint is characterized by the fact that it can only limit the radial movement of the object in any direction, and cannot limit the rotation of the object around the axis of the cylindrical pin and the movement parallel to the axis of the cylindrical pin. The constraint reaction force should therefore be normal to the contact surface and perpendicular to the axis through the center of the cylindrical pin, as shown in Figure. 1-9c; Because the contact position cannot be predetermined, the direction of the reactive force cannot be predetermined, but the reactive force of the smooth pin must act in the plane perpendicular to the pin and through the center of the shaft and the center of the hole. Therefore, we usually use two perpendicular components $\boldsymbol{F}_x$ and $\boldsymbol{F}_y$ through the axis to represent the constraint reaction force of the pin. Once the magnitude of $\boldsymbol{F}_x$ and $\boldsymbol{F}_y$ is determined, the parallelogram law of force can be used to determine the magnitude and direction of the constraint reaction force, as shown in Fig.1-9d.

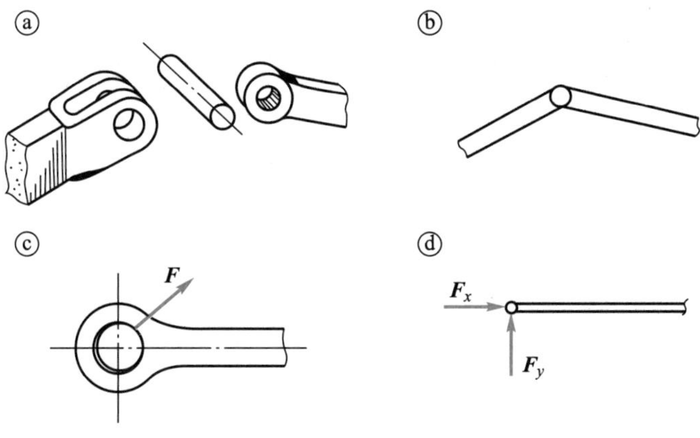

Fig.1-9

The following is a description of several types of supports composed of cylindrical pin constraints, namely fixed-hinge support and roller support.

As shown in Fig.1-10a and 1-10b, a smooth cylindrical pin is used to connect the structure or member to the base, and the base is fixed on top of the support. The cylindrical hinge constraint is called fixed-hinge support. Fig. 1 - 10c shows a sketch of the fixed-hinge support. This type of constraint is characterized by the fact that the member can only rotate around the pin axis without radial movement in any direction perpendicular to the pin axis, so the constraint reaction force of the fixed-hinge support is in the plane perpendicular to the pin axis, through the center of the pin axis, in a variable direction, usually expressed as two mutually perpendicular components, as shown in Fig.1-10d.

As shown in Fig.1-11a, a row of rollers is installed at the bottom of the fixed-hinge support, which can make the support roll along the fixed support surface, called roller support, and Fig.1-11b is its sketch. When the member is deformed, this support constraint allows both a small rotation and a small movement, but restricts the object to move in the direction perpendicular to the support surface. Therefore, the constraint reaction force is normal to the support surface, through the center of the pin and directed toward the object, as shown in Fig.1-11c.

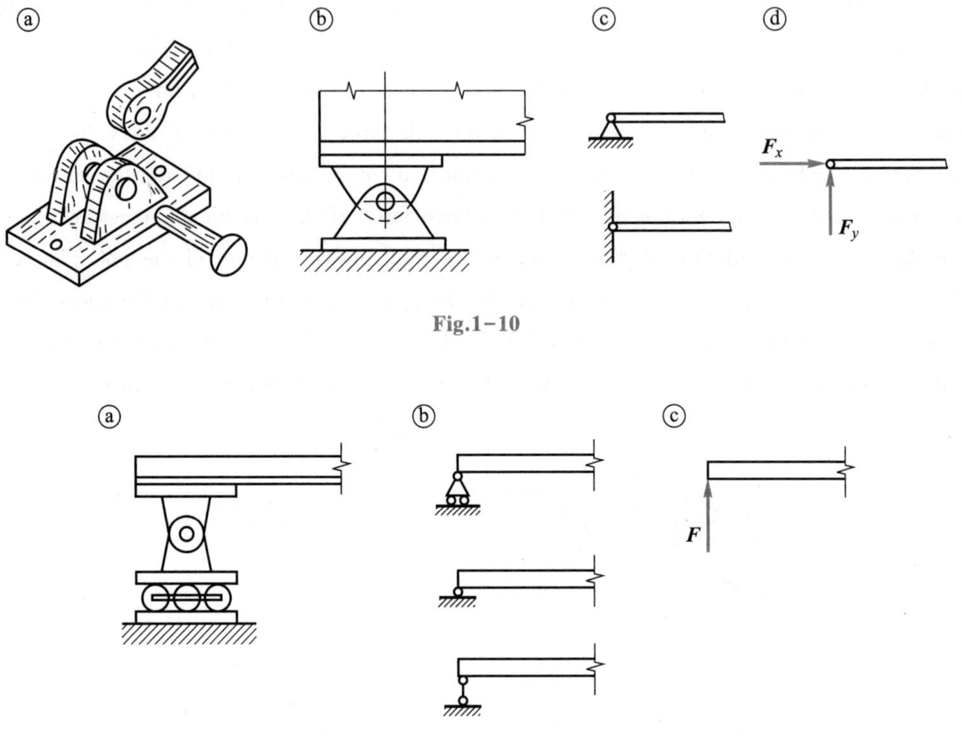

Fig.1–10

Fig.1–11

4. Other types of constraints

(1) Bearings

This is shown in Fig.1–12a. These bearings allow the shaft to rotate, but limit the displacement of the shaft perpendicular to the axis, and Fig.1–12b shows a sketch of this bearing. The characteristics of the bearing constraint are the same with the smooth pin support, as shown in Fig.1–12c.

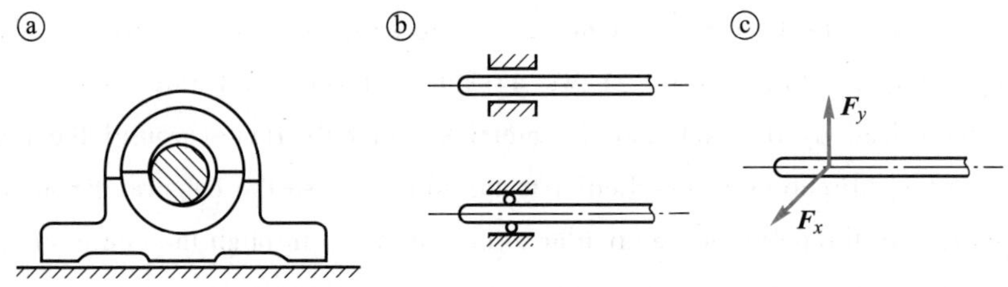

Fig.1–12

(2) Two-force member

Objects in engineering structures are sometimes supported by a rigid bar with hinge connections at both ends, such as the bar *AB* in Fig.1–13a. The rigid body acts with force at the two hinge points. If the weight of the rigid bar is neglected, this kind of member is in equilibrium only at two points of force, which is called two-force member. According to the two-force equilibrium principle, the lines of action of the constraint reactions acting on the two ends of the two-force member must pass through the two-hinged points, such as the lines of action of $\boldsymbol{F}_A$ and $\boldsymbol{F}_B$ in Fig.1−13b pass through points A and B.

The member does not have to be a straight bar, and it might be a curved bar (Fig.1−14a), the line of action of the constraint reaction force acting on the object passes through the two-hinged points, as shown in Fig.1−14b.

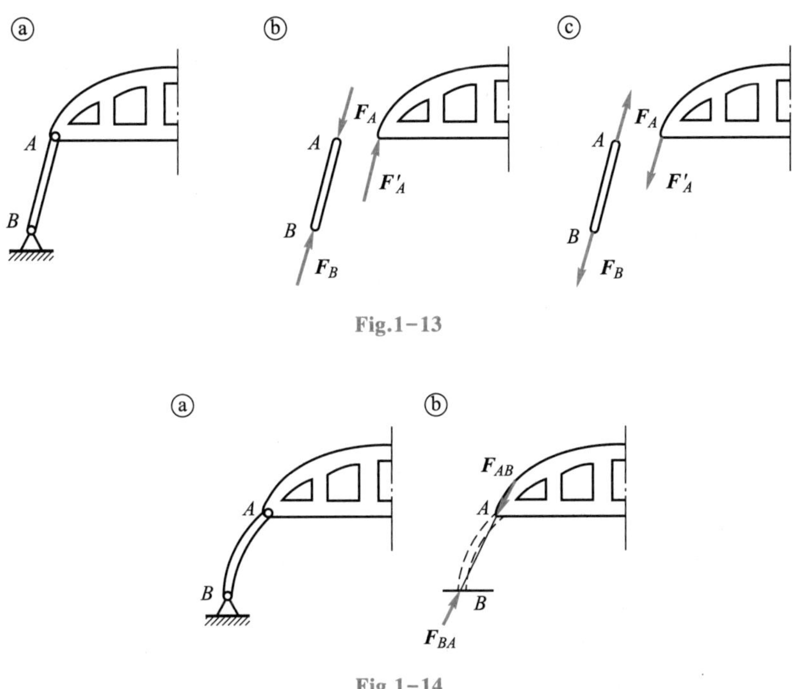

Fig.1−13

Fig.1−14

1.4 Force analysis and free-body diagram

In order to carry out force analysis of the object, it is necessary to separate the object from the surrounding objects with which it is associated. We need to draw all the applied forces and constraint reaction force acting on the body. The steps for drawing free-body diagram are usually as follows:

① According to the problem, the object of study is selected appropriately and separated from the whole structure.

② Draw all the applied forces acting on the object of study (generally known forces).

③ The constraint reaction forces are drawn one by one for each location where the object is in direct contact with each constraint. The line of action and direction of the constraint reaction forces are determined by the nature of the constraint. When the line of action of some constraint reaction forces can be determined, but its direction cannot be determined, it can be assumed along its line of action.

④ All forces on the free-body diagram should be marked according to the nature of the force, the type of constraint, and the location of the point of action. For example, the applied force is expressed by $\boldsymbol{W}$, $\boldsymbol{G}$, $\boldsymbol{P}$, etc.; the flexible cable tension is expressed by $\boldsymbol{F}_{\mathrm{T}}$; the reaction force at fixed-hinge A is expressed by $\boldsymbol{F}_{Ax}$ and $\boldsymbol{F}_{Ay}$; the letters labeled for the action and reaction forces should be expressed as $\boldsymbol{F}$ and $\boldsymbol{F}'$, respectively.

Example 1−1 A structure is shown in Fig.1−15a, the bars AB and BC have equal length and are hinged at B. It is the fixed-hinge support at A and C. A horizontal force is applied at the midpoint D of the bar AB. The weight of the two bars are negligible, try to draw a free-body diagram of the bars AB and BC.

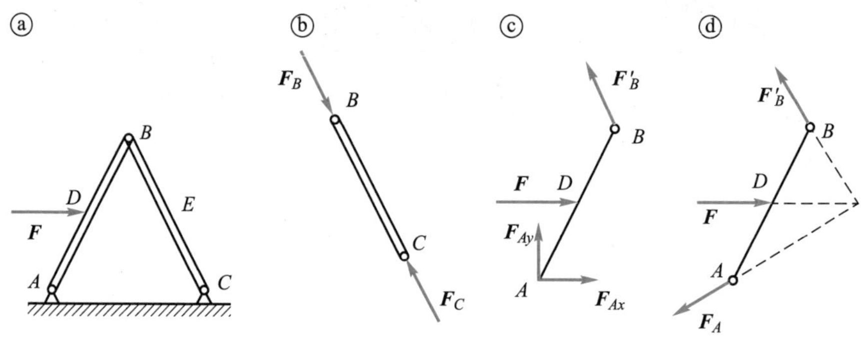

Fig.1−15

Solution Taking the bar BC as a free body, it is balanced by two forces, so it is a two-force member, and the line of action of these two forces should be along the line between points A and B, the direction is opposite, as shown in Fig.1−15b.

Taking the bar AB as a free body, the end A is constrained by the fixed-hinged support, and the constraint reaction force is usually expressed by two perpendicular components, as shown in Fig.1−15c for $\boldsymbol{F}_{Ax}$ and $\boldsymbol{F}_{Ay}$, the direction of these two forces can be set to the positive directions. If the value determined by the equilibrium equation is negative, it means that the direction of the actual force is opposite to the direction shown in the diagram.

Further analysis of the bar AB, the actual reaction force at A is a resultant force. The bar AB is in equilibrium under the action of three forces, so these three forces should be in the same plane and intersect at common point, as shown in Fig.1−15d.

Example 1−2 As shown in Fig.1−16a, a homogeneous ball A weighing $\boldsymbol{G}_1$ is resting on a smooth inclined plane with angle α. A rope through the pulley C is used to connect the ball with another object B weighing $\boldsymbol{G}_2$. Draw the free-body diagram of each equilibrium object without considering the weight of pulley C and the rope.

Solution Taking object B as a free body, as shown in Fig.1−16b, it is subjected to two forces, including the gravity $\boldsymbol{G}_2$ and the tension $\boldsymbol{F}_D$ of the rope DG.

Then taking the ball A as a free body, as shown in Fig.1−16c, it is subjected to three forces, namely, gravity $\boldsymbol{G}_1$, the pulling force $\boldsymbol{F}_E$ of the rope EH and the reaction force $\boldsymbol{F}_F$ of the smooth contact surface.

Taking pulley C as a free body, as shown in Fig.1−16d, subjecting to the tension $\boldsymbol{F}_H$ of the rope EH and the tension $\boldsymbol{F}_G$ of the rope CD, noting that $\boldsymbol{F}_H$ should be equal, opposite and colinear with $\boldsymbol{F}_E$, $\boldsymbol{F}_G$ should be equal, opposite and colinear with $\boldsymbol{F}_D$, and $\boldsymbol{F}_C$ is the constraint reaction force at the fixed-

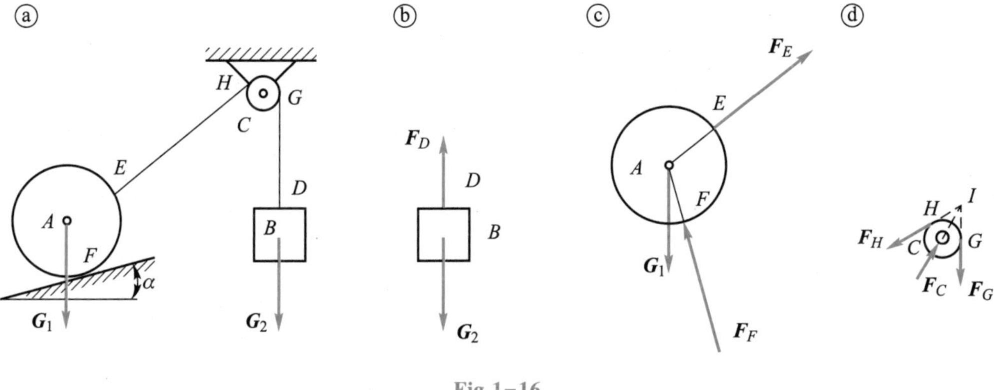

Fig.1–16

hinge support C. Since pulley C is equilibrated by three forces, the three forces acting on pulley C should be in the same plane and intersect at a point I.

Example 1–3 As shown in Fig.1–17a, the bar AB and AC have the same length and are hinged at A, and connected by a rope at D and E two points, placed on a smooth horizontal surface, a force $\boldsymbol{F}$ acts on the bar AB at the midpoint, regardless of the weight of the two bars, draw the free-body diagram of each bar and the whole structure.

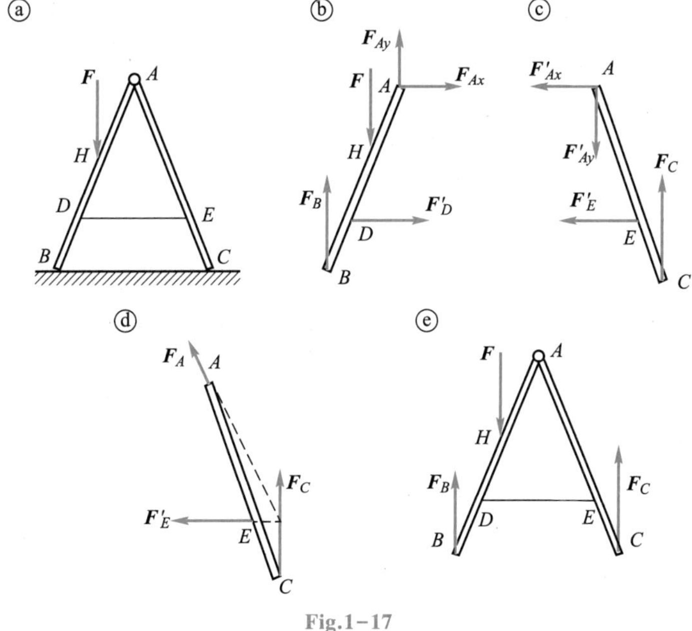

Fig.1–17

Solution Taking the bar AB with pin A as a free-body, as shown in Fig.1–17b, $\boldsymbol{F}_B$ is the constraint reaction force from the smooth contact surface, which is vertical to the contact surface; $\boldsymbol{F}$ is the applied force; $\boldsymbol{F}_{Ax}$, $\boldsymbol{F}_{Ay}$ are two perpendicular constraint reaction forces acting on the pin A, set the direction of the force as shown in Fig.1–17b; $\boldsymbol{F}'_D$ is the tension force acting on the bar AB by the rope DE.

Taking the bar AC as a free-body, $\boldsymbol{F}'_{Ax}$, $\boldsymbol{F}'_{Ay}$ are two perpendicular constraint reaction forces acting on the end A of the bar AC by the pin A, as shown in Fig.1-17c, and $\boldsymbol{F}_{Ax}$, $\boldsymbol{F}_{Ay}$ in Fig.1-17b are the action and reaction forces, equal in magnitude and opposite in direction; $\boldsymbol{F}_C$ is the constraint reaction force from the smooth contact surface, perpendicular to the contact surface; $\boldsymbol{F}'_E$ is the tension force acting on the bar AC by the rope DE. For the bar AC, three-force equilibrium confluence theorem can be used to make sure the position of action line of $\boldsymbol{F}_A$, as shown in Fig.1-17d.

Taking the whole for force analysis, as shown in Fig.1-17e, $\boldsymbol{F}_B$ and $\boldsymbol{F}_C$ are the same as in Fig.1-17b and Fig.1-17c, $\boldsymbol{F}_{Ax}$, $\boldsymbol{F}_{Ay}$ and $\boldsymbol{F}'_{Ax}$, $\boldsymbol{F}'_{Ay}$ or $\boldsymbol{F}'_D$, $\boldsymbol{F}'_E$ do not appear in the free-body diagram of the whole structure (not drawn), these forces are internal forces for the whole.

Example 1-4 As shown in Fig.1-18a, an object weighing 20 kN is connected to a pulley D by a rope wrapped around pulley B. Draw free-body diagram of two bars and pulley B, the weight of the two bars and pulley and friction are negligible.

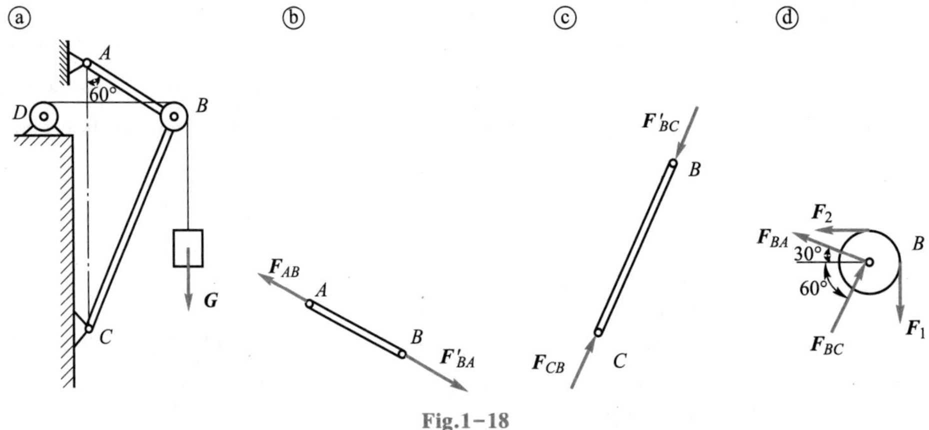

Fig.1-18

Solution it can be seen that the bar AB is a two-force member, which is equilibrated by two forces at A and B, and these two forces are equal, opposite, and colinear, as shown in Fig.1-18b.

As shown in Fig.1-18c, the bar BC is also a two-force member, so the action lines of $\boldsymbol{F}_{CB}$ and $\boldsymbol{F}'_{BC}$ are both along the BC in opposite directions.

The free-body diagram of the pulley with pin is shown in Fig.1-18d, $\boldsymbol{F}_2$ and $\boldsymbol{F}_1$ are the tensile forces of the rope acting on the edge of the wheel; $\boldsymbol{F}_{BA}$ and $\boldsymbol{F}_{BC}$ act along AB and BC, respectively, and are equal in magnitude and opposite in direction to $\boldsymbol{F}'_{BA}$ and $\boldsymbol{F}'_{BC}$ in Fig.1-18b and Fig.1-18c.

Problems

1-1 Draw the free-body diagram for object A or member AB in Fig.P1-1. The weight of each object is negligible, and all contacts are smooth.

Answer: Omit

1-2 Draw a free-body diagram for each of the objects labeled in Fig.P1-2. The weight of each

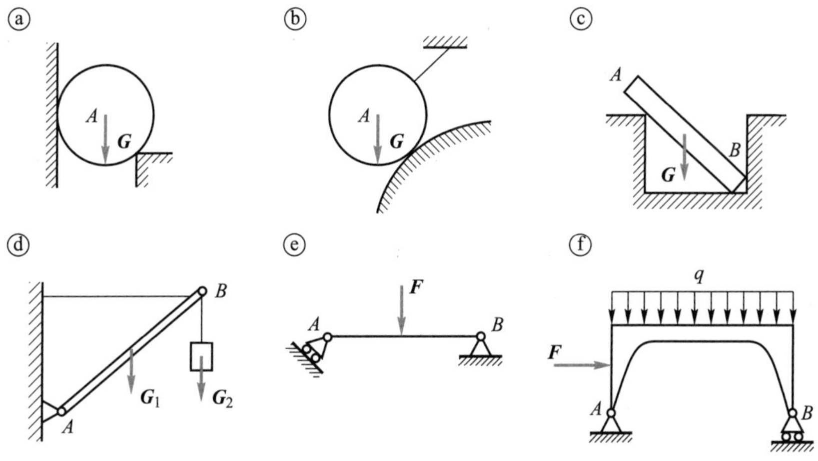

Fig.P1-1

object is negligible, and all contacts are smooth.

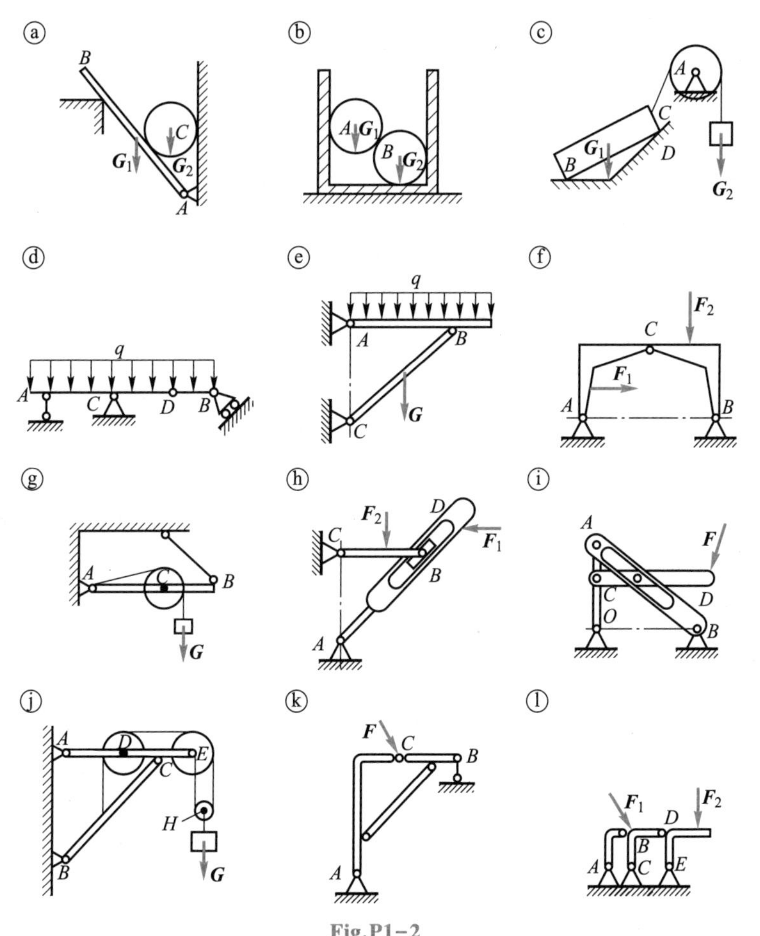

Fig.P1-2

Answer: Omit

Chapter 2 Coplanar Concurrent Force System and Couple System

Teaching Scheme of Chapter 2

According to the orientation of the action lines of force in a force system in space, the force system can be divided into a coplanar force system and a space force system. Coplanar force system can be divided into coplanar concurrent force system (or coplanar common point force system), coplanar force couple system, coplanar parallel force system and coplanar arbitrary force system (or coplanar general force system). Space force system can be divided into space concurrent force system, space force couple system, space parallel force system and space arbitrary force system (space general force system).

The two simplest force systems are the coplanar concurrent force system and the coplanar force couple system. This chapter focuses on the reduction and equilibrium conditions of these two force systems, which are the basis for the study of complex force system.

2.1 Reduction and equilibrium of coplanar concurrent force system

The action lines of all forces intersect at a point and locate in the same plane, which are called the coplanar concurrent force system. There are two methods to study plane concurrent force system, namely, geometric method and analytical method. In this section, the reduction and equilibrium conditions of the plane concurrent force system will be studied by these two methods, respectively.

2.1.1 Reduction of coplanar concurrent force system

1. geometric method—force polygon law

Let a rigid body act as a coplanar concurrent force system consisting of four forces that intersect at point A, as shown in Fig.2-1a. According to the transmissibility of the forces, the forces can be moved to point A along their respective lines of action to form a common point force system. We can use the parallelogram law of force to reduce these four forces to obtain a resultant force acting at point A:

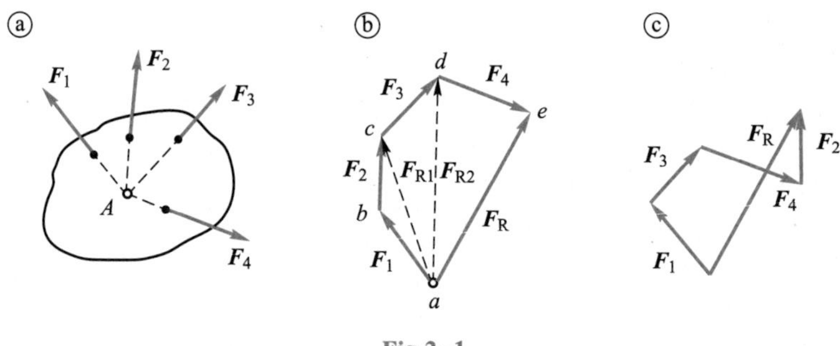

Fig.2-1

$$F_R = \sum F_i$$

Or use the triangle law of force to add the forces in sequence:

$$F_{R1} = F_1 + F_2, \quad F_{R2} = F_{R1} + F_3 = F_1 + F_2 + F_3, \quad F_R = F_{R2} + F_4 = F_1 + F_2 + F_3 + F_4$$

Also get the resultant force:

$$F_R = \sum F_i$$

Delete the F_{R1} and F_{R2} to obtain a polygon.

This polygon, formed by the vector of component forces and the vector of resultant forces, is called a force polygon (Fig. 2-1b). The resultant force vector F_R is called the closed side of the force polygon, and this method is called the geometric method. Fig.2-1b shows the magnitude and direction of each force vector, and illustrates the relationship between each force vector and the resultant force vector, but each force vector does not indicate its position of action. In the case of force polygons, if the forces are combined in a different order, the shape of the resulting force polygon is different, but the resultant force vector is identical, as shown in Fig.2-1c.

Through the above analysis, the following conclusions can be obtained: the plane concurrent force system can be reduced into a resultant force, the action line of the resultant force passes through the intersection point, the magnitude and direction of the resultant force is expressed by the closed side of the force polygon, that is, the resultant force vector is equal to the vector sum of the component force vectors. A vector expression as

$$F_R = F_1 + F_2 + \cdots + F_n = \sum F_i \tag{2-1}$$

2. Analytical method — projection method

(1) Projection of the force on the rectangular coordinate system

As shown in Fig. 2-2, there is a force F acting on point A. Take a rectangular coordinate system Oxy in the plane where the force F is located, and make vertical lines from the ends of the force vector to the x-axis, then the distance between the a_1 and b_1 is given an appropriate plus or minus sign to represent the projection of the force vector F on the x-axis, and note it as F_x. The projection F_x is positive if the direction from the a_1 to b_1 is the same as the positive direction of the x-axis, and negative if the opposite is true. Similarly, if a vertical line is drawn from the ends of the force vector to the y-axis, the line a_2b_2 is given an appropriate plus or minus sign to represent the projection of the force vector F on the y-axis, denoted as F_y. If the angle between the force vector F and the positive x-axis is α, and the angle with the positive y-axis is β, then from Fig.2-2, we can see that

$$F_x = F\cos\alpha, \quad F_y = F\cos\beta \tag{2-2}$$

That is, the projection of the force on an axis is equal to the magnitude of the force multiplied by the cosine of the angle between the force and the positive direction of the axis. Here, F is the magnitude of the force, usually take a positive value.

(2) Analytical expressions of forces

Based on the parallelogram law of force, the force F in Fig.2-2 can be decomposed orthogonally along the coordinate axes x and y into two mutually perpendicular components F_x and F_y, which are vectors and represent the components of the

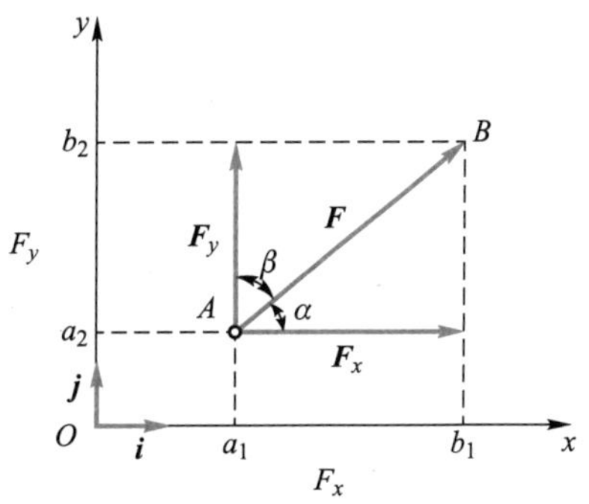

Fig.2-2

force along that direction, i.e.

$$\boldsymbol{F}=\boldsymbol{F}_x+\boldsymbol{F}_y$$

It is easy to see that these components of the force $\boldsymbol{F}$ are related to the projection of the force on the axis as follows, i.e.

$$\boldsymbol{F}_x=F_x\boldsymbol{i},\quad \boldsymbol{F}_y=F_y\boldsymbol{j}$$

where $\boldsymbol{i}$ and $\boldsymbol{j}$ are unit vectors in the positive direction along the x and y coordinate axes. Thus, the force $\boldsymbol{F}$ can be expressed as its projection on the axes as

$$\boldsymbol{F}=F_x\boldsymbol{i}+F_y\boldsymbol{j}$$

If the projections F_x and F_y of the force $\boldsymbol{F}$ on the coordinate axes x and y are known, the magnitude and direction of the force $\boldsymbol{F}$ can be obtained from the following equation, i.e.

$$F=\sqrt{F_x^2+F_y^2}$$

$$\cos\alpha=\frac{F_x}{F},\quad \cos\beta=\frac{F_y}{F}$$

(3) Analytical method of reduction of coplanar concurrent force system

As shown in Fig.2-3, firstly, the projection theorem of the resultant force of the coplanar concurrent force system is introduced, which is the basis of the analytical method of reduction of coplanar concurrent force system. The coplanar concurrent force system acting on the rigid body is $\boldsymbol{F}_1$, $\boldsymbol{F}_2$, $\boldsymbol{F}_3$ and $\boldsymbol{F}_4$, and from any point a force polygon $abcde$ is made, then the closed edge ae represents the resultant force vector $\boldsymbol{F}_{\mathrm{R}}$ of the force system, take the coordinate system Oxy, project all force vectors to the x-axis and y-axis, we can obtain

$$\left.\begin{aligned}a_1e_1&=a_1b_1+b_1c_1+c_1d_1+d_1e_1\\a_2e_2&=a_2b_2+b_2c_2+c_2d_2+d_2e_2\end{aligned}\right\}\tag{2-3}$$

$$\left.\begin{aligned}F_{\mathrm{R}x}&=F_{1x}+F_{2x}+F_{3x}+F_{4x}\\F_{\mathrm{R}y}&=F_{1y}+F_{2y}+F_{3y}+F_{4y}\end{aligned}\right\}$$

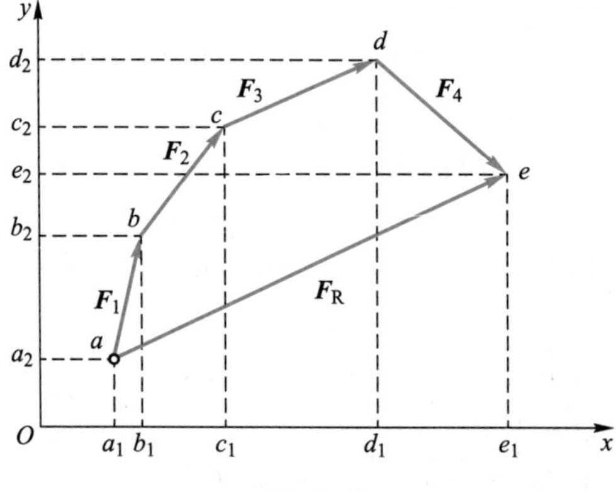

Fig.2-3

Extending the above relationship between the resultant force projection and the component force projection to a coplanar concurrent force system consisting of n forces, we can get

$$
\left.\begin{aligned}
F_{Rx} &= F_{1x} + F_{2x} + \cdots + F_{nx} = \sum F_{ix} \\
F_{Ry} &= F_{1y} + F_{2y} + \cdots + F_{ny} = \sum F_{iy}
\end{aligned}\right\} \tag{2-4}
$$

That is, the projection of the resultant force of the coplanar concurrent force system on a certain axis is equal to the algebraic sum of the projections of the component forces on the same axis, called the resultant force projection theorem.

After calculating the projections F_{Rx} and F_{Ry} of the resultant force $\boldsymbol{F}_R$, the magnitude and direction of the resultant force can be calculated as

$$
\left.\begin{aligned}
F_R = \sqrt{F_{Rx}^2 + F_{Ry}^2} = \sqrt{\left(\sum F_{ix}\right)^2 + \left(\sum F_{iy}\right)^2} \\
\cos\alpha = \frac{F_{Rx}}{F_R}, \quad \cos\beta = \frac{F_{Ry}}{F_R}
\end{aligned}\right\} \tag{2-5}
$$

The method of calculating the magnitude and direction of the resultant force by applying equation (2-4) and equation (2-5) is called the analytical method or projection method.

2.1.2 Equilibrium conditions of the coplanar concurrent force system

1. Geometric conditions for the equilibrium of the coplanar concurrent force system

The resultant force vector of a coplanar concurrent force system is the closed edge of a force polygon with each component force vector as its side. If the resultant force vector is equal to zero, it means that the ending of the last force vector in the force polygon coincides with the beginning O of the first force vector, i.e., the force polygon closes itself. This is the geometric condition for the equilibrium of the coplanar concurrent force system.

For example, a coplanar concurrent force system ($\boldsymbol{F}_1$, $\boldsymbol{F}_2$, $\boldsymbol{F}_3$, $\boldsymbol{F}_4$) in equilibrium (Fig.2-4a), its polygon of forces is self-closing (Fig.2-4b).

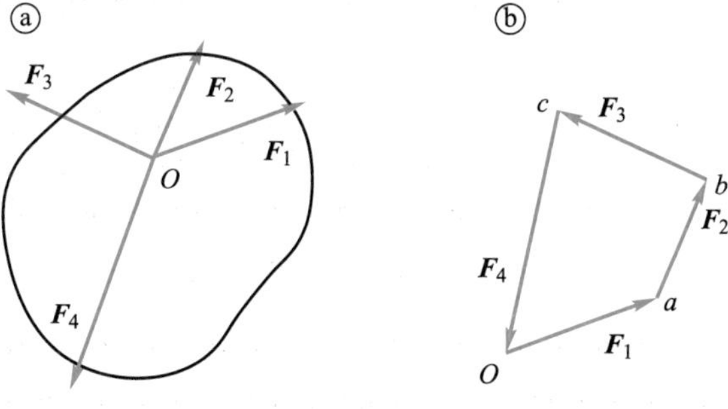

Fig.2-4

Therefore, the necessary and sufficient condition for a coplanar concurrent force system in equilibrium is that the polygon of forces constituted by each force vector in the force system is closed by itself, or the vector sum of each force is equal to zero.

$$
\boldsymbol{F}_R = \boldsymbol{F}_1 + \boldsymbol{F}_2 + \cdots + \boldsymbol{F}_n = \sum \boldsymbol{F}_i = \boldsymbol{0} \tag{2-6}
$$

Applying the geometric conditions for the equilibrium of the coplanar concurrent force system, the two unknown quantities can be solved by the characteristics of the force polygon closed by itself, which determines the magnitude and direction of the unknown force.

Example 2-1 Fig. 2-5a shows a brake mechanism with a force $F = 212$ N and $\alpha = 45°$. Determine the force acting on the bar BC without considering the weight of the components.

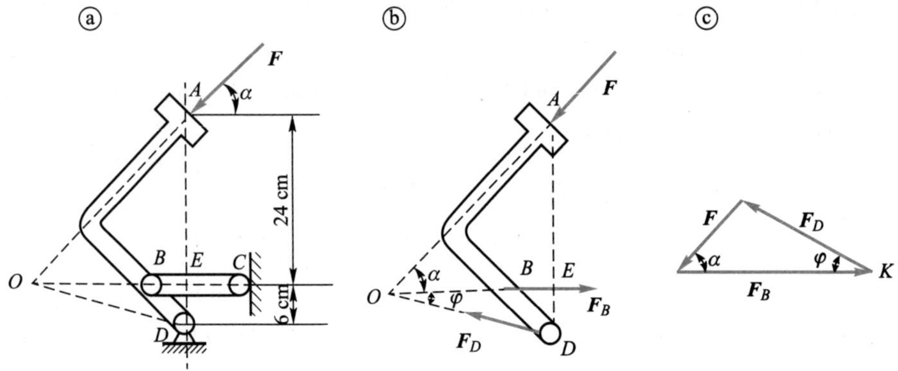

Fig.2-5

Solution Take the curved bar ABD as the free-body and do the force analysis (Fig.2-5b), make a closed force polygon (Fig.2-5c), from the geometric relationship of Fig.2-5b we have

$$OE = EA = 24\ \text{cm},\quad \tan\varphi = \frac{DE}{OE} = \frac{6}{24} = \frac{1}{4},\quad \varphi = \arctan\frac{1}{4} = 14.01°$$

From the triangular relationship in Fig.2-5c, we have

$$F_B = 751\ \text{N}$$

2. Analytical conditions for the equilibrium of the coplanar concurrent force system

The necessary and sufficient condition for the coplanar concurrent force system in equilibrium is $\boldsymbol{F}_{\text{R}} = \sum \boldsymbol{F}_i = \boldsymbol{0}$, that is, the magnitude of the resultant force F_{R} should be equal to zero, from the equation (2-6), we can get

$$F_{\text{R}} = \sqrt{F_{\text{R}x}^2 + F_{\text{R}y}^2} = \sqrt{\left(\sum F_{ix}\right)^2 + \left(\sum F_{iy}\right)^2} = 0$$

In the equation, the two terms in the root must be zero at the same time, then we have

$$\sum F_{ix} = 0,\quad \sum F_{iy} = 0$$

This is the analytical condition for the equilibrium of a coplanar concurrent force system, i.e., the algebraic sum of the projections of each force in the force system on the two axes is equal to zero, respectively. These two equations are called the equilibrium equations of the coplanar concurrent force system. These two equations establish the interrelationship between the forces in equilibrium, and they are independent of each other to solve the two unknown quantities.

Example 2-2 Fig.2-6 shows a coplanar concurrent force system acting on an object at point O. $F_1 = 200$ N, $F_2 = 300$ N, $F_3 = 100$ N, and $F_4 = 250$ N. Determine the magnitude and direction of the resultant force of this force system.

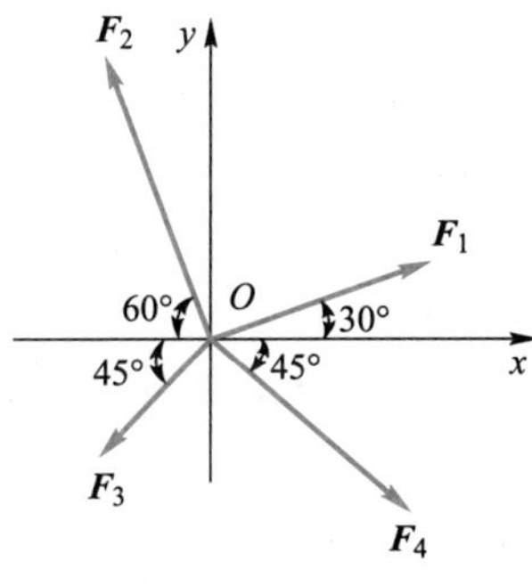

Fig.2-6

Solution Based on the resultant force projection theorem, we can write

$$\begin{aligned} F_{Rx} &= \sum F_{ix} \\ &= F_1\cos 30° - F_2\cos 60° - F_3\cos 45° + F_4\cos 45° \\ &= 129.3 \text{ N} \end{aligned}$$

$$\begin{aligned} F_{Ry} &= \sum F_{iy} \\ &= F_1\cos 60° + F_2\cos 30° - F_3\cos 45° - F_4\cos 45° \\ &= 112.3 \text{ N} \end{aligned}$$

The magnitude of the resultant force can be expressed as

$$F_R = \sqrt{F_{Rx}^2 + F_{Ry}^2} = 171.3 \text{ N}$$

The direction of the resultant force can be expressed as

$$\cos\alpha = \frac{F_{Rx}}{F_R} = 0.754, \quad \cos\beta = \frac{F_{Ry}}{F_R} = 0.656$$

$$\alpha = 40.99°, \quad \beta = 49.01°$$

2.2 Coplanar couple system

2.2.1 Moment of a force about a point in the plane

The force on the object may produce the movement effect and the rotation effect, or both of these two effects. The moving effect of a force depends on the magnitude and direction of the force, while the effect of a force to rotate an object around a point depends on the moment of the force about that point.

As shown in Fig.2-7.The rotation effect of a force is not only related to the magnitude of the force, but also to the vertical distance d from the point to the action line of the force, the greater the force or distance, the more significant the rotation effect.

As we known, the rotation effect of a force $\boldsymbol{F}$ depends on two factors, namely, the product of the magnitude of the force F and the distance d, and the direction in which the force rotates the object.

In order to measure the effect of a force to rotate an object around a point, we define the moment of the force about the point, let a force $\boldsymbol{F}$ act on the object at point A, and take a point O in the plane where the force $\boldsymbol{F}$ is located, then the moment of the force about point O can be expressed as

$$M_O(\boldsymbol{F}) = \pm Fh$$

where point O is called the center of moment; h is the vertical distance from the center of moment O to the action line of the force $\boldsymbol{F}$, called the force arm. And the positive and negative signs are usually specified as the moment of the force about the point is positive when the force makes the object rotate counterclockwise around the center of the moment; and vice versa is negative. From Fig.2-8, it can be seen that the magnitude of the moment of the force F about point O can also be expressed as two times the area of $\triangle OAB$, i.e.

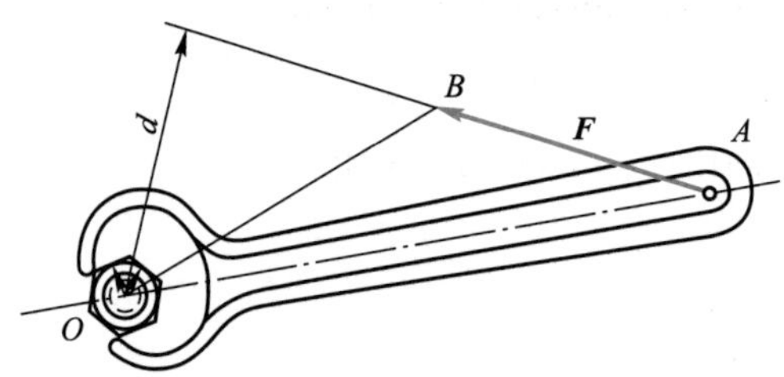

Fig.2-7

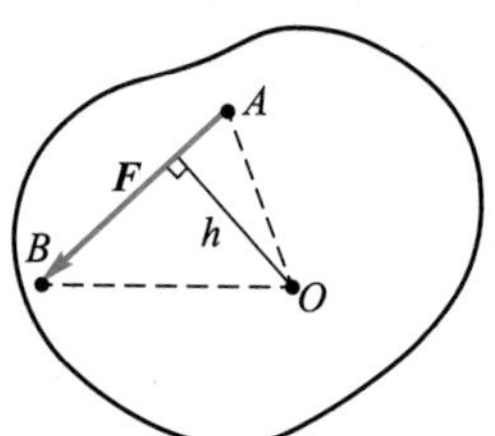

Fig.2-8

$$M_O(\boldsymbol{F}) = \pm 2A_{\triangle OAB}$$

If the unit of force is Newton (N) and the unit of force arm is meter (m), then the unit of force moment is the Newton · meter (N · m).

The moment is equal to zero in the following two cases: ① $F=0$; ② $h=0$ (the action line of the force passes through the center of the moment). In addition, if $\boldsymbol{F}_R$ is the resultant force of the coplanar concurrent force system, the moment of the resultant force about any point O is equal to the algebraic sum of the moments of the component force in the force system about the same point, i.e.

$$M_O(\boldsymbol{F}_R) = \sum M_O(\boldsymbol{F}_i)$$

This relationship is called the resultant moment theorem for coplanar concurrent force system.

2.2.2 Coplanar couple

1. Couple

A force system consisting of two opposing parallel forces with equal magnitude is called a couple (Fig.2-9a). The couple can be written as $(\boldsymbol{F}, \boldsymbol{F}')$. The vertical distance d between the two forces in the couple is called the moment arm, and the plane where the couple is located is called the couple's plane of action. For example, the driver of a car turns the steering wheel (Fig.2-9b) and rotates the reamer with both hands in order to tap the threads on the workpiece (Fig.2-9c), etc.

2. Properties of couple

(1) The couple can not be equivalent to a force (couple has no resultant force), at the same time the couple itself can not equilibrate, also can not equilibrate with a force, which is another basic

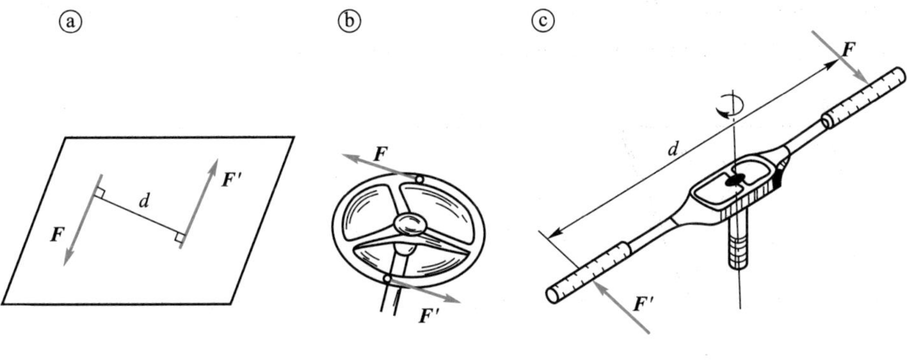

Fig.2-9

element in statics.

(2) The effect of the couple on the object can only produce the effect of rotation, but never the effect of movement, the effect of the couple on the rotation of the object can be measured by the moment of couple.

The couple moment is an algebraic quantity whose magnitude is equal to the product of the magnitude of a force in the couple and the moment arm, independent of the position of the moment center, and is denoted by the symbol M or M (F, F'), i.e.

$$M = \pm Fd$$

In the formula, the positive and negative signs are the same as the moment of force, i.e., the positive sign is taken when the couple makes the object rotate counterclockwise, and the negative sign is taken vice versa.

(3) Two couples acting in the same plane, if their moments are equal, the two couples are equivalent to each other, that is, they have the same effect on the rigid body, which is the equivalence theorem of plane couples.

The nature of the equivalent transformation of the couple can be described as:

① The couple can move arbitrarily within its plane of action without changing its effect on the rigid body.

② As long as the magnitude and direction of the couple moment remain unchanged, the magnitude of the force in the couple and the length of the moment arm can be changed simultaneously without changing the effect of the couple on the rigid body.

The couple moment (Fig. 2 - 10) is represented by M indicating the magnitude of the couple moment and the arc with an arrow indicating the direction of the couple moment.

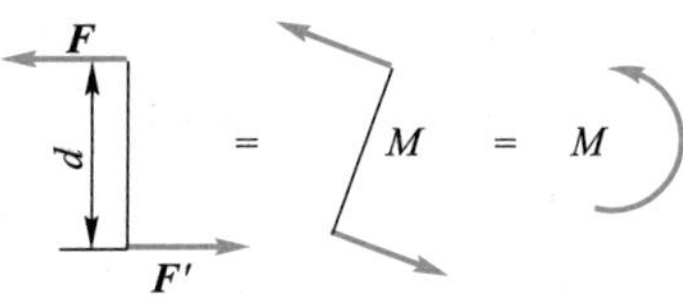

Fig.2-10

3. Resultant and equilibrium of coplanar couples

A system consisting of several couples acting on an object in the same plane is called planar couples. Because the couple itself cannot be reduced a resultant force, resultant of the planar couples must be a couple.

Let there be two couples $(\boldsymbol{F}_1, \boldsymbol{F}_1')$ and $(\boldsymbol{F}_2, \boldsymbol{F}_2')$ in the same plane with moment arms d_1 and d_2, as shown in Fig.2-11a, then the moments of each couple are

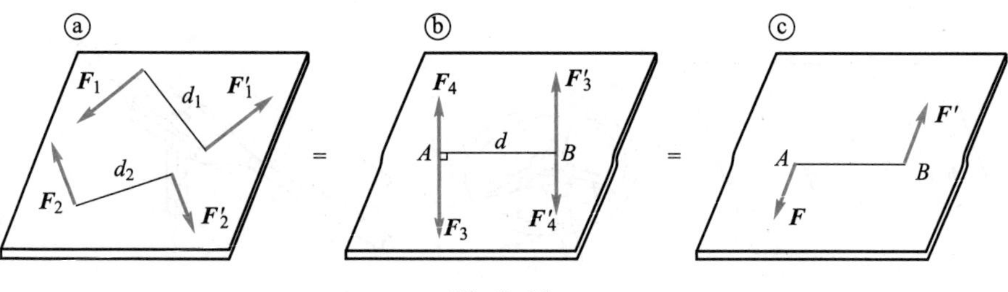

Fig.2-11

$$M_1 = F_1 d_1, \quad M_2 = F_2 d_2$$

Take any line $AB = d$ in the action plane of the couple, and make the moment arm of each couple as d while keeping the couple moment unchanged, so the magnitude of each force in the couple should be changed to

$$F_3 = \frac{F_1 d_1}{d}, \quad F_4 = \frac{F_2 d_2}{d}$$

Then move each couple so that their moment arms coincide with AB, and the original planar couples is transformed into two common line force systems acting at points A and B (Fig.2-11b). Then the two common linear force systems are reduced separately to obtain the two forces $\boldsymbol{F}$ and $\boldsymbol{F}_1'$ shown in Fig.2-11c, whose magnitudes are

$$F = F_3 - F_4, \quad F' = F_3' - F_4'$$

It can be seen that the forces $\boldsymbol{F}$ and $\boldsymbol{F}'$ are equal in magnitude, opposite in direction and not in the same line, and they form a resultant couple $(\boldsymbol{F}, \boldsymbol{F}')$ equivalent to the original couples, whose couple moment is

$$M = Fd = (F_3 - F_4)d = F_1 d_1 - F_2 d_2$$

so

$$M = M_1 + M_2$$

If there are n couples acting in the same plane, the above equation can be extended to

$$M = M_1 + M_2 + \cdots + M_n = \sum M_i$$

It can be seen that any couple in the same plane can be reduced a resultant couple, and the resultant couple moment is equal to the algebraic sum of each couple moment.

The result shows that the resultant couple moment should be equal to zero when the couples is in equilibrium. Therefore, a sufficient necessary condition for the equilibrium of the planar couples is that the algebraic sum of all couple moments is equal to zero, i.e.

$$\sum M_i = 0$$

This equation is the equilibrium equation for planar couples, and there is only one independent equation that can be solved for only one unknown quantity.

Example 2-3 As shown in Fig.2-12a, length of a beam l, a fixed-hinge support at B, and a bar AD connected by a cylindrical hinge at A. The bar AD is a two-force member, regardless of the weight of the beam, try to determine the reaction forces at A and B.

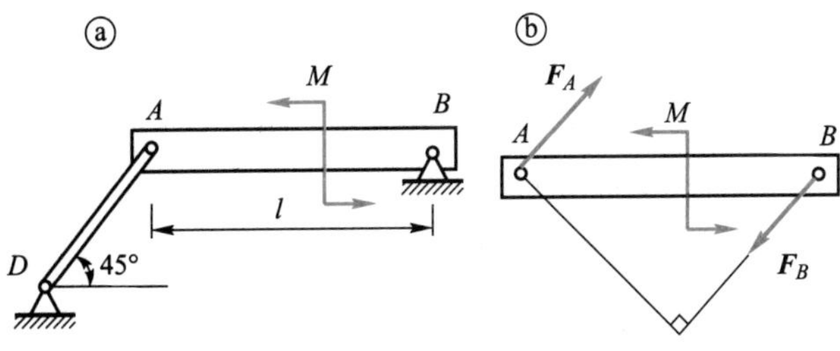

Fig.2-12

Solution Since the bar AD is a two-force member, the line of action of the force on beam AB is along the line DA, in the direction shown in Fig.2-12b. Since a couple can only be equilibrated by another couple, the constraint reaction force at B should be parallel to the action line of the force at A, moreover, $\boldsymbol{F}_A$ and $\boldsymbol{F}_B$ constitutes a couple in equilibrium with the couple M, list the equilibrium equation of the planar couples, there are

$$\sum M_B = 0, \quad M - F_A l\cos 45° = 0$$

$$F_A = F_B = \frac{M}{l\cos 45°} = \frac{\sqrt{2}M}{l}$$

Problems

2-1 A uniform sphere of weight W is supported by a soft rope and a smooth inclined surface as shown in Fig.P2-1.Knowing the angle α and angle β, determine the magnitude of the tension on the rope and the compression force on the inclined surface.

Answer: $F_T = \dfrac{\sin\beta}{\sin(\alpha+\beta)}W$, $F_N = \dfrac{-\sin\alpha}{\sin(\alpha+\beta)}W$

2-2 As shown in Fig.P2-2, a horizontal force $F = 20$ kN acts on the plane steel frame $ABCD$ at point B, the weight of the steel frame is negligible, determine the reaction force at points A and D.

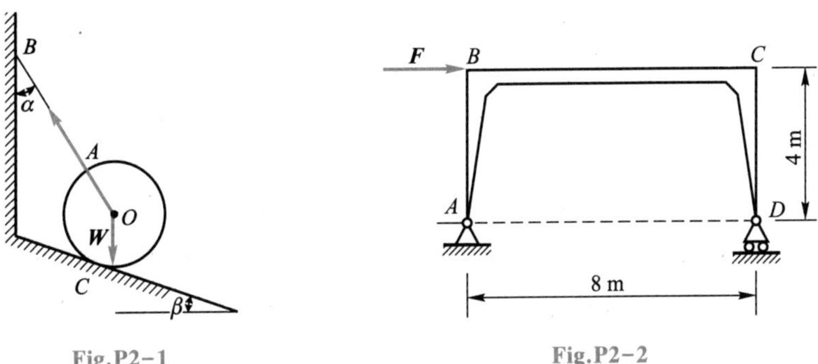

Fig.P2-1 Fig.P2-2

Answer: $F_A = -22.36$ kN, $F_D = 10$ kN

2-3 As shown in Fig.P2-3, a couple acts on beam AB with moment M, beam length l, and the weight of beam is negligible, determine the reactive forces at points A and B.

Answer: $F_A = F_B = \dfrac{M}{l\cos\theta}$

2-4 The CD side of a hinged four-bar mechanism $CABD$ is fixed and forces $\boldsymbol{F}_1$ and $\boldsymbol{F}_2$ act at hinges A and B as shown in Fig.P2-4. The mechanism is in equilibrium, and all bars weight are negligible, determine the relationship between the forces $\boldsymbol{F}_1$ and $\boldsymbol{F}_2$.

Answer: $\dfrac{F_1}{F_2} = 0.644$

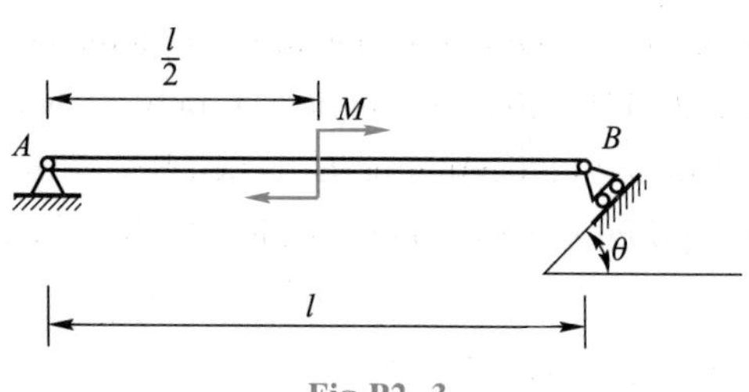

Fig.P2-3

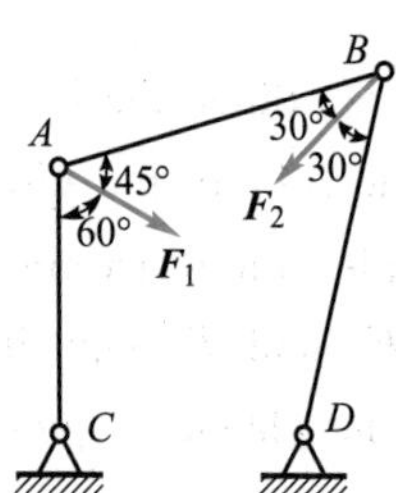

Fig.P2-4

2-5 As shown in Fig.P2-5, a couple is known to act on beam AB with moment M, beam length l, and the weight of beam is negligible. Determine the reaction forces at supports A and B in the three cases shown in Fig.P2-5.

Answer: (a) $F_A = F_B = \dfrac{M}{l}$; (b) $F_A = F_B = \dfrac{M}{l}$; (c) $F_A = F_B = \dfrac{M}{l\cos\theta}$

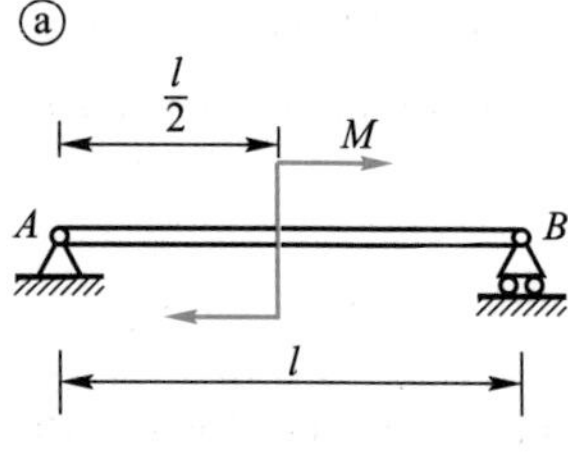

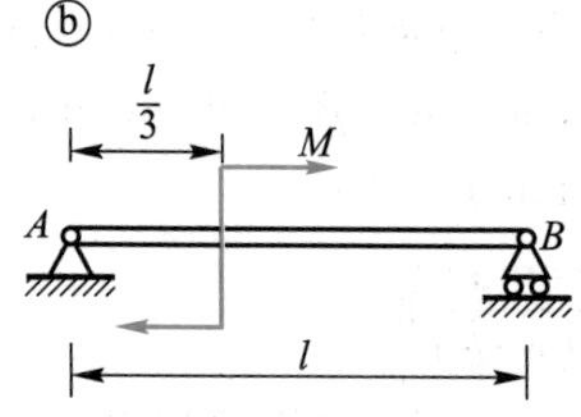

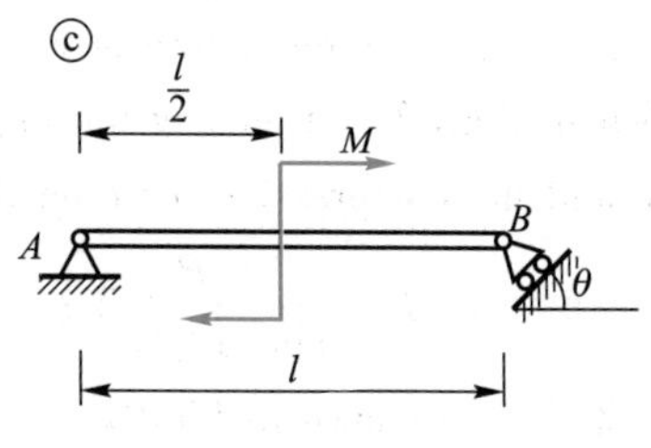

Fig.P2-5

2-6 The structure is shown in Fig.P2-6, the weight of each member is negligible and a couple with moment M is applied to member AB, determine the reaction force at supports A and C.

Answer: $F_A = F_C = \dfrac{M}{2\sqrt{2}a}$

2-7 The structure is shown in Fig.P2-7, the weight of each member is negligible and a couple with moment M is applied to member BC, determine the reaction force at support A and C.

Answer: $F_C = \dfrac{M}{l}$, $F_A = \dfrac{\sqrt{2}M}{l}$

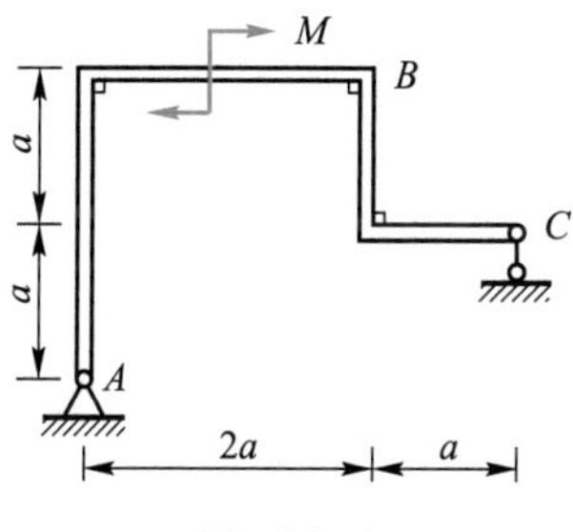

Fig.P2-6

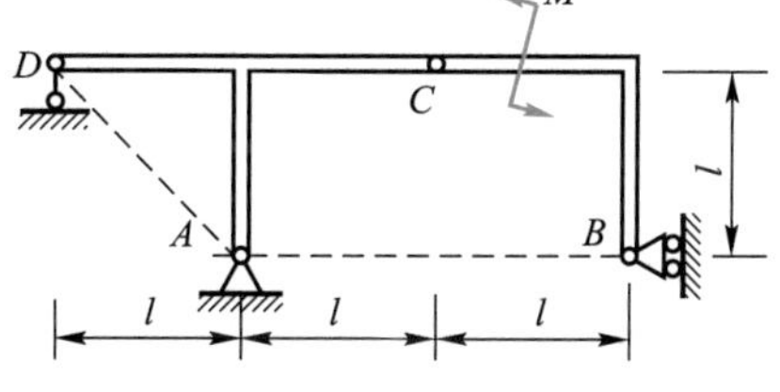

Fig.P2-7

2-8 As shown in Fig.P2-8, a planar four-link articulated mechanism $OABD$ has two couples with moments M_1 and M_2 acting in the plane on the bar OA and the bar BD, respectively, and the structure is in equilibrium at the position shown in Fig.P2-8. Knowing that $OA=r$, $BD=2r$, $\alpha=30°$, and the weight of the bar is negligible, determine the relationship between M_1 and M_2.

Answer: $M_2=2M_1$

2-9 The structure shown in Fig.P2-9 has no self-weight and a smooth surface contact at A. Determine the reaction force at D under the action of a known couple M.

Answer: $F_D=\dfrac{2M}{a}$

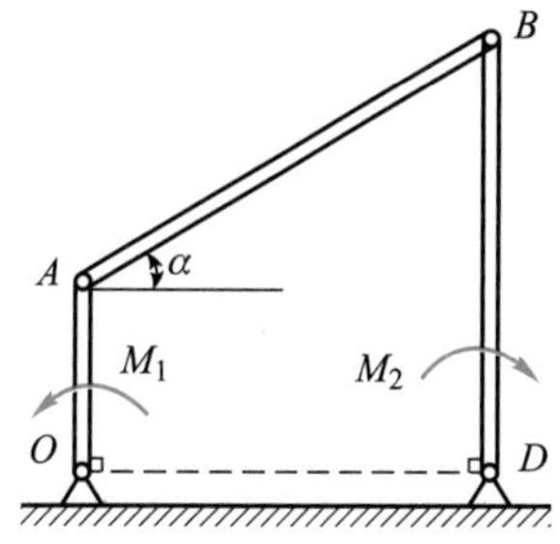

Fig.P2-8

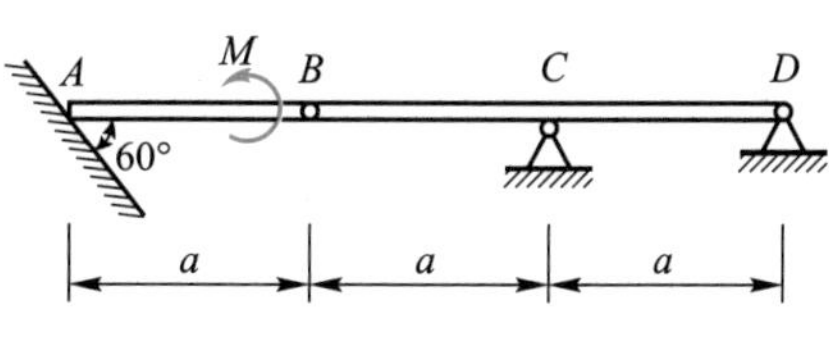

Fig.P2-9

Chapter 3 Coplanar General Force System

Teaching Scheme of Chapter 3

The so-called coplanar general force system means that the lines of action of the forces acting on the object are distributed arbitrarily in the same plane.

3.1 The translation theorem of forces

The point of action of a force acting on a rigid body can be shifted along its line of action without changing the effect of the force on the rigid body. However, if the force is moved parallel to any point beyond the original line of action of the force, it will change its effect on the rigid body.

Let the force $\boldsymbol{F}$ act on the rigid body at point B, and now translate it to any point A beyond the line of action. The condition is that the effect of the primary force $\boldsymbol{F}$ on the rigid body cannot be changed (Fig. 3 - 1a). For this purpose, a pair of forces $\boldsymbol{F}'$ and $\boldsymbol{F}''$ of equal magnitude, opposite direction and parallel to the force $\boldsymbol{F}$, $\boldsymbol{F}=\boldsymbol{F}'=-\boldsymbol{F}''$, act at point A. It is obvious that a new equilibrium force system is added, based on the principle of plus or minus equilibrium force system, this does not change the effect of the original force on the rigid body (Fig.3-1b). $\boldsymbol{F}''$ and $\boldsymbol{F}$ exactly form a couple, and the couple moment is $M = Fd$, d is the vertical distance between the two lines of force action (Fig.3-1c). At the same time the couple moment is equal to the moment of the original force $\boldsymbol{F}$ about point A, i.e. $M = M_A(\boldsymbol{F})$. Since $\boldsymbol{F}' = \boldsymbol{F}$, it can be considered that the force $\boldsymbol{F}$ is translated to the point A, but as an equivalent force system, there is an additional couple, and the moment of the additional couple is equal to the moment of the original force about the new point A.

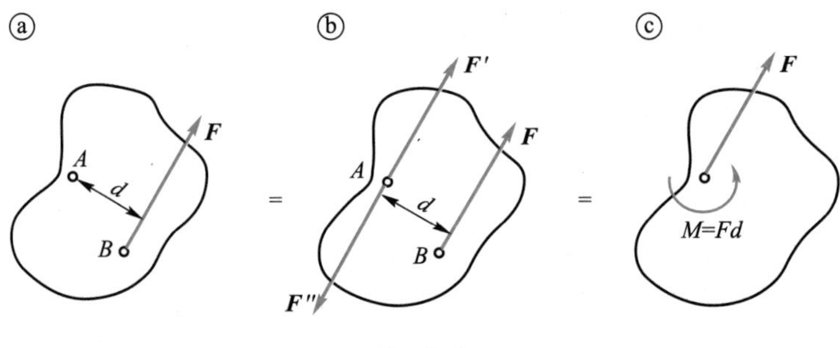

Fig.3-1

The translation theorem of forces: The force acting on a rigid body can be translated to any point on the rigid body, but at the same time a couple must be added, and the moment of the additional couple is equal to the moment of the original force about the new point of action.

The translation theorem of forces shows that a force before translation is equivalent to a force and a couple after translation. Conversely, a force and a couple acting on a point can also be reduced into a force.

3.2 Reduction of the coplanar general force system

In the following, the translation theorem of forces is applied to study the reduction of the coplanar general force system.

Let a coplanar general force system ($\boldsymbol{F}_1$, $\boldsymbol{F}_2$, $\cdots$, $\boldsymbol{F}_n$) consisting of n forces act on a rigid body (Fig.3-2a), and take any point O in the plane of action of the force system, which is called the center of simplification. Based on the translation theorem of force, each force in the force system is translated to point O to obtain a coplanar concurrent force system ($\boldsymbol{F}_1'$, $\boldsymbol{F}_2'$, $\cdots$, $\boldsymbol{F}_n'$) with all lines of action passing through point O and a planar couple system (M_1, M_2, $\cdots$, M_n) consisting of the corresponding additional couple (Fig.3-2b). In this way, the coplanar general force system is equivalent to two basic force systems, the coplanar concurrent force system and the planar couple system. Then, the two force systems are reduced, respectively.

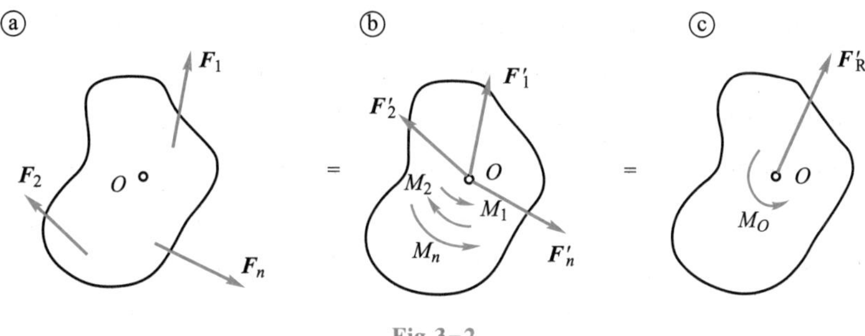

Fig.3-2

The coplanar concurrent force system ($\boldsymbol{F}_1'$, $\boldsymbol{F}_2'$, $\cdots$, $\boldsymbol{F}_n'$) can be reduced into a resultant force through the simplified center O point, the resultant force vector is

$$\boldsymbol{F}_R' = \boldsymbol{F}_1' + \boldsymbol{F}_2' + \cdots + \boldsymbol{F}_n' \tag{3-1}$$

Noting that $\boldsymbol{F}_1', \boldsymbol{F}_2', \cdots, \boldsymbol{F}_n'$ are equal in magnitude and direction to the forces $\boldsymbol{F}_1, \boldsymbol{F}_2, \cdots, \boldsymbol{F}_n$ of the original force system, i.e. $\boldsymbol{F}_i = \boldsymbol{F}_i'$, so we have

$$\boldsymbol{F}_R' = \boldsymbol{F}_1 + \boldsymbol{F}_2 + \cdots + \boldsymbol{F}_n = \sum \boldsymbol{F}_i \tag{3-2}$$

Equation (3-2) shows that the resultant force vector of the coplanar concurrent force system is equal to the vector sum of each force in the original force system, and we call $\boldsymbol{F}_R'$ as the principal vector of the original force system. Obviously, the principal vector is independent of the choice of the simplification center.

The planar couple system can be reduced into a resultant couple, and the resultant couple moment is equal to the algebraic sum of the additional couple moments, and also equal to the algebraic sum of the moments of each force in the original force system about the simplified center O point, i.e.

$$M_O = M_1 + M_2 + \cdots + M_n = M_O(\boldsymbol{F}_1) + M_O(\boldsymbol{F}_2) + \cdots + M_O(\boldsymbol{F}_n) = \sum M_O(\boldsymbol{F}_i) \tag{3-3}$$

Equation (3-3) shows that the resultant moment of the planar couple system is equal to the algebraic sum of the moments of each force in the original force system about the simplified center O, and M_O is called the principal moment of the original force system about the simplified center. Obviously, in general, the principal moment is related to the choice of the simplification center, so it must be stated to which point it is the principal moment. The principal moment is noted as M_O, and the subscript O indicates that it is the principal moment about the simplification center O point.

In general, the coplanar general force system is reduced toward a point in the plane of action to obtain a force and couple. As shown in Fig.3-2c, the action line of this force passes through the center of the simplification and its force vector is called the principal vector of the force system and is equal to the vector sum of each force in the force system. This couple acts in the original plane and its couple moment is called the principal moment of the force system about the simplified center, which is equal to the algebraic sum of the moments of each force in the force system about the simplified center.

It should be noted that the principal vector of the force system is not a force, but only a vector without point of action or effect, and is independent of the choice of the simplification center. The principal moment is also not a couple, which is an algebraic quantity without action effect. Choosing different simplification centers will have different principal moments.

In order to represent the principal vector $\boldsymbol{F}'_{\mathrm{R}}$ analytically, the orthogonal x and y axes can be built in the plane of action of the force system, and the unit vectors $\boldsymbol{i}$ and $\boldsymbol{j}$ can be introduced, then

$$\boldsymbol{F}'_{\mathrm{R}}=F'_{\mathrm{R}x}\boldsymbol{i}+F'_{\mathrm{R}y}\boldsymbol{j} \tag{3-4}$$

where $F'_{\mathrm{R}x}$ and $F'_{\mathrm{R}y}$ represent the projections of the principal vector on the x and y axes, respectively. Projecting equation (3-2) onto the x-axis and y-axis, respectively, we get

$$\left.\begin{aligned} F'_{\mathrm{R}x} &= \sum F_{ix} \\ F'_{\mathrm{R}y} &= \sum F_{iy} \end{aligned}\right\} \tag{3-5}$$

Therefore, the magnitude and direction of the principal vector is

$$F'_{\mathrm{R}} = \sqrt{F'^2_{\mathrm{R}x} + F'^2_{\mathrm{R}y}} = \sqrt{\left(\sum F_{ix}\right)^2 + \left(\sum F_{iy}\right)^2} \tag{3-6}$$

$$\cos(\boldsymbol{F}'_{\mathrm{R}},\boldsymbol{i})=\frac{F'_{\mathrm{R}x}}{F_{\mathrm{R}}},\quad \cos(\boldsymbol{F}'_{\mathrm{R}},\boldsymbol{j})=\frac{F'_{\mathrm{R}y}}{F_{\mathrm{R}}} \tag{3-7}$$

The analytical expression of the principal moment for the force system about point O is

$$M_O = \sum M_O(\boldsymbol{F}_i) \tag{3-8}$$

For fixed-end support at the end A, the reaction forces are arbitrarily distributed (Fig.3-3a) and are more complex. However, when the applied force is a coplanar force system, the force system composed of these constraint reaction forces can be identified as a coplanar general force system, which is reduced toward point A to obtain a reaction force $\boldsymbol{F}_A$ and a reaction couple M_A(Fig.3-3b). From the reduced results of the coplanar general force system, it is known that both the direction of this reaction force and the sense of rotation of this reaction coupling cannot be determined in advance.

Therefore, the reaction force $\boldsymbol{F}_A$ can be expressed as two perpendicular component forces $\boldsymbol{F}_{Ax}$ and $\boldsymbol{F}_{Ay}$, while the moment of the reaction couple is denoted as M_A (Fig.3-3c), and its sense can be set as counterclockwise. If the calculated $\boldsymbol{F}_{Ax}$, $\boldsymbol{F}_{Ay}$ and M_A are negative values from the equations, it means that the direction of the actual reaction force and reaction couple is opposite to that shown in Fig.3-3c. The reduced results of the fixed-end reaction force toward point A show that it not only limits the movement of the object, but also limits the rotation of the object.

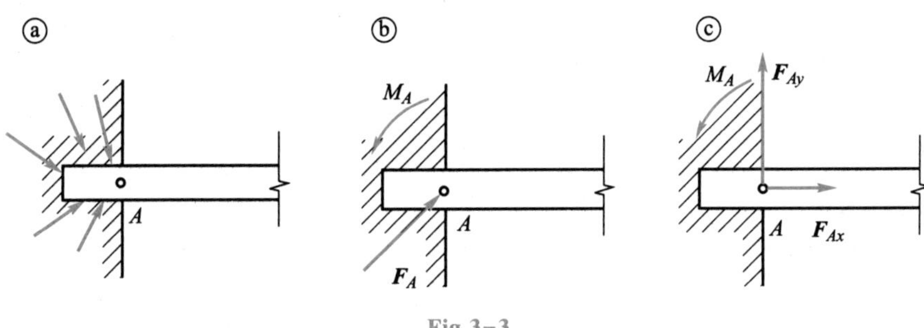

Fig.3-3

3.3 Reduced resultant of coplanar general force system and resultant moment theorem

There are four possible cases for reduced resultant of the coplanar general force system, and each of these cases is discussed below.

1. $F'_R=0, M_O=0$

Both the principal vector and principal moment are equal to zero, which means that the reduced coplanar concurrent force system and planar couple system are equilibrium. Therefore, the original coplanar general force system is also equilibrium.

2. $F'_R=0, M_O\neq 0$

The principal vector is equal to zero and the principal moment is not equal to zero, which means that the original coplanar general force system is equivalent to a planar couple system, that is, the original coplanar general force system can be reduced into a resultant couple, and the resultant couple moment is equal to the principal moment of the original coplanar general force system. In the case the principal moment of the force system is not relevant to the choice of the simplification center.

3. $F'_R\neq 0, M_O=0$

The principal vector is not equal to zero, while the principal moment is equal to zero, indicating that the original coplanar general force system is equivalent to a resultant force whose line of action passes through the simplified center, and the magnitude and direction of the resultant force are determined by the principal vector $\boldsymbol{F}'_R$. This case is related to the position of the simplification center, i.e., the additional couple system is just equilibrium at this time.

4. $F'_R \neq 0, M_O \neq 0$

Both the principal vector and principal moment are not equal to zero, as shown in Fig.3–4a. The couple with principal moment M_O is now represented by two forces F''_R and F_R, such that $F_R = F'_R = -F''_R$ (Fig.3–4b), then the result can be further reduced into a resultant force F_R. The magnitude and direction of the resultant force F_R is the same with F'_R, but the resultant force does not act at the simplified center O point, but deviates from the distance d, i.e.

$$d = M_O / F_R \tag{3-9}$$

In this case, the moment of the resultant force F_R about point O in Fig.3–4c is

$$M_O(F_R) = F_R d$$

Substituting equation (3–9) into the above equation, we have

$$M_O(F_R) = F_R d = M_O$$

M_O is the principal moment of the force system, and $M_O = \sum M_O(F_i)$, then we have

$$M_O(F_R) = \sum M_O(F_i) \tag{3-10}$$

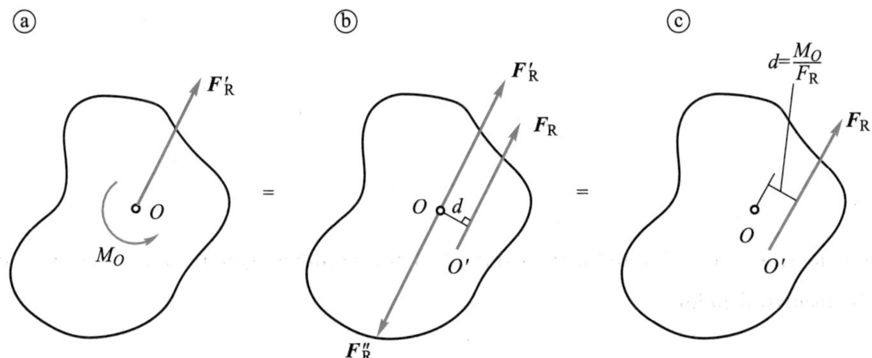

Fig.3–4

Equation (3–10) shows that if a coplanar general force system can be reduced to a resultant force, then the moment of the resultant force about any point in the plane is equal to the algebraic sum of the moments of the forces in the system about the same point, which is the resultant moment theorem of the coplanar general force system.

By applying the resultant moment theorem, the analytic expression of the moment of the force F about the coordinate origin O can be calculated. As shown in Fig.3–5, let the force F be decomposed into two component forces F_x and F_y along the coordinate axes, then by the resultant moment theorem we have

$$M_O(F) = M_O(F_x) + M_O(F_y) = xF_y - yF_x \tag{3-11}$$

where F_x and F_y are the projections of the force F on the coordinate axes; x and y are the coordinates of any point on the line of action of the force F.

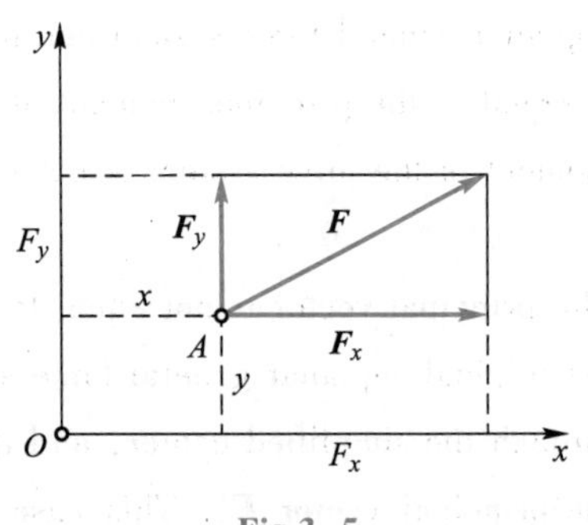

Fig.3–5

Example 3-1 As shown in Fig.3-6, the gravity dam is subjected to a plane general force system, $G_1 = 450$ kN, $G_2 = 200$ kN, $F_1 = 300$ kN, $F_2 = 70$ kN, determine the principal vector and principal moment of the force system simplified to point O.

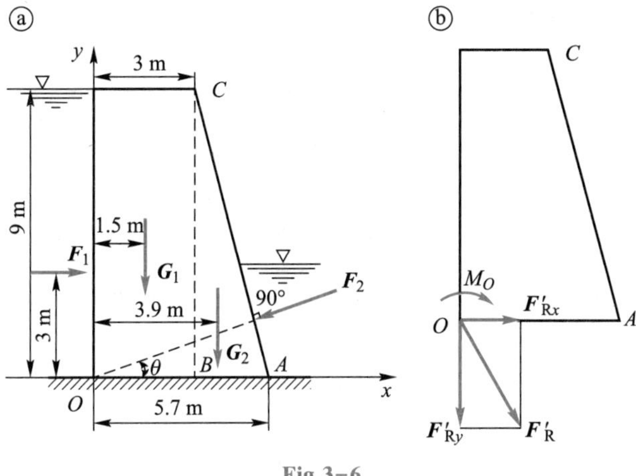

Fig.3-6

Solution Calculate the principal vector:

$$F'_{Rx} = \sum F_{ix} = F_1 - F_2\cos\theta = 232.9\ \text{kN}$$

$$F'_{Ry} = \sum F_{iy} = -G_1 - G_2 - F_2\sin\theta = -670.1\ \text{kN}$$

$$F'_R = \sqrt{232.9^2 + (-670.1)^2}\ \text{kN} = 709.4\ \text{kN}$$

$$\cos(\boldsymbol{F}'_R, \boldsymbol{i}) = \frac{F'_{Rx}}{F'_R} = 0.328, \quad \cos(\boldsymbol{F}'_R, \boldsymbol{j}) = \frac{F'_{Ry}}{F'_R} = -0.945$$

Calculate the principal moment:

$$M_O = \sum M_O(\boldsymbol{F}_i) = -F_1 \times 3\ \text{m} - G_1 \times 1.5\ \text{m} - G_2 \times 3.9\ \text{m} = -2\ 355\ \text{kN}\cdot\text{m}$$

Example 3-2 The forces $\boldsymbol{F}_1$, $\boldsymbol{F}_2$, $\boldsymbol{F}_3$, and $\boldsymbol{F}_4$ acting at points O, A, B, and C of the rectangular plate shown in Fig.3-7 are a coplanar general force system with $\boldsymbol{F}_1 = 1$ kN, $F_2 = 2$ kN, and $F_3 = F_4 = 3$ kN, determine the principal vector and principal moment of this force system simplified to point O in the plane and the final reduced result.

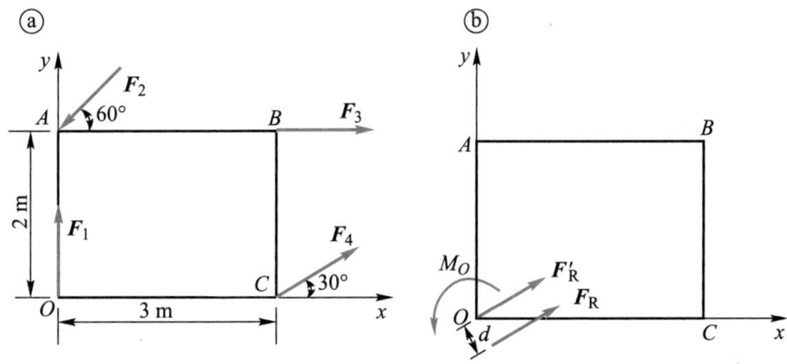

Fig.3-7

Solution Calculate the principal vector:

$$F'_{Rx}=\sum F_{ix}=-F_2\cos 60°+F_3+F_4\cos 30°=0.598\ \text{kN}$$

$$F'_{Ry}=\sum F_{iy}=F_1-F_2\sin 60°+F_4\sin 30°=0.768\ \text{kN}$$

$$F'_R=\sqrt{0.598^2+0.768^2}\ \text{kN}=0.794\ \text{kN}$$

$$\cos(\boldsymbol{F}'_R,\boldsymbol{i})=\frac{F'_{Rx}}{F'_R}=0.614,\quad \cos(\boldsymbol{F}'_R,\boldsymbol{j})=\frac{F'_{Ry}}{F'_R}=0.789$$

Calculate the principal moment:

$$M_O=\sum M_O(\boldsymbol{F}_i)=2F_2\cos 60°-2F_3+3F_4\sin 30°=0.5\ \text{kN}\cdot\text{m}$$

We can reduce the principal vector and moment to a resultant force, see Fig.3-7.

$$F_R=F'_R=0.794\ \text{kN}$$

$$d=\frac{M_O}{F_R}=\frac{0.5}{0.794}=0.63\ \text{m}$$

3.4 Equilibrium of coplanar general force system

A sufficient necessary condition for the equilibrium of a coplanar general force system is that both the principal vector and principle moment of the force system are equal to zero, i.e.

$$\boldsymbol{F}'_R=0,\quad M_O=0 \tag{3-12}$$

These equilibrium conditions can be expressed analytically. Substituting equations (3-5) and (3-8) into equation (3-12), we get

$$\left.\begin{aligned}\sum F_{ix}&=0\\ \sum F_{iy}&=0\\ \sum M_O(\boldsymbol{F}_i)&=0\end{aligned}\right\} \tag{3-13}$$

Equation (3-13) is the basic form of the equilibrium equation of the coplanar general force system. It has two projection equations and one moment equation, there are three independent equilibrium equations, and three unknown quantities can be solved.

The equilibrium equation of a coplanar general force system has two alternative sets, which are equivalent to the basic form of the equilibrium equation, as discussed below.

1. Two-moment equilibrium equation

$$\left.\begin{aligned}\sum F_{ix}&=0\\ \sum M_A(\boldsymbol{F}_i)&=0\\ \sum M_B(\boldsymbol{F}_i)&=0\end{aligned}\right\} \tag{3-14}$$

where A and B are any two points in the plane and the line between A and B cannot be perpendicular to the x-axis.

2. Three moment equilibrium equation

$$\left.\begin{aligned}\sum M_A(\boldsymbol{F}_i) = 0\\ \sum M_B(\boldsymbol{F}_i) = 0\\ \sum M_C(\boldsymbol{F}_i) = 0\end{aligned}\right\} \tag{3-15}$$

where A, B and C are any three points in the plane that do not lie on the same line.

These three different forms of equilibrium equations for the coplanar general force system can be appropriately selected according to the specific conditions when solving specific practical problems.

3.5 Reduction and equilibrium of coplanar parallel force system

A force system in which the action lines of all forces are in the same plane and parallel to each other is called a coplanar parallel force system. It is a special case of the coplanar general force system. Based on the reduction and equilibrium equations of the coplanar general force system, the reduction and equilibrium equations of the coplanar parallel force system are given below.

As shown in Fig.3-8, there is a coplanar parallel force system ($\boldsymbol{F}_1$, $\boldsymbol{F}_2$, $\cdots$, $\boldsymbol{F}_n$) acting on the rigid body, and the points of action of each force are A_1, A_2, $\cdots$, A_n, and the force system is reduced to any point in the plane. Take a point O in the plane as the center of simplification, and establish a rectangular coordinate system with point O as the origin, and make the y-axis parallel to the line of action of each force. Based on the reduction method of coplanar general force system, it is known that the coplanar parallel force system can be reduced to O point to obtain a force and couple, this force passes through the simplified center, its force vector is equal to the principal vector of the force system, from equation (3-4) and equation (3-5) have

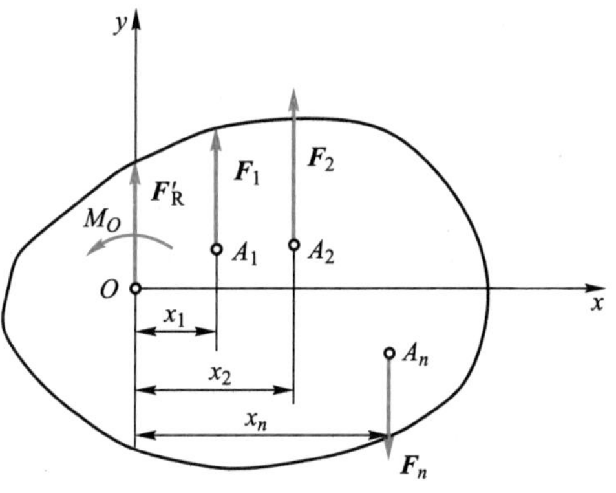

Fig.3-8

$$\boldsymbol{F}'_{\mathrm{R}} = F'_{\mathrm{R}x}\boldsymbol{i} + F'_{\mathrm{R}y}\boldsymbol{j}$$

$$\left.\begin{aligned}F'_{\mathrm{R}x} = \sum F_{ix}\\ F'_{\mathrm{R}y} = \sum F_{iy}\end{aligned}\right\}$$

Obviously the magnitude of this force F'_{R} is equal to the algebraic sum of the projections of each force on the y-axis, the direction is parallel to the direction of each force, and the pointing is determined by the sign of the algebraic sum, i.e.

$$F'_{\mathrm{R}y} = \sum F_{iy}$$

The couple moment is calculated by equation (3-8) and is equal to the algebraic sum of the moments of each force about point O, i.e.

$$M_O = \sum M_O(\boldsymbol{F}_i) = F_1x_1 + F_2x_2 + \cdots + F_nx_n = \sum F_ix_i$$

where F_i is the algebraic value of $\boldsymbol{F}_i$; x_i is the horizontal coordinate of the action point A_i of force $\boldsymbol{F}_i$.

If both the principal vector and principal moment are not equal to zero, the force system can further reduce to a resultant force $\boldsymbol{F}_{\mathrm{R}}$, whose magnitude and direction should be the same as $\boldsymbol{F}'_{\mathrm{R}}$, but the action line of the resultant force $\boldsymbol{F}_{\mathrm{R}}$ does not pass through the simplified center, but deviates from a distance d, i.e.

$$d = \frac{M_O}{F'_{\mathrm{R}}} = \frac{\sum F_ix_i}{\sum F_i}$$

When both the principal vector and the principal moment of the plane parallel force system are equal to zero, the force system is in equilibrium, and the equilibrium equation of the coplanar parallel force system is

$$\sum F_{iy} = 0, \quad \sum M_O(\boldsymbol{F}_i) = 0 \tag{3-16}$$

A sufficient necessary condition for the equilibrium of a coplanar parallel force system is that the algebraic sum of all the forces in the force system is equal to zero, and that the algebraic sum of the moments of each force about any point in the plane is equal to zero.

The equilibrium equation of the coplanar parallel force system can also be expressed using another set, that is

$$\left.\begin{aligned}\sum M_A(\boldsymbol{F}_i) = 0\\ \sum M_B(\boldsymbol{F}_i) = 0\end{aligned}\right\} \tag{3-17}$$

In these two equations, the line AB can not be parallel to the action line of the forces. The coplanar parallel force system has two independent equilibrium equations, which can be solved for two unknown forces.

The horizontal beam AB, as shown in Fig. 3 – 9, is exerted a triangularly distributed load with a maximum set of loads q, and its length is l.

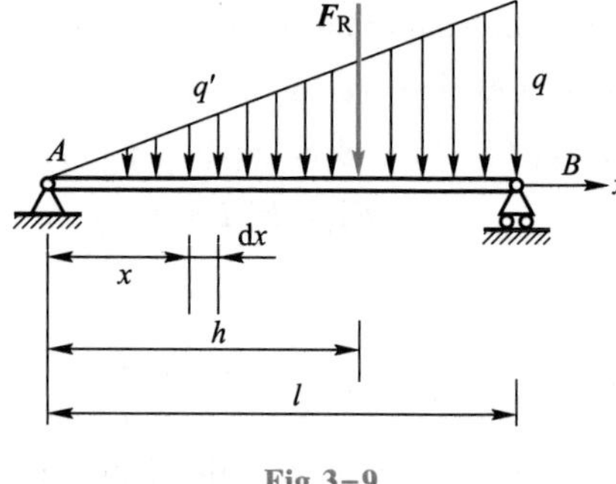

Fig.3-9

For the micro-segment dx of the beam, the magnitude of the force is $q'\mathrm{d}x$, where q' is the set of loads at that point, and according to the similar triangle relationship, we can get

$$q' = \frac{x}{l}q \tag{3-18}$$

Therefore, the magnitude of the resultant force of the distributed load is

$$F_{\mathrm{R}} = \int_0^l q'\mathrm{d}x = \frac{1}{2}ql \tag{3-19}$$

The distance between the action line of the resultant force F_R and the end A is h. According to the resultant moment theorem, we have $F_R h = \int_0^l q'x\mathrm{d}x$. Substituting the values of q' and F_R, we can get

$$h = \frac{2}{3}l \tag{3-20}$$

Example 3-3 A horizontal homogeneous beam as shown in Fig.3-10a is supported at A on a fixed-hinge support and at C on a roller support. A force $\boldsymbol{F}_P$ is applied at the end D of the beam, and a uniform load q is applied at the section AC of the beam. $F_P = 2qa$. Determine the reaction force at fixed-hinge support A and roller support C.

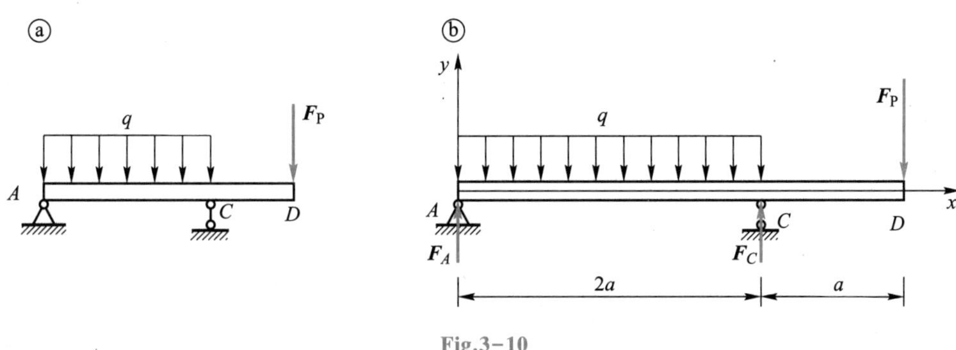

Fig.3-10

Solution Since all the forces acting on the beam are in the xy plane and the action lines of the forces are parallel to the y-axis, the action lines of the reaction forces at A and C are also parallel to the y-axis, as shown in Fig.3-10b. For this system, two equilibrium equations can be presented. There are two unknown forces $\boldsymbol{F}_A$ and $\boldsymbol{F}_C$, the resultant force of the uniform forces is $2qa$, acting at the middle of the AC section.

Firstly, the moment equilibrium equation about point A is listed, and $\boldsymbol{F}_C$ can be determined directly. Second, the projection equation on the y-axis is used to determine the $\boldsymbol{F}_A$.

$$\sum M_A(\boldsymbol{F}_i) = 0, \quad F_C \times 2a - q \times 2a \times a - F_P \times 3a = 0$$

$$F_C = 4qa$$

$$\sum F_{iy} = 0, \quad F_A + F_C - q \times 2a - F_P = 0$$

$$F_A = 0$$

Example 3-4 As shown in Fig.3-11a, the curved bar A is supported by a fixed-hinge, and the end B is a smooth surface support. Try to determine the reaction force at point A.

Solution The force analysis of the crank is shown in Fig.3-11b, the reaction force $\boldsymbol{F}_B$ at B is perpendicular to AB, the two perpendicular forces at A is $\boldsymbol{F}_{Ax}$ and $\boldsymbol{F}_{Ay}$, respectively. The moment equilibrium equation about point A can be used to directly solve $\boldsymbol{F}_B$, that is

$$\sum M_A(\boldsymbol{F}_i) = 0, \quad -90\ \mathrm{N\cdot m} - 60\ \mathrm{N} \times 1\ \mathrm{m} + F_B \times 0.75\ \mathrm{m} = 0$$

$$F_B = 200\ \mathrm{N}$$

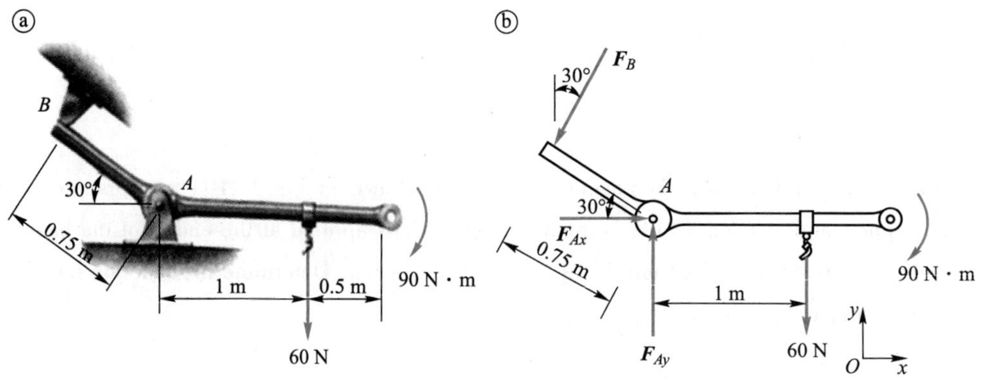

Fig.3-11

The equilibrium equation along x and y direction can written as

$$\sum F_{ix} = 0, \quad F_{Ax} - F_B \sin 30° = 0$$

$$F_{Ax} = 100 \text{ N}$$

$$\sum F_{iy} = 0, \quad F_{Ay} - F_B \cos 30° - 60 \text{ N} = 0$$

$$F_{Ay} = 233 \text{ N}$$

Example 3-5 The cantilever beam shown in Fig.3-12a has a fixed-end support at A, a force $\boldsymbol{F}$ and a couple with moment M at point B, and a uniform load with load set q. Determine the reaction force at the fixed-end A.

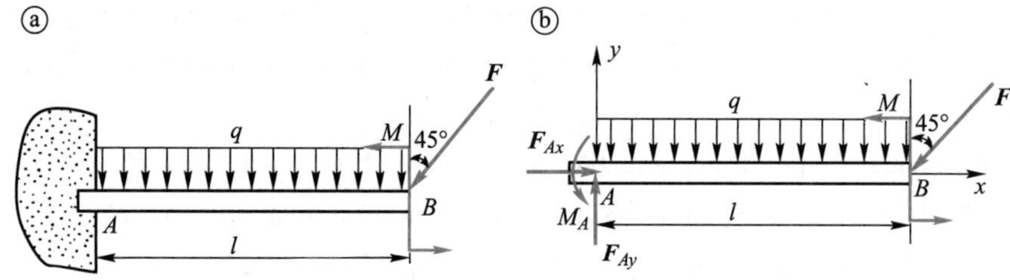

Fig.3-12

Solution The force analysis of the beam is shown in Fig.3-12b. For the fixed-end support, there are two perpendicular components and one couple, namely $\boldsymbol{F}_{Ax}$, $\boldsymbol{F}_{Ay}$ and M_A.

Build the coordinate system as shown in Fig.3-12b, and list the projection equation along the x-axis direction, so that the equation contains only one unknown force F_{Ax}; and then, list the projection equation along the y-axis direction and the equation of moment about point A, that is

$$\sum F_{ix} = 0, \quad F_{Ax} - F\cos 45° = 0$$

$$F_{Ax} = F\cos 45° = 0.707F$$

$$\sum F_{iy} = 0, \quad F_{Ay} - ql - F\sin 45° = 0$$

$$F_{Ay} = ql + 0.707F$$

$$\sum M_A(\boldsymbol{F}_i) = 0, \quad M_A - ql \times \frac{l}{2} - F\cos 45^\circ \times l + M = 0$$

$$M_A = \frac{1}{2}ql^2 + 0.707Fl - M$$

3.6 Equilibrium of composite bodies

A system consisting of two or more rigid bodies interacting with each other is called composite bodies.

When a composite body is in equilibrium, every body of the body system is also in equilibrium. Therefore, when studying the equilibrium of a composite body, the whole composite body can be taken as the object of study, and each body composed of the composite body can also be taken as the object of study. The object of study is selected to be analyzed in terms of known and unknown forces.

When performing force analysis of a composite body, it is important to distinguish between internal and external forces. A system composed of two or more bodies, the interaction between the bodies is called internal forces in the system. Since the internal forces always appear in pairs, equal in magnitude and opposite in direction, the lines of action coincide, action and reaction forces, therefore, the sum of their projections on any axis and the moment about any point are equal to zero, respectively. When taking the whole or part as the object of study, the internal force need not be drawn in the free-body diagram, and the external force must all be drawn.

It must be noted that the division between internal and external forces is not absolute, but for a certain object of study. For example, a force is internal for the whole system, but becomes external for some object.

The following is an example about how to solve the equilibrium problem of composite bodies.

Example 3-6 C is a fixed-end support, B is a pin support, and A is a roller support, as shown in Fig.3-13a. It is known that the uniform load set is $q = 15$ kN/m and the moment of couple is $M = 20$ kN · m. Determine the reaction forces at points A, C and B.

Solution The system consists of two rigid bodies. Based on the force analysis of the whole structure, we can know that there are four unknown forces ($\boldsymbol{F}_A$, $\boldsymbol{F}_{Cx}$, $\boldsymbol{F}_{Cy}$ and M_C), but only three equilibrium equations, so you can not determine these four unknown forces.

Comparing beams AB and BC, the free-body diagrams are shown in Fig.3-13b and Fig.3-13c, respectively. It can be seen that there are five unknown forces on beam BC and only three unknown forces on beam AB, and the force system is coplanar general force system. Therefore, the beam AB should be chosen as the object of study, and the equation for taking the moment about point B should be listed to determine $\boldsymbol{F}_A$ firstly. The second object of study can be either beam BC or whole. If the beam BC is chosen as the object of study, F'_{Bx} and F'_{By} need to be solved firstly; if the whole object of study is chosen, the whole structure has only three unknown forces, which can all be solved.

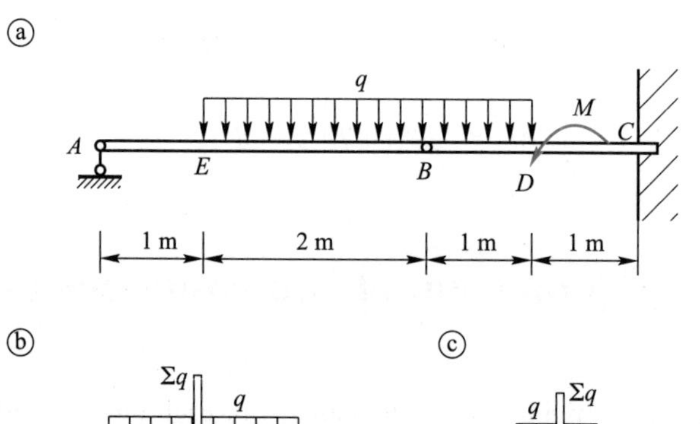

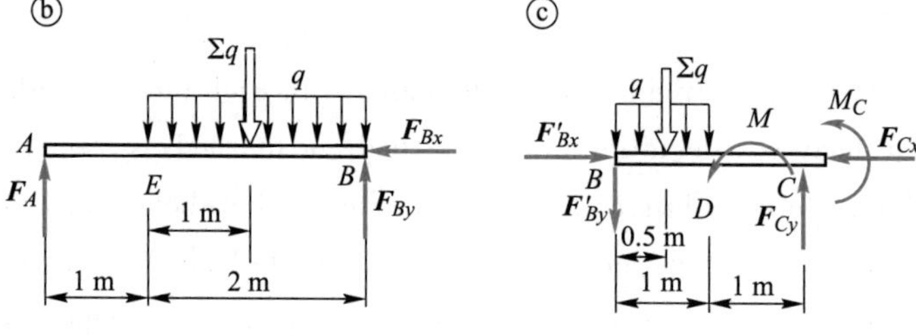

Fig.3-13

(1) Firstly, select the beam AB as the object of study and draw the free-body diagram as shown in Fig.3-13b. The moment equilibrium equation is

$$\sum M_B(\boldsymbol{F}_i) = 0\ ,\quad F_A\times 3\ \mathrm{m} - q\times 2\ \mathrm{m}\times 1\ \mathrm{m} = 0 \tag{3-21}$$

From equation (3-21), we have

$$F_A = 10\ \mathrm{kN}$$

From $\sum F_{ix} = 0$, we have

$$F_{Bx} = 0$$

From $\sum F_{iy} = 0$, we have

$$\sum F_{iy} = 0\ ,\quad F_A + F_{By} - q\times 2\ \mathrm{m} = 0 \tag{3-22}$$

From equation (3-22), we have

$$F_{By} = 20\ \mathrm{kN}$$

(2) For the beam BC, its free-body diagram is shown in Fig.3-13c. Note that $\boldsymbol{F}'_{Bx}$, $\boldsymbol{F}'_{By}$ and $\boldsymbol{F}_{Bx}$, $\boldsymbol{F}_{By}$ are the action and reaction forces, equal in magnitude and opposite in direction, i.e., $F'_{Bx} = F_{Bx}$, $F'_{By} = F_{By}$.

The moment equilibrium equation is

$$\sum M_C(\boldsymbol{F}_i) = 0,\quad M_C + M + F'_{By}\times 2\ \mathrm{m} + q\times 1\ \mathrm{m}\times 1.5\ \mathrm{m} = 0 \tag{3-23}$$

From equation (3-23), we have

$$M_C = -82.5\ \mathrm{kN\cdot m}$$

From $\sum F_{ix} = 0$, we have

$$\sum F_{ix} = 0,\quad F'_{Bx} - F_{Cx} = 0 \tag{3-24}$$

The sign "–" indicates that the actual direction of M_C is opposite to the direction shown in the figure. From equation (3-24), we have

$$F_{Cx}=0$$

From $\sum F_{iy}=0$, we have

$$\sum F_{iy}=0, \quad F_{Cy}-F'_{By}-q\times 1\ \mathrm{m}=0 \tag{3-25}$$

$$F_{Cy}=35\ \mathrm{kN}$$

Example 3-7 As shown in Fig.3-14a, a frame structure is formed by the bars AB, BC and CO. The weight of the object is $G=12$ kN, which is connected to point E on the wall by a rope around a pulley, $AD=BD=2$ m, $CD=DO=1.5$ m, the weight of the bar and pulley are negligible, try to calculate the reaction force at A, B and the internal force of the bar BC.

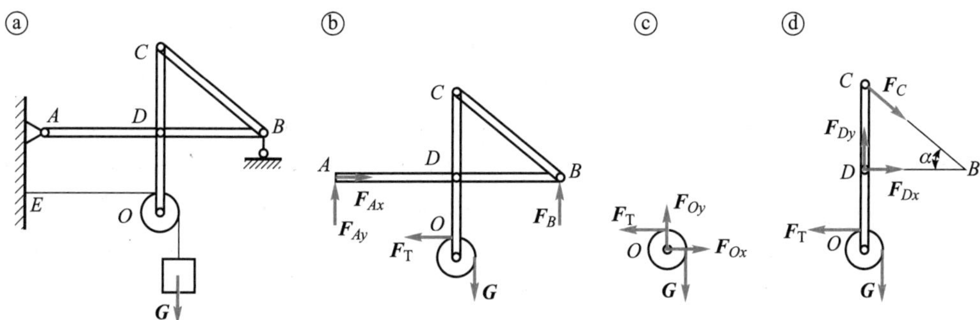

Fig.3-14

Solution The force analysis of the whole structure is shown in Fig.3-14b. There are five forces: the object gravity $\boldsymbol{G}$ and four unknown forces ($\boldsymbol{F}_{Ax}$, $\boldsymbol{F}_{Ay}$, $\boldsymbol{F}_B$, $\boldsymbol{F}_T$). For the coplanar general force system, there are only three independent equilibrium equations, so the four unknown quantities cannot be determined.

Take pulley O as the object of study, the force analysis is shown in Fig.3-14c, which contains three unknown forces ($\boldsymbol{F}_T$, $\boldsymbol{F}_{Ox}$ and $\boldsymbol{F}_{Oy}$). The moment equation about point O is

$$\sum M_O(\boldsymbol{F}_i)=0, \quad F_T\cdot r-G\cdot r=0 \tag{3-26}$$

From equation (3-26), we have

$$F_T=G$$

Now there are only three unknown forces, which can be determined by three independent equilibrium equations, namely

$$\sum M_A(\boldsymbol{F}_i)=0, \quad 2\cdot AD\cdot F_B-AD\cdot G-OD\cdot F_T=0 \tag{3-27}$$

$$\sum F_{ix}=0, \quad F_{Ax}-F_T=0 \tag{3-28}$$

$$\sum F_{iy}=0, \quad F_{Ay}+F_B-G=0 \tag{3-29}$$

From equation (3-27), we have

$$F_B=\frac{(AD+OD)G}{2\cdot AD}=10.5\ \mathrm{kN}$$

From equation (3−28), we have

$$F_{Ax}=F_{T}=12\ \text{kN}$$

Substituting F_B into equation (3−29) yields

$$F_{Ay}=G-F_{B}=1.5\ \text{kN}$$

For the internal force of the BC bar, note that the bar BC is a two-force member, take the bar OC and the pulley O as the object of study, force analysis is shown in Fig.3−14d. The action line of force F_C is along the CB line, as shown in Fig.3−14d, take the moment about point D, that is

$$\sum M_D(\boldsymbol{F}_i)=0,\quad F_{T}\cdot OD+F_C\cdot CD\cdot\cos\alpha=0 \tag{3-30}$$

From equation (3−30), we have

$$F_C=-15\ \text{kN}$$

The actual direction is opposite to the assumed direction.

Example 3−8 A combined beam consists of the beam AC and the beam CD (Fig.3−15a). A fixed end supports at A point, a pin supports at the middle C, and a two-force member supports at B with an angle of 60° to the horizontal. A couple with a moment of $M=20$ kN · m acts on the beam AC, and a force $\boldsymbol{F}$ at point D is at an angle of 30° to the plumb line. The combined beam is subjected to a uniform load with a load set of $q=10$ kN/m and $l=1$ m. Try to determine all the forces acting on the combined beam.

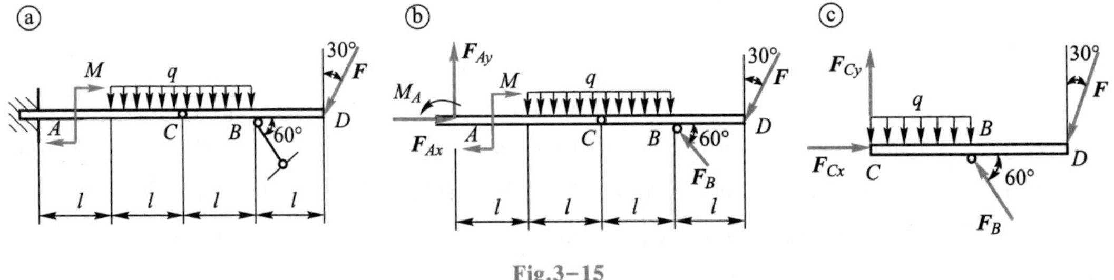

Fig.3−15

Solution Taking the whole as the object of study, the force analysis is shown in Fig.3−15b, containing four unknown forces ($\boldsymbol{F}_{Ax}$, $\boldsymbol{F}_{Ay}$, M_A and $\boldsymbol{F}_B$), and three equilibrium equations are listed, namely

$$\sum F_{ix}=0,\quad F_{Ax}-F_B\cos 60^\circ-F\sin 30^\circ=0 \tag{3-31}$$

$$\sum F_{iy}=0,\quad F_{Ay}+F_B\sin 60^\circ-2ql-F\cos 30^\circ=0 \tag{3-32}$$

$$\sum M_A(\boldsymbol{F}_i)=0,\quad M_A-M-2ql\times 2l+F_B\sin 60^\circ\times 3l-F\cos 30^\circ\times 4l=0 \tag{3-33}$$

From these three equations, we can not determine the four unknown forces. And then take the bar CD as the object of study, force analysis is shown in Fig.3−15c. To take the moment about point C, the equation does not contain the unknown force $\boldsymbol{F}_{Cx}$, $\boldsymbol{F}_{Cy}$, that is

$$\sum M_C(\boldsymbol{F}_i)=0,\quad F_B\sin 60^\circ\times l-ql\times\frac{l}{2}-F\cos 30^\circ\times 2l=0 \tag{3-34}$$

Combining equations (3−31) ~ (3−34), the solution is

$$F_B=45.77\ \text{kN},\quad F_{Ax}=32.89\ \text{kN},\quad F_{Ay}=-2.32\ \text{kN},\quad M_A=10.37\ \text{kN}\cdot\text{m}$$

Example 3-9 As shown in Fig.3-16a, a structure consisting of the bars AB, DC and pulley D, A and C are fixed-hinge support, and B is a pin support. The weight of the object is $\boldsymbol{G}$, which connected to point E on the bar AB through the rope around the pulley, ignoring the weight of all components, try to calculate the reaction force at B.

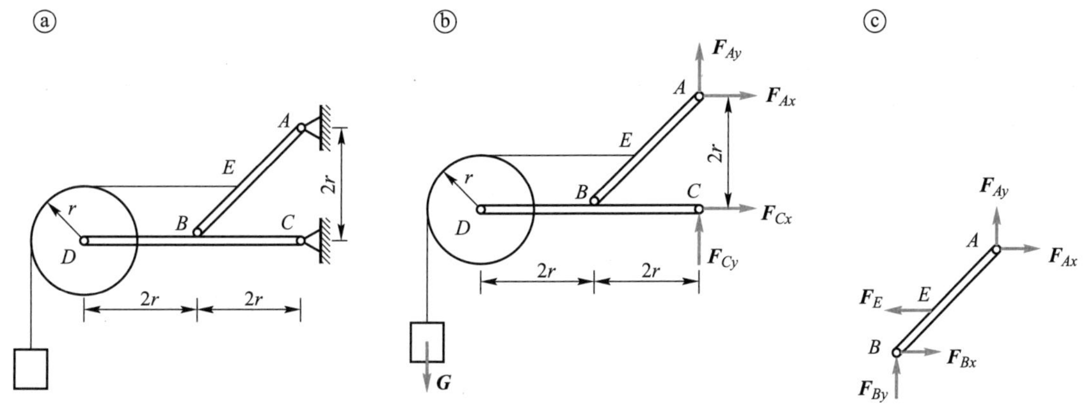

Fig.3-16

Solution Taking the whole as the object of study, the force analysis is shown in Fig.3-16b, containing four unknown forces, and the magnitude of $\boldsymbol{F}_{Ax}$ is obtained by taking the moment equation about point C, that is

$$\sum M_C(\boldsymbol{F}_i) = 0, \quad 5r \times G - 2r \times F_{Ax} = 0 \tag{3-35}$$

$$F_{Ax} = 2.5G$$

Then take the bar AB as the object of study, force analysis is shown in Fig.3-16c, list two equilibrium equations, that is

$$\sum F_{ix} = 0, \quad F_{Ax} + F_{Bx} - F_E = 0 \tag{3-36}$$

$$\sum M_A(\boldsymbol{F}) = 0, \quad 2r \times F_{Bx} - 2r \times F_{By} - r \times F_E = 0 \tag{3-37}$$

Solving equation (3-36) and equation (3-37), we get

$$F_{Bx} = -1.5G, \quad F_{By} = -2G$$

The actual direction is opposite to the assumed direction.

3.7 Coplanar truss analysis

A truss is an engineering structure composed of some bars joined together at their end points, and its structure remains unchanged after being subjected to forces. A truss in which the bars are in the same plane is called a coplanar truss. The connection points of the bars in a truss are called joints. For example, roof frames of houses, arches of bridges, crane cantilevers and high-voltage transmission line towers are all truss structures. The plant roof frame shown in Fig.3-17 is a typical truss. Truss structure has wide application in engineering, mainly because it has the characteristics of economical and reasonable use of materials and small mass.

ⓐ ⓑ

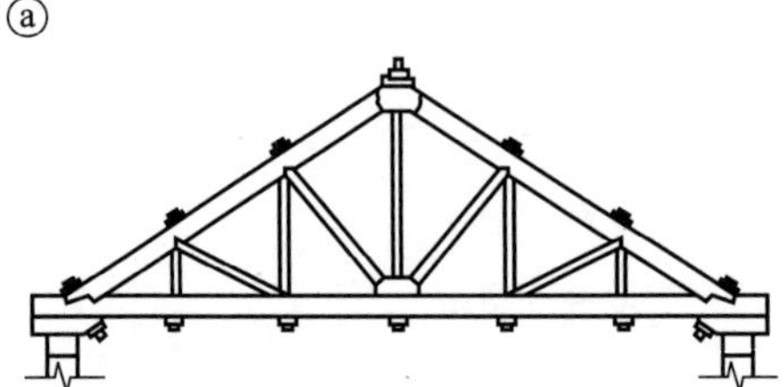

Fig.3-17

The following is the basic method of calculating the internal force of coplanar truss. In engineering, the following basic assumptions are made for the calculation of coplanar trusses in order to simplify the calculation.

① All bar weights are negligible or equally distributed at the joints at the ends of the bars. Each bar can be regarded as a two-force member.

② All the bars are connected with ideal smooth pins.

③ All external forces act on the joints and are located in the plane of the truss.

Based on the above assumptions, the truss shown in Fig. 3 - 17a can be simplified to the mechanical model shown in Fig.3-17b. The truss satisfying the above assumptions is called ideal truss.

The three basic assumptions have proven to induce little error in the calculation of the internal forces of the coplanar truss and to be safe, and make the calculation of the internal forces much simpler.

There are two methods for calculating the internal force of each bar in a truss, the joint method and the section method.

1. Joint method

From the basic assumption of truss calculation, it is known that each joint in the truss is subjected to the action of coplanar concurrent force system, and the internal force of the bar can be determined by listing the equilibrium equation of each joint in turn. This method is called the joint method.

2. Section method

An imaginary section is used to cut the truss into two parts. The external force on this part and the internal force of the truncated bar form a coplanar general force system, and the equilibrium equation is listed, so that the internal force of the truncated bar can be determined. This method is called the section method.

When applying the joint method and section method to solve the internal force of the truss, it is generally necessary to find out the reaction force of the truss firstly. For the joint method, it is necessary to start from a joint containing only two unknown forces; for the section method, try to select a section that, in general, passes through not more than three members in which the forces are unknown. Usually each bar is assumed to be subject to tension, if the calculated internal force of the bar is positive then it is a tension bar, and vice versa for the compression bar. Moreover, the line of action of each member force is specified from the geometry of the truss, since the force in a member passes along its axis. Also, the member forces acting on one part of the truss are equal but opposite to those acting on the other part.

The application of the joint and section methods is illustrated by the following examples.

Example 3-10 As shown in Fig.3-18a, the internal force of each bar of the truss is calculated by the joint method, $F = 10$ kN.

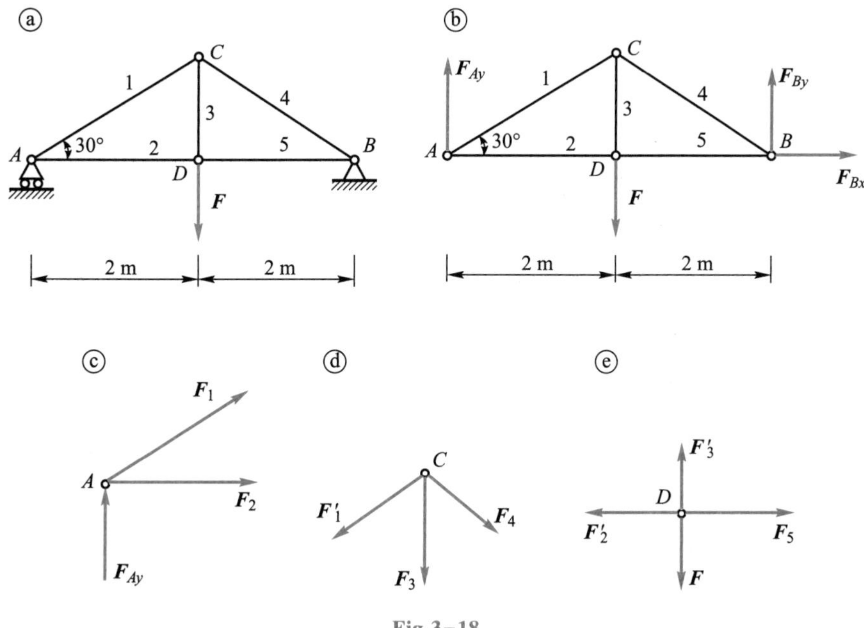

Fig.3-18

Solution Taking the whole as the object of study, the force analysis is shown in Fig.3-18b, and three equilibrium equations are listed to determine the three unknown forces $\boldsymbol{F}_{Ay}$, $\boldsymbol{F}_{Bx}$ and $\boldsymbol{F}_{By}$, i.e.

$$\sum F_{ix} = 0, \quad F_{Bx} = 0$$

$$\sum M_A(\boldsymbol{F}_i) = 0, \quad F_{By} \times 4 \text{ m} - F \times 2 \text{ m} = 0$$

$$F_{By} = 5 \text{ kN}$$

$$\sum M_B(\boldsymbol{F}_i) = 0, \quad F \times 2 \text{ m} - F_{Ay} \times 4 \text{ m} = 0$$

$$F_{Ay} = 5 \text{ kN}$$

The internal force of each bar will be calculated, usually assume that each bar is subject to tension, if the calculation result is positive, the bar is subject to tension; if negative, the bar is subject to compression. The force system acting on each joint is a coplanar concurrent force system.

Taking pin A as the object of study, the force analysis is shown in Fig.3-18c, which contains two unknown forces $\boldsymbol{F}_1$ and $\boldsymbol{F}_2$. Equilibrium equations can be listed to solve these two forces, namely

$$\sum F_{ix} = 0, \quad F_2 + F_1 \cos 30° = 0$$

$$\sum F_{iy}=0,\quad F_{Ay}+F_1\sin 30^\circ=0$$

The solution is

$$F_1=-10\ \text{kN},\quad F_2=8.66\ \text{kN}$$

Then take the pin C as the object of study, the force analysis as shown in Fig.3-18d, including the forces $\boldsymbol{F}_1'$, $\boldsymbol{F}_3$ and $\boldsymbol{F}_4$, and $\boldsymbol{F}_1'$ has been determined, that is, $F_1'=F_1=-10$ kN. Two equilibrium equations, namely

$$\sum F_{ix}=0,\quad F_4\cos 30^\circ-F_1'\cos 30^\circ=0$$

$$\sum F_{iy}=0,\quad -F_3-(F_1'+F_4)\sin 30^\circ=0$$

Solving the above two equations

$$F_3=10\ \text{kN},\quad F_4=-10\ \text{kN}$$

The pin D is the object of study, and the force analysis is shown in Fig.3-18e. Here, $\boldsymbol{F}_2'$ and $\boldsymbol{F}_3'$ have been determined before, that is, $F_2'=F_2$, $F_3'=F_3$. Equilibrium equation is listed, that is

$$\sum F_{ix}=0,\quad F_5-F_2'=0$$

The solution is

$$F_5=8.66\ \text{kN}$$

Example 3-11 Calculate the internal forces of the bar FE, CE and CD shown in Fig.3-19a by the section method, where the force $F_C=4$ kN and $F_E=2$ kN.

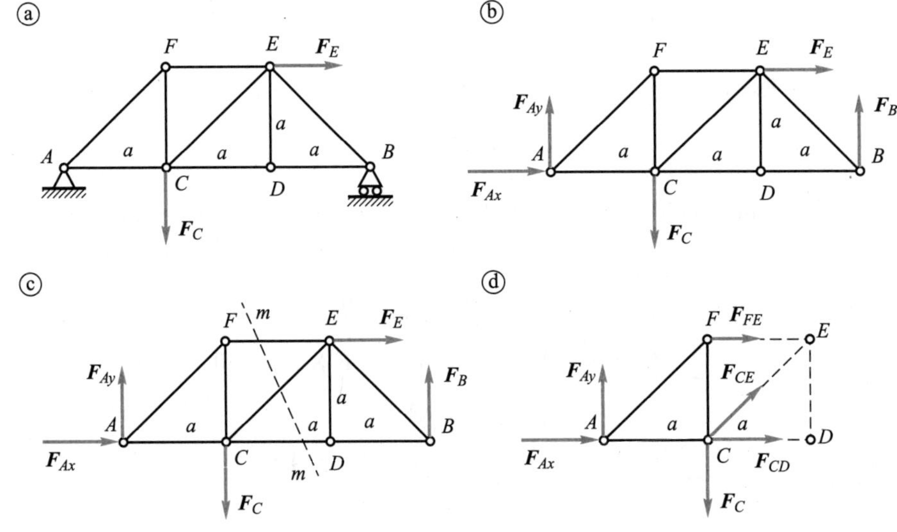

Fig.3-19

Solution Calculating the internal force of a bar by the section method is a relatively simple method. The internal force of the truncated bar becomes the external force, the force system acting on the truncated part of the truss is the coplanar general force system, through the three equilibrium equations, the three unknown forces can be determined.

Taking the whole as the object of study, the force analysis is shown in Fig.3-19b, for the reaction forces $\boldsymbol{F}_{Ax}$, $\boldsymbol{F}_{Ay}$ and $\boldsymbol{F}_B$, three equilibrium equations can be listed to determine, namely

$$\sum F_{ix} = 0, \quad F_{Ax} + F_E = 0$$

$$\sum F_{iy} = 0, \quad F_B + F_{Ay} - F_C = 0$$

$$\sum M_A(\boldsymbol{F}_i) = 0, \quad -F_C \times a - F_E \times a + F_B \times 3a = 0$$

Solving the above three equations

$$F_{Ax} = -2\ \text{kN}, \quad F_{Ay} = 2\ \text{kN}, \quad F_B = 2\ \text{kN}$$

Then a section m—m is used to cut the bars FE, CE and CD (Fig.3-19c), and the left part of the structure is chosen as the object of study. Force analysis as shown in Fig.3-19d, all forces form a coplanar general force system, list three equilibrium equations:

$$\sum F_{ix} = 0, \quad F_{CD} + F_{Ax} + F_{FE} + F_{CE}\cos 45° = 0$$

$$\sum F_{iy} = 0, \quad F_{Ay} - F_C + F_{CE}\cos 45° = 0$$

$$\sum M_C(\boldsymbol{F}_i) = 0, \quad -F_{FE} \times a - F_{Ay} \times a = 0$$

Solving the above three equations:

$$F_{CE} = -2\sqrt{2}\ \text{kN}, \quad F_{CD} = 2\ \text{kN}, \quad F_{FE} = -2\ \text{kN}$$

The actual direction of $\boldsymbol{F}_{CE}$ and $\boldsymbol{F}_{FE}$ are opposite to the assumed direction.

Problems

3-1 A coplanar force system consists of three forces and two couples, as shown in Fig.P3-1. Determine the resultant force $\boldsymbol{F}_R$ of this system and the coordinates of the intersection of the line of action of $\boldsymbol{F}_R$ with the x-axis.

Answer: $\boldsymbol{F}_R = (-1.5\boldsymbol{i} - 2\boldsymbol{j})$ kN, $x = 0.29$ m

3-2 As shown in Fig.P3-2, the square has side a, A is the center, B is the midpoint of the side, and $F_1 = F_2 = F_3 = F_4$. Find the reduction of this force system toward point A.

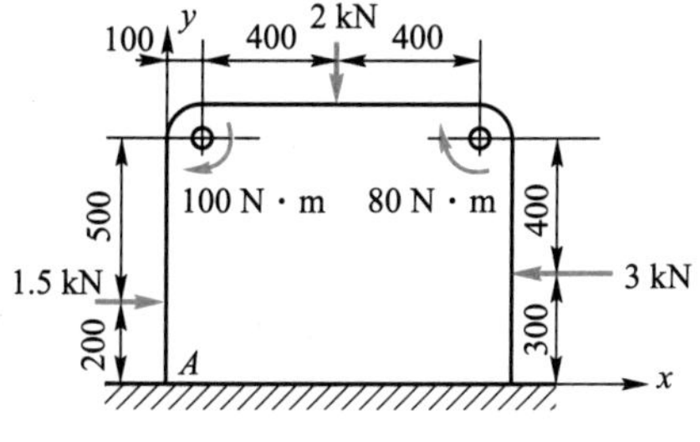

Fig.P3-1

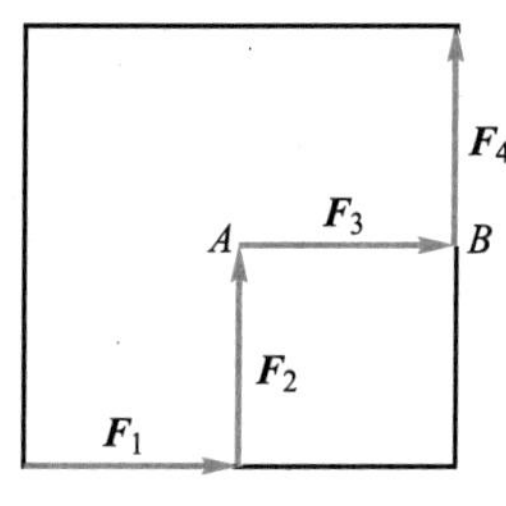

Fig.P3-2

Answer: $\boldsymbol{F}_R = 2F\boldsymbol{i} + 2F\boldsymbol{j}$, $M_A = Fa$

3-3 A is the fixed-end, the whole structure is subjected to the load shown in Fig. P3-3.

Determine the reaction force at the fixed-end A.

Answer: $F_{Ax}=-12.2$ kN, $F_{Ay}=-1.25$ kN, $M_A=16.28$ kN · m

3-4 As shown in Fig.P3-4, a couple is applied to a horizontal beam with a moment $M=60$ kN · m. A load $F_P=20$ kN is applied at point C. Determine the reaction forces of supports A and B.

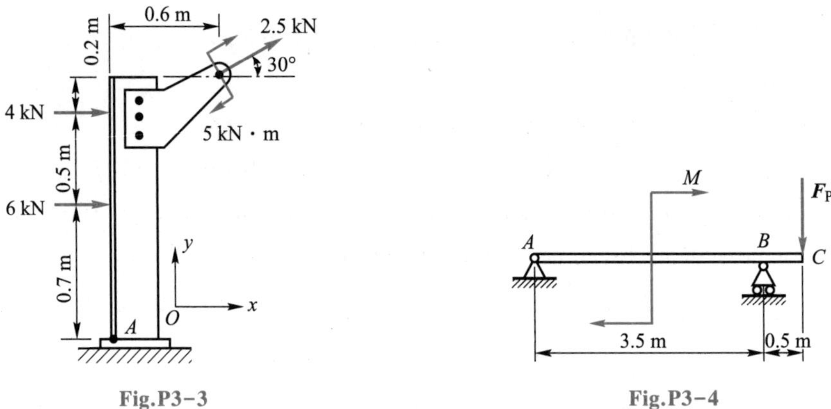

Fig.P3-3 Fig.P3-4

Answer: $F_{Ax}=0$, $F_{Ay}=-20$ kN, $F_B=40$ kN

3-5 As shown in Fig.P3-5, there is a couple ($\boldsymbol{F}$, $\boldsymbol{F}'$) acting on the horizontal beam, a uniformly distributed load with load set q on the left arm, and a load $\boldsymbol{F}_D$ at the end point of the right arm. It is known that $F=10$ kN, $F_D=20$ kN, $q=20$ kN/m, $a=0.8$ m, try to find the reaction force of supports A and B.

Answer: $F_{Ax}=0$, $F_{Ay}=15$ kN, $F_B=21$ kN

3-6 In the structure shown in Fig.P3-6, the self-weight of each member is negligible. The moment of the couple $M=800$ N · m is applied to the member AB. Determine the reaction forces at points A and C.

Answer: $F_C=2.7$ kN, along the $\overrightarrow{BC}$ direction; $F_A=2.7$ kN, along the $\overrightarrow{CB}$ direction

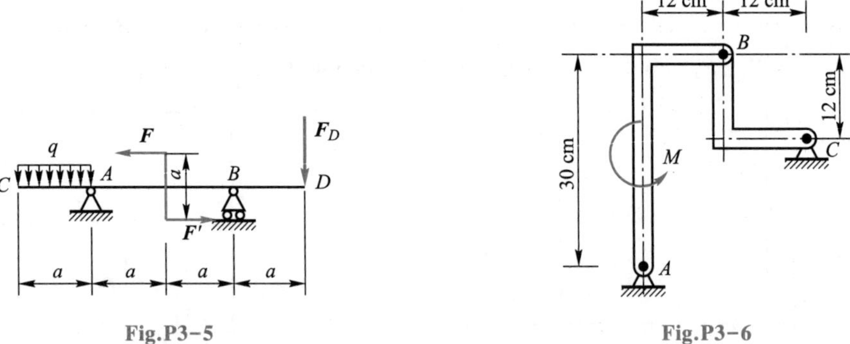

Fig.P3-5 Fig.P3-6

3-7 The structure ABC consists of the bars AB, AC and DG, as shown in Fig.P3-7.

The pin E on the bar DG slides in the slot of the bar AC. Determine the forces at points A, D and B on the bar AB when a force $\boldsymbol{F}$ is applied to the horizontal bar DG.

Answer: Point A: $F_{Ax}=F(\leftarrow)$, $F_{Ay}=F(\downarrow)$; Point D: $F_{Dx}=2F(\rightarrow)$, $F_{Dy}=F(\uparrow)$; Point B: $F_{Bx}=F(\leftarrow)$, $F_{By}=0$

3-8 Find the reaction forces at points A, B and C for each of the beams shown in Fig.P3-8, $AB=BC=a$. a, q and M are known.

Answer: (a) $F_{Ax}=0$, $F_{Ay}=2qa$, $M_A=2qa^2$, $F_B=F_C=0$

(b) $F_{Ax}=0$, $F_{Ay}=qa$, $M_A=2qa^2$, $F_{Bx}=0$, $F_{By}=qa$, $F_C=qa$

(c) $F_{Ax}=0$, $F_{Ay}=\dfrac{7}{4}qa$, $M_A=3qa^2$, $F_{Bx}=0$, $F_{By}=\dfrac{3}{4}qa$, $F_C=\dfrac{q}{4}a$

(d) $F_{Ax}=0$, $F_{Ay}=-\dfrac{M}{2a}$, $M_A=-M$, $F_{Bx}=0$, $F_{By}=-\dfrac{M}{2a}$, $F_C=\dfrac{M}{2a}$

(e) $F_{Ax}=F_{Ay}=F_{Bx}=F_{By}=F_C=0$, $M_A=-M$

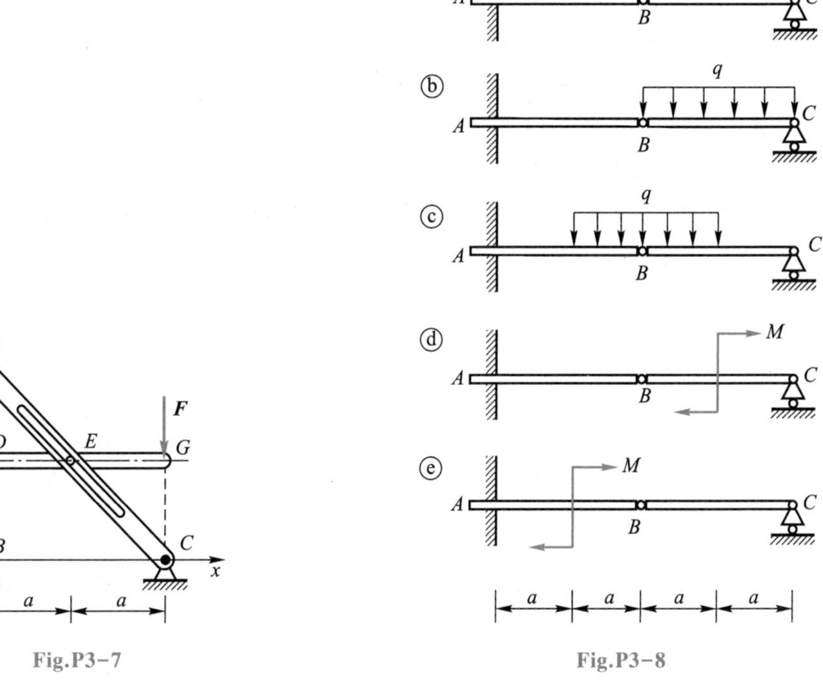

Fig.P3-7

Fig.P3-8

3-9 The system is supported as shown in Fig.P3-9.It is known that $\angle ABC=60°$, $\angle BAC=30°$, $AB=12r$, $EC=CD=2r$, the radii of pulleys D and E are r, the diameter of pulley H is r, and the object M weighs G. If the weights of the bar and pulley are negligible, determine the reaction forces at supports A and B.

Answer: $F_{Ax}=\dfrac{8\sqrt{3}+1}{24}G$, $F_{Ay}=\dfrac{18+\sqrt{3}}{24}G$, $F_{Bx}=\dfrac{8\sqrt{3}+1}{24}G$, $F_{By}=\dfrac{6-\sqrt{3}}{24}G$

3-10 Determine the internal forces of coplanar truss shown in Fig.P3-10.

Answer: $F_{CD}=-\sqrt{5}F$, $F_{ED}=2F$, $F_{EC}=F$, $F_{BC}=-2F$, $F_{AE}=\sqrt{5}F$, $F_{BE}=0$

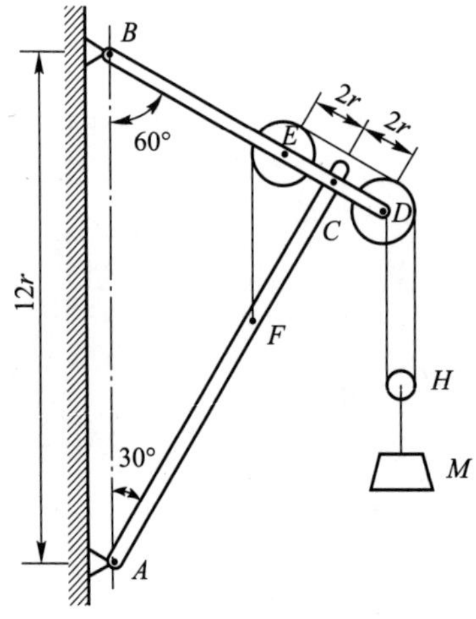

Fig.P3–9

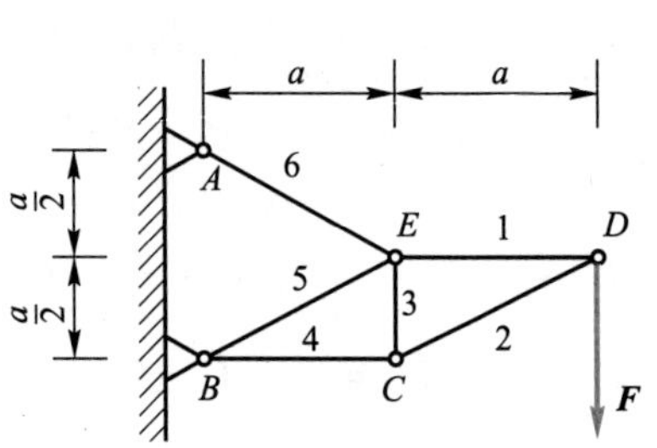

Fig.P3–10

3–11 The loads on the coplanar truss are shown in Fig.P3–11. Use the section method to determine the internal forces of member 1, member 2 and member 3.

Answer: $F_1=-5.333F$, $F_2=2F$, $F_3=-1.667F$

3–12 The support and load of a coplanar truss are shown in Fig.P3–12. *ABC* is an equilateral triangle, *E* and *F* are the midpoints of the two sides, and *AD*=*DB*. Find the internal force of the bar *CD*.

Answer: $F_{CD}=-\dfrac{\sqrt{3}}{2}F$

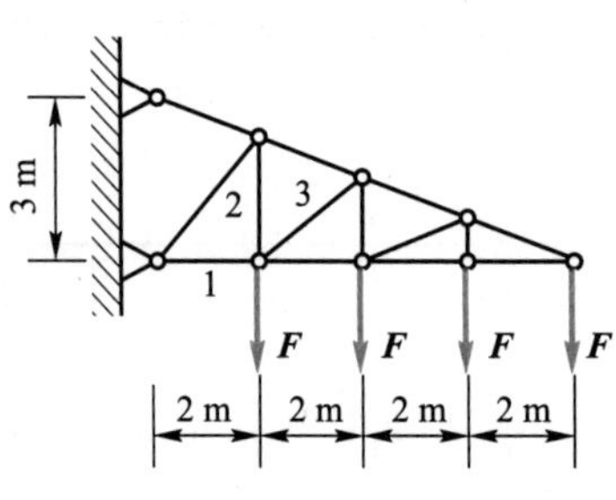

Fig.P3–11

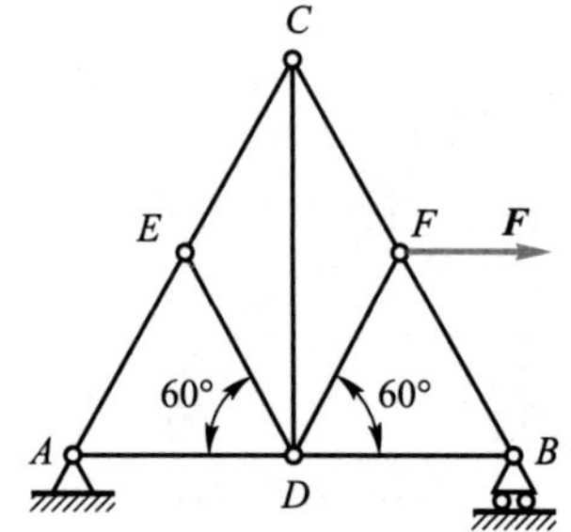

Fig.P3–12

3–13 The steel frame *ABC* and beam *CD*, supported and loaded as shown in Fig.P3–13. It is known that $F=5$ kN, $q=200$ N/m, $q_0=300$ N/m. Find the reaction forces of supports *A* and *B*.

Answer: $F_{Ax}=0.3$ kN, $F_{Ay}=-0.537$ kN, $F_{By}=3.537\ 5$ kN

3–14 As shown in Fig. P3–14, an automatic ridge breaker is supported by lever *AB* on a trapezoidal frame consisting of legs *CD* and *EF*. A force $\boldsymbol{F}_A$ is applied at the end of lever *A* to lift a stone *G* at the end *B*. If $AB=2.8$ m, $BC=0.7$ m, $CD=1.2$ m, $EH=HD=0.85$ m, $HI=HJ=0.5$ m, rope length $IJ=0.5$ m, $G=1\ 000$ N, $F_1=100$ N, $F_2=60$ N, $F_3=40$ N. $\boldsymbol{F}_1$, $\boldsymbol{F}_2$, and $\boldsymbol{F}_3$ act on the midpoints of the bars *AB*, *EH*, and *CD*, respectively. Determine the reaction force at *H*.

Answer: $F_{Hx} = 983.24$ N, $F_{Hy} = 994.49$ N

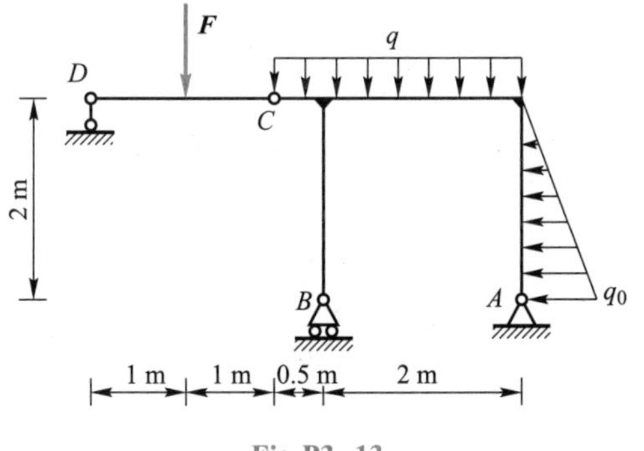

Fig.P3-13

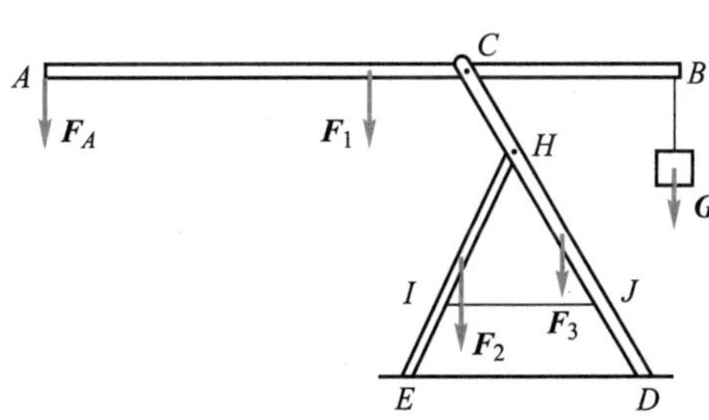

Fig.P3-14

3-15 The combined beam consisting of AC and CD is connected by the pin C, as shown in Fig.P3-15.It is known that the load strength $q = 10$ kN/m, the moment of couple is $M = 40$ kN · m and the weight of the beam is negligible. Find the reaction forces at supports A, B, C and D.

Answer: $F_{Cy} = 5$ kN, $F_D = 15$ kN, $F_{Ay} = -15$ kN, $F_B = 40$ kN

3-16 The frame are shown in Fig.P3-16, the force $F = 60$ kN, and the self-weight of the bars are negligible. Find the reaction forces at points A and E, and the internal forces of the bars BD and BC.

Answer: $F_{Ax} = 60$ kN(←), $F_{Ay} = 30$ kN(↑), $F_{Bx} = 60$ kN(→), $F_{By} = 30$ kN(↑), $F_{BD} = 30$ kN (press), $F_{BC} = 50$ kN(pull)

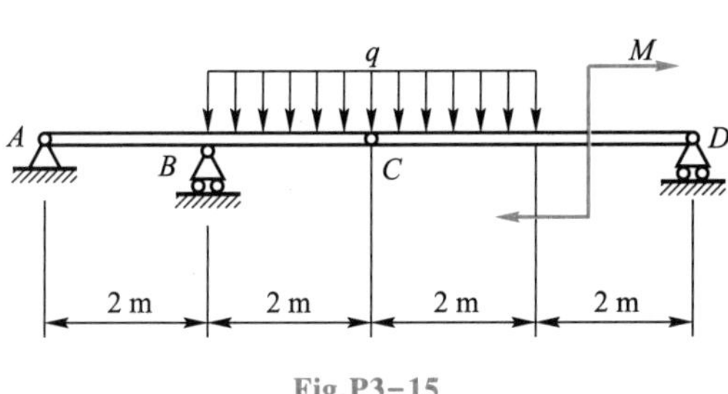

Fig.P3-15

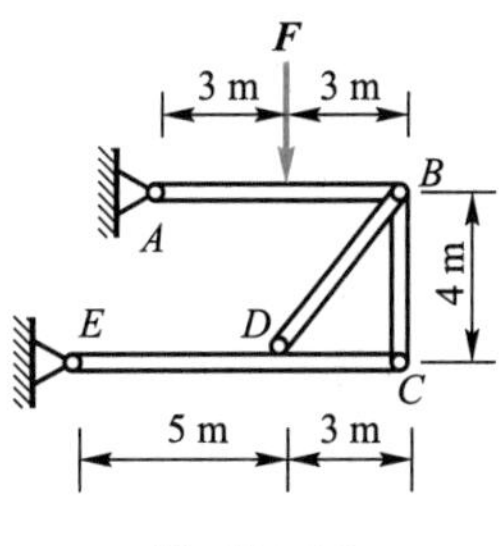

Fig.P3-16

3-17 The system shown in Fig.3-17, we know that $q = 5$ kN/m, $F = 20$ kN, and $M = 20$ kN · m. Find the reaction forces of supports A and B, and the internal forces of the bars CE and DE (bar weights are negligible, and C, D, and E are pin supports).

Answer: $F_{Ax} = -2.5$ kN, $F_{Ay} = 15$ kN, $F_{Bx} = 2.5$ kN, $F_{By} = 10$ kN, $F_{CE} = 30$ kN, $F_{DE} = -\frac{5\sqrt{2}}{2}$ kN = -3.53 kN

3-18 The block is suspended as shown in Fig.P3-18, its weight $G = 1.8$ kN, and the weight of rope, bar and pulley are negligible. Find the reaction force of point A, and the force on the bar BC.

Answer: $F_{Ax} = 2.4$ kN, $F_{Ay} = 1.2$ kN, $F_{BC} = 0.848$ kN

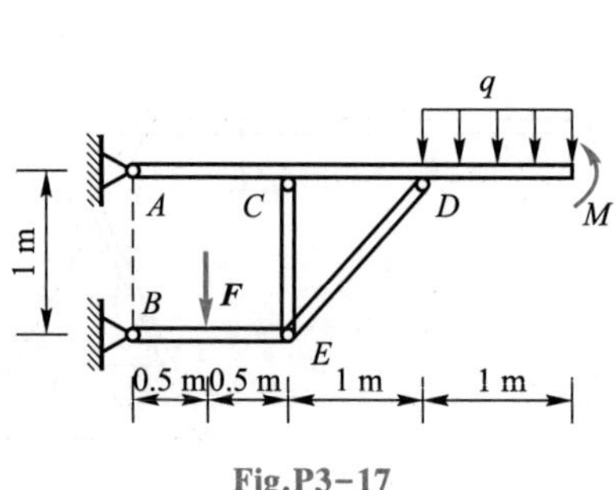

Fig.P3-17

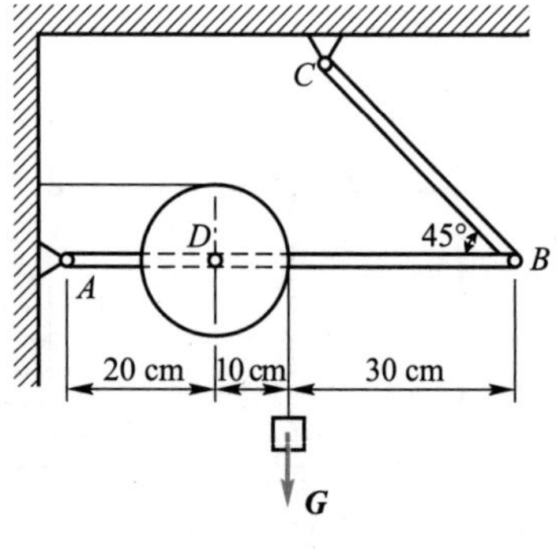

Fig.P3-18

3-19 The three-hinged arch is subjected to a uniform load q, and the structure are shown in Fig. P3-19, determine the reaction force at supports A and B (the self-weight of each member is negligible).

Answer: $F_{Ax}=F_{Bx}=\dfrac{ql^2}{8h}$, $F_{Ay}=F_{By}=\dfrac{ql}{2}$

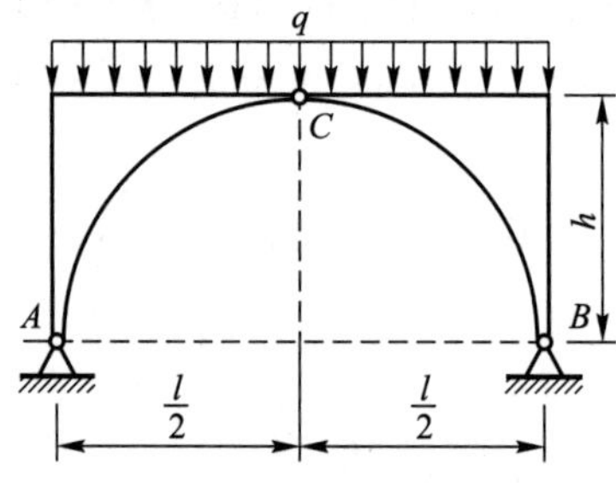

Fig.P3-19

Chapter 4 Space Force System

Teaching Scheme of Chapter 4

This chapter focuses on the reduction and equilibrium of space force system. The force system in which the lines of action of the forces acting on the object are distributed in space is called the space force system, which can be divided into the space concurrent force system, the space couple system and the space general force system.

4.1 Decomposition and projection of space force

A right-angle parallelepiped is made with the force vector $\boldsymbol{F}$ as the diagonal and its three edges are parallel to the coordinate axes, as shown in Fig.4-1.By applying the parallelogram law of force, the force $\boldsymbol{F}$ is first decomposed into $\boldsymbol{F}_{xy}$ and $\boldsymbol{F}_z$, and then $\boldsymbol{F}_{xy}$ is decomposed into $\boldsymbol{F}_x$ and $\boldsymbol{F}_y$ by applying the parallelogram law of force. As a result, the force is decomposed into three orthogonal components $\boldsymbol{F}_x$, $\boldsymbol{F}_y$, and $\boldsymbol{F}_z$ along the coordinate axes by decomposing twice, as shown in Fig.4-1a.

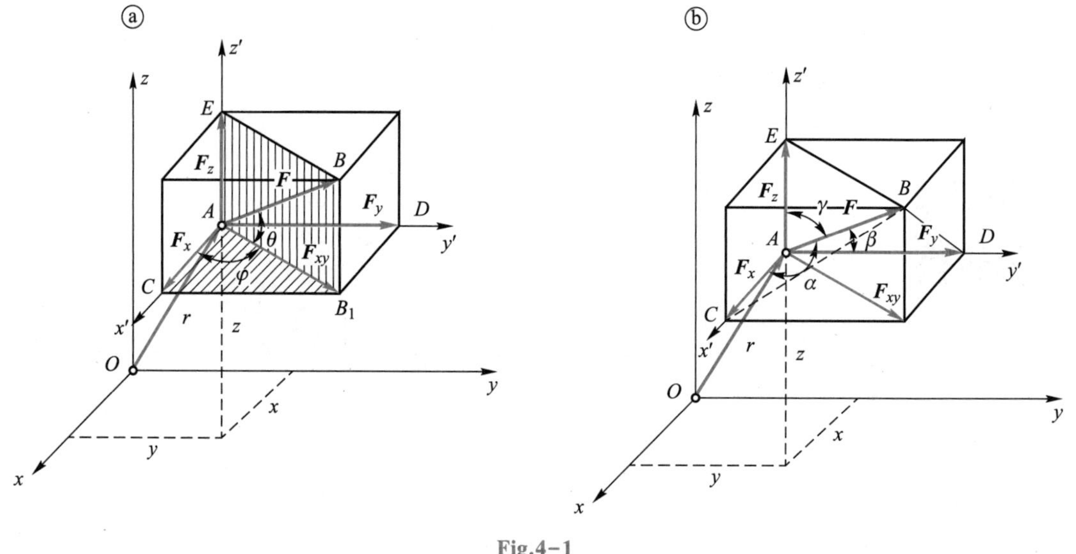

Fig.4-1

The projection of the force $\boldsymbol{F}$ on each axis of the coordinate system $Oxyz$ is denoted by F_x, F_y, and F_z, and the angles between the force and the positive direction of the coordinate axes are α, β, and γ, As shown in Fig.4-1b. Respectively, then the projection of the force on the coordinate axes is

$$F_x = F\cos\alpha, \quad F_y = F\cos\beta, \quad F_z = F\cos\gamma \tag{4-1}$$

This projection method is called the direct projection method.

The method of projecting the force onto the corresponding coordinate plane first and then onto the corresponding coordinate axis, which is called the two-step projection method. As shown in Fig.4-1a, the force $\boldsymbol{F}$ is firstly projected onto the xy-plane $\boldsymbol{F}_{xy}$ and z-axis $\boldsymbol{F}_z$, and then the $\boldsymbol{F}_{xy}$ is projected onto the x-axis and y-axis, respectively, we can obtain

$$F_x = F\cos\theta\cos\varphi,\quad F_y = F\cos\theta\sin\varphi,\quad F_z = F\sin\theta \tag{4-2}$$

It should be noted that the projection of the force on the axis is a scalar quantity, while the projection of the force on the plane is a vector quantity.

If the projections F_x, F_y, and F_z of the force $\boldsymbol{F}$ on the coordinate axes are known, the magnitude and direction of the force is

$$\left.\begin{array}{c} F = \sqrt{F_x^2 + F_y^2 + F_z^2} \\ \cos\alpha = \dfrac{F_x}{F},\quad \cos\beta = \dfrac{F_y}{F},\quad \cos\gamma = \dfrac{F_z}{F} \end{array}\right\} \tag{4-3}$$

Based on the relationship between the vector components along an axis and their projection on that axis, the analytical expression for the decomposition of the force $\boldsymbol{F}$ along a rectangular coordinate in space is

$$\boldsymbol{F} = F_x\boldsymbol{i} + F_y\boldsymbol{j} + F_z\boldsymbol{k}$$

where $\boldsymbol{i}$, $\boldsymbol{j}$, and $\boldsymbol{k}$ are unit vectors along the positive direction of each coordinate axis.

Example 4-1 The cube with side length a shown in Fig.4-2 has forces $\boldsymbol{F}_1$ and $\boldsymbol{F}_2$ at points A and B, respectively, and $F_1 = F_2 = F$. Try to calculate the projection of each force on the x, y, and z axes.

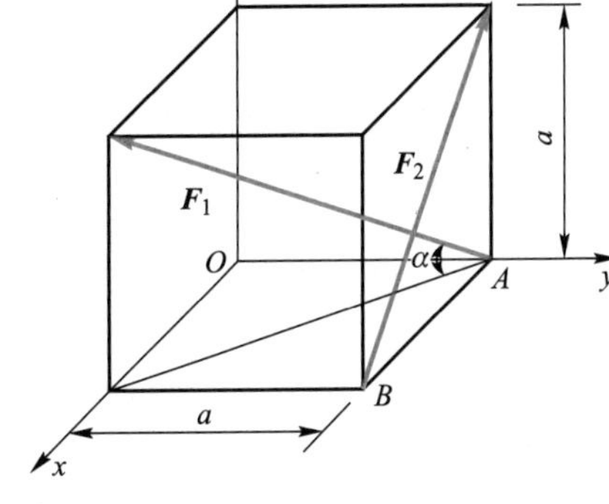

Fig.4-2

Solution Firstly, apply the two-step projection method to calculate the projection of $\boldsymbol{F}_1$ on each axis, the angle between $\boldsymbol{F}_1$ and xy plane is α. From the geometric relationship in Fig.4-2, we have

$$\cos\alpha = \frac{\sqrt{2}a}{\sqrt{3}a} = \frac{\sqrt{2}}{\sqrt{3}},\quad \sin\alpha = \frac{a}{\sqrt{3}a} = \frac{1}{\sqrt{3}}$$

The projection of $\boldsymbol{F}_1$ onto the xy plane is

$$F_{xy} = F\cos\alpha = F\frac{\sqrt{2}}{\sqrt{3}}$$

Then projecting $\boldsymbol{F}_{xy}$ to the x and y axes, we have

$$F_x = F_{xy}\cos 45^\circ = F\frac{\sqrt{2}}{\sqrt{3}}\frac{\sqrt{2}}{2} = \frac{F}{\sqrt{3}}$$

$$F_y = -F_{xy}\cos 45^\circ = -F\frac{\sqrt{2}}{\sqrt{3}}\frac{\sqrt{2}}{2} = -\frac{F}{\sqrt{3}}$$

The projection of $\boldsymbol{F}_1$ on the z-axis has

$$F_z = F\sin\alpha = \frac{F}{\sqrt{3}}$$

Then apply the direct projection method to calculate the projection of $\boldsymbol{F}_2$ on each axis, the angle between the positive direction of $\boldsymbol{F}_2$ and the positive direction of x-axis, y-axis and z-axis are 135°,

0° and 45° respectively. Then we have

$$F_x = -F\cos 45° = -\frac{\sqrt{2}F}{2}$$

$$F_y = 0$$

$$F_z = F\cos 45° = \frac{\sqrt{2}F}{2}$$

4.2 Reduction and equilibrium of space concurrent force system

As shown in Fig.4-3, there is a space concurrent force system ($\boldsymbol{F}_1$, $\boldsymbol{F}_2$, ⋯, $\boldsymbol{F}_n$) on the rigid body, the concurrent point is O. Take point O as the coordinate origin to establish a rectangular coordinate system $Oxyz$, the forces will be expressed by the analytical expression, that is

$$\boldsymbol{F}_i = F_{ix}\boldsymbol{i} + F_{iy}\boldsymbol{j} + F_{iz}\boldsymbol{k} \tag{4-4}$$

The equation (4-4) can be expressed as

$$\boldsymbol{F}_\mathrm{R} = \sum \boldsymbol{F}_i = \sum F_{ix}\boldsymbol{i} + \sum F_{iy}\boldsymbol{j} + \sum F_{iz}\boldsymbol{k} \tag{4-5}$$

where the coefficients of $\boldsymbol{i}$, $\boldsymbol{j}$ and $\boldsymbol{k}$ should be the projection of the resultant force $\boldsymbol{F}_\mathrm{R}$ on each coordinate axis, respectively, so we get

$$F_{\mathrm{R}x} = \sum F_{ix},\quad F_{\mathrm{R}y} = \sum F_{iy},\quad F_{\mathrm{R}z} = \sum F_{iz} \tag{4-6}$$

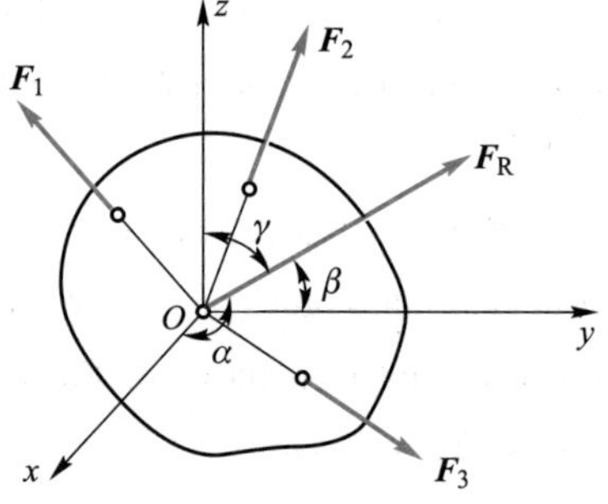

Fig.4-3

In other words, the projection of the resultant force on an axis is equal to the algebraic sum of the projections of the components of the force system on the same axis, which is the projection theorem of the resultant force for space concurrent force system.

From equation (4-6), the projection of the resultant force of space concurrent force system on the three axes can be obtained, and the magnitude and direction of the resultant force $\boldsymbol{F}_\mathrm{R}$ is

$$F_\mathrm{R} = \sqrt{F_{\mathrm{R}x}^2 + F_{\mathrm{R}y}^2 + F_{\mathrm{R}z}^2} = \sqrt{\left(\sum F_{ix}\right)^2 + \left(\sum F_{iy}\right)^2 + \left(\sum F_{iz}\right)^2}$$

$$\cos\alpha = \frac{F_{\mathrm{R}x}}{F_\mathrm{R}},\quad \cos\beta = \frac{F_{\mathrm{R}y}}{F_\mathrm{R}},\quad \cos\gamma = \frac{F_{\mathrm{R}z}}{F_\mathrm{R}}$$

Since a space concurrent force system can be reduced a resultant force, the sufficient and necessary condition for the equilibrium of a space concurrent force system is that the resultant force is equal to zero, i.e.

$$\boldsymbol{F}_\mathrm{R} = \boldsymbol{F}_1 + \boldsymbol{F}_2 + \cdots + \boldsymbol{F}_n = \sum \boldsymbol{F}_i = 0 \tag{4-7}$$

By projecting equation (4-7) onto the three axes, the equilibrium condition can be expressed as

$$\sum F_{ix} = 0,\quad \sum F_{iy} = 0,\quad \sum F_{iz} = 0 \tag{4-8}$$

These are the three independent equilibrium equations for the space concurrent force system.

Example 4-2 Fig.4-4 shows a suspension node O on an aerodynamic balance which is used to measure the resistance of a model, and a force $\boldsymbol{F}$ acts on the node O. The plane formed by wires OA and OB is perpendicular to the plane Oyz and intersects the plane at OD, while wire OC is in the same direction with the horizontal axis y. It is known that the angle between OD and z axis is β, and $\angle AOD = \angle BOD = \alpha$, try to find the tension in each wire.

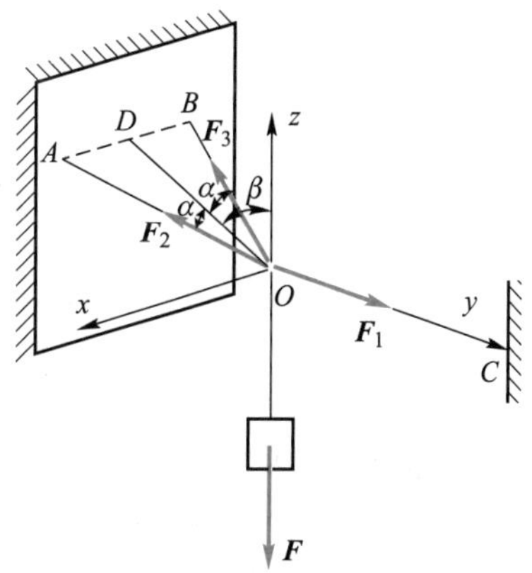

Fig.4-4

Solution The force analysis of node O is shown in Fig.4-4, and the force system acting at node O is the space concurrent force system. The projections of $\boldsymbol{F}_1$, $\boldsymbol{F}_2$ and $\boldsymbol{F}_3$ on the coordinate axes are

$$F_{1x}=0,\quad F_{1y}=F_1,\quad F_{1z}=0$$

$$F_{2x}=F_2\sin\alpha,\quad F_{2y}=-F_2\cos\alpha\sin\beta,\quad F_{2z}=F_2\cos\alpha\cos\beta$$

$$F_{3x}=F_3\sin\alpha,\quad F_{3y}=-F_3\cos\alpha\sin\beta,\quad F_{3z}=F_3\cos\alpha\cos\beta$$

From the equilibrium equation of the spatial concurrent force system, we have

$$\sum F_{ix}=0,\quad F_2\sin\alpha - F_3\sin\alpha = 0$$

$$\sum F_{iy}=0,\quad F_1 - F_2\cos\alpha\sin\beta - F_3\cos\alpha\sin\beta = 0$$

$$\sum F_{iz}=0,\quad F_2\cos\alpha\cos\beta + F_3\cos\alpha\cos\beta - F = 0$$

Solving the above three equations together, yields

$$F_1 = F\tan\beta,\quad F_2 = F_3 = \frac{F}{2\cos\alpha\cos\beta}$$

4.3 Space couple system

The effect of couple on the rigid body depends on three elements: the magnitude of the couple moment, the direction of couple and the orientation of the couple plane in space. Therefore, a vector is used to represent a space couple whose direction is perpendicular to the plane of action of the couple, whose pointing is in accordance with the right-handed screw rule of steering of the couple, and whose length represents the magnitude of the couple moment, so that this vector completely represents the three elements of the couple, which is called the couple moment vector (Fig.4-5), denoted as $\boldsymbol{M}$. The resultant of the couple moment vectors obeys the parallelogram law of vector.

The couple moment vector is free to slip along its line of action, and the couple is free to rotate in its plane of action, so the couple moment vector is a free vector without a fixed line and point of action.

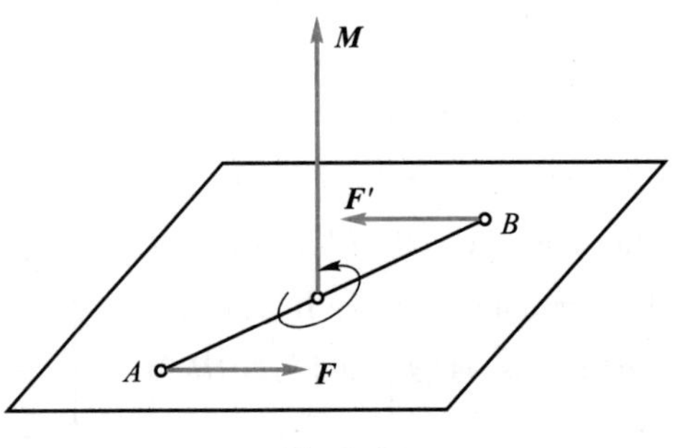

Fig.4-5

According to the above analysis, it can be seen that the condition for the equivalence of two space couples is that the couple moment vectors of the two couples are equal.

The following study investigates the reduction and equilibrium of the space couple system.

There is n space couples acting on the object, and these couples form a space couple system. Let the each couple moment vector in the system be $\boldsymbol{M}_1$, $\boldsymbol{M}_2$, $\cdots$, $\boldsymbol{M}_n$. According to the characteristic of the couple moment vector, it is possible to slip and translate them so that the couple moment vectors intersect at a point and then reduce them. The space couple system can be reduced into a resultant couple, the resultant couple moment vector is equal to the vector sum of each couple moment vector in the couple system, that is

$$\boldsymbol{M} = \boldsymbol{M}_1 + \boldsymbol{M}_2 + \cdots + \boldsymbol{M}_n = \sum \boldsymbol{M}_i \tag{4-9}$$

The magnitude and direction of the resultant couple moment vector can be obtained analytically, establish the rectangular coordinates $Oxyz$, and then project equation (4-9) onto the rectangular coordinate axes, we have

$$M_x = \sum M_{ix}, \quad M_y = \sum M_{iy}, \quad M_z = \sum M_{iz} \tag{4-10}$$

Equation (4-10) shows that the projection of the couple moment vector on each axis is equal to the algebraic sum of the projection of each couple moment vector on the same axis. Then, the magnitude and direction of the couple moment vector are

$$M = \sqrt{M_x^2 + M_y^2 + M_z^2} = \sqrt{\left(\sum M_{ix}\right)^2 + \left(\sum M_{iy}\right)^2 + \left(\sum M_{iz}\right)^2}$$

$$\cos\alpha = \frac{M_x}{M}, \quad \cos\beta = \frac{M_y}{M}, \quad \cos\gamma = \frac{M_z}{M}$$

where α, β and γ are the angles between the resultant couple moment vector $\boldsymbol{M}$ and the positive x, y and z axes, respectively.

A sufficient necessary condition for the equilibrium of the space couple system is that the resultant couple moment vector is equal to zero, i.e.

$$\sum \boldsymbol{M}_i = \boldsymbol{0} \tag{4-11}$$

The projection of equation (4-11) onto the coordinate axes gives

$$\sum M_{ix} = 0, \quad \sum M_{iy} = 0, \quad \sum M_{iz} = 0$$

This is the equilibrium equation of the space couple system.

Example 4-3 The triangular column rigid body shown in Fig.4-6a is half of a square body. A couple acts on each of the three sides. The couple moment $(\boldsymbol{F}_1, \boldsymbol{F}_1')$ is known to be $M_1 = 20$ N · m; the couple moment $(\boldsymbol{F}_2, \boldsymbol{F}_2')$ is $M_2 = 20$ N · m; and the couple moment $(\boldsymbol{F}_3, \boldsymbol{F}_3')$ is $M_3 = 20$ N · m. Determine the resultant couple moment vector $\boldsymbol{M}$, and if the rigid body is in equilibrium, what kind of couple needs to be introduced?

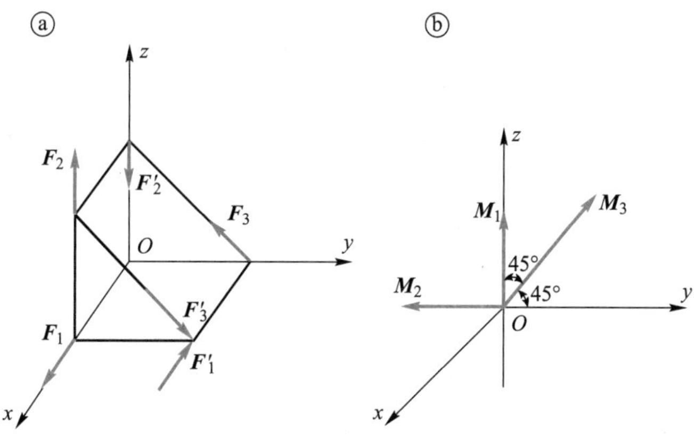

Fig.4-6

Solution Applying the right-handed screw rule, the three couples are expressed as the couple moment vectors and translated to the origin of the coordinate system, the resultant couple moment vector is equal to the vector sum of each couple moment vector, and the projection of the resultant couple moment vector on each coordinate axis is equal to the algebraic sum of the projection of each couple moment vector on the same coordinate axis, i.e.

$$M_x = \sum M_{ix} = 0$$

$$M_y = \sum M_{iy} = 11.2 \text{ N} \cdot \text{m}$$

$$M_z = \sum M_{iz} = 41.2 \text{ N} \cdot \text{m}$$

Then the resultant couple moment vector is

$$\boldsymbol{M} = 11.2\boldsymbol{j} + 41.2\boldsymbol{k}$$

The magnitude of the resultant couple moment vector is

$$M = \sqrt{M_x^2 + M_y^2 + M_z^2} = 42.7 \text{ N} \cdot \text{m}$$

The direction of the resultant couple moment vector is

$$\cos(\boldsymbol{M}, \boldsymbol{i}) = \frac{M_x}{M} = 0, \quad \angle(\boldsymbol{M}, \boldsymbol{i}) = 90^\circ$$

$$\cos(\boldsymbol{M}, \boldsymbol{j}) = \frac{M_y}{M} = 0.262, \quad \angle(\boldsymbol{M}, \boldsymbol{j}) = 74.8^\circ$$

$$\cos(\boldsymbol{M},\ \boldsymbol{k})=\frac{M_z}{M}=0.965,\quad \angle(\boldsymbol{M},\ \boldsymbol{k})=15.2°$$

To equilibrate this rigid body, it is necessary to add a couple with the couple moment vector $\boldsymbol{M}_4=-\boldsymbol{M}$.

4.4 Moment of a space force about a point and a specified axis

1. Moment of a space force about a point

The moment of a space force about a point needs to be represented by a vector, which is called the moment vector, noted as $\boldsymbol{M}_O(\boldsymbol{F})$ (Fig.4-7). The moment vector is perpendicular to the moment plane through the moment center O, and its pointing is determined by the right-handed screw rule, i.e., the moment turns counterclockwise when viewed from the end of the vector; the length of the vector indicates the magnitude of the moment, i.e.

$$M_O(\boldsymbol{F})=Fd=2S_{\triangle OAB} \tag{4-12}$$

When the position of the moment center changes, the magnitude and direction of $\boldsymbol{M}_O(\boldsymbol{F})$ also change, so the beginning of the moment vector must be drawn at the moment center, and cannot be moved arbitrarily. Moment vector is a positioning vector.

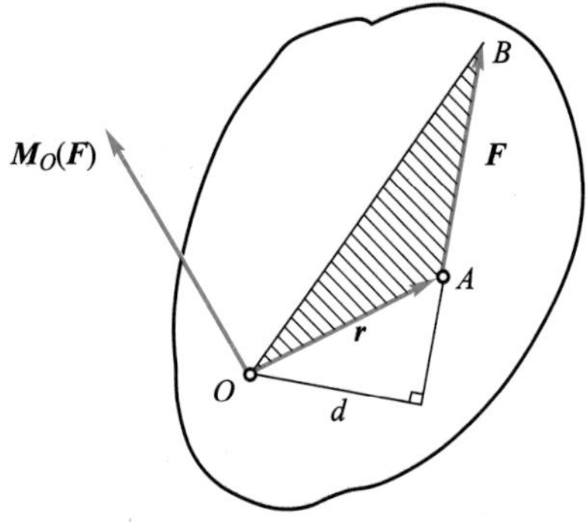

Fig.4-7

If $\boldsymbol{r}$ is the radius vector from the center of moment O to the point of action A of force $\boldsymbol{F}$, the vector product $\boldsymbol{r}\times\boldsymbol{F}$ is also a vector, and its magnitude is equal to twice the area of $\triangle OAB$, and its direction is the same with the direction of the moment vector. Thus it is obtained that

$$\boldsymbol{M}_O(\boldsymbol{F})=\boldsymbol{r}\times\boldsymbol{F} \tag{4-13}$$

That is, the moment of the force about any point is a vector, equal to the vector product of the radius vector from the center of the moment to the point of action of the force and the force. Equation (4-13) is called the expression for the vector product of the moments of the force about the point.

How to write the analytical expression of the moment of the force about the point? The rectangular coordinate system $Oxyz$ is chosen, the projections of force $\boldsymbol{F}$ on the coordinate axes are F_x, F_y, F_z, the coordinates of the action point A of the force are $(x,\ y,\ z)$ (Fig.4-8), and the unit vectors of the coordinate axes are $\boldsymbol{i}$, $\boldsymbol{j}$, $\boldsymbol{k}$. Then the radius vector $\boldsymbol{r}$ and the force vector $\boldsymbol{F}$ can be expressed as, respectively

$$\boldsymbol{r}=x\boldsymbol{i}+y\boldsymbol{j}+z\boldsymbol{k}$$
$$\boldsymbol{F}=F_x\boldsymbol{i}+F_y\boldsymbol{j}+F_z\boldsymbol{k}$$

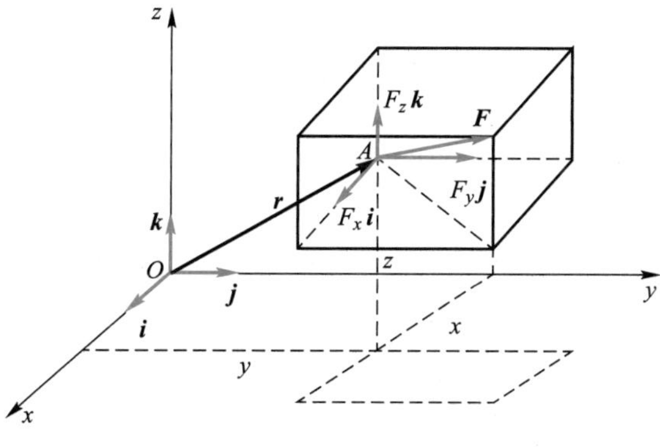

Fig.4-8

The moment vector of the force about the point O can be expressed as

$$\boldsymbol{M}_O(\boldsymbol{F})=\boldsymbol{r}\times\boldsymbol{F}=(x\boldsymbol{i}+y\boldsymbol{j}+z\boldsymbol{k})\times(F_x\boldsymbol{i}+F_y\boldsymbol{j}+F_z\boldsymbol{k})$$

$$=\begin{vmatrix}\boldsymbol{i} & \boldsymbol{j} & \boldsymbol{k}\\ x & y & z\\ F_x & F_y & F_z\end{vmatrix}=(yF_z-zF_y)\boldsymbol{i}+(zF_x-xF_z)\boldsymbol{j}+(xF_y-yF_x)\boldsymbol{k} \tag{4-14}$$

This is the analytical expression for the moment of the force about the point O, it can be known that the coefficients of the unit moments $\boldsymbol{i}$, $\boldsymbol{j}$, and $\boldsymbol{k}$ should represent the projection of the moment vector $\boldsymbol{M}_O(\boldsymbol{F})$ on the three coordinate axes, respectively, i.e.

$$[\boldsymbol{M}_O(\boldsymbol{F})]_x=yF_z-zF_y,\quad [\boldsymbol{M}_O(\boldsymbol{F})]_y=zF_x-xF_z,\quad [\boldsymbol{M}_O(\boldsymbol{F})]_z=xF_y-yF_x \tag{4-15}$$

2. Moment of a space force about a specified axis

The moment of a force about an axis is the measure of the effect of rotation about that axis.

There is an axis Oz on the rigid body, the line of action of the force $\boldsymbol{F}$ is neither parallel nor perpendicular to this axis (Fig.4-9). The force $\boldsymbol{F}$ is decomposed into two components, $\boldsymbol{F}_z$ and $\boldsymbol{F}_{xy}$, where the component $\boldsymbol{F}_z$ is parallel to the z-axis, which has no effect on the rotation of the rigid body around the axis Oz; only the component $\boldsymbol{F}_{xy}$ perpendicular to the axis Oz can make the rigid body rotate. Thus, the effect of a force rotating a rigid body about the axis Oz can be measured by the moment of the force component $\boldsymbol{F}_{xy}$ about point O, and the force component $\boldsymbol{F}_{xy}$ is the projection of the force $\boldsymbol{F}$ on the Oxy plane. The moment of the space force about the axis can be defined as follows.

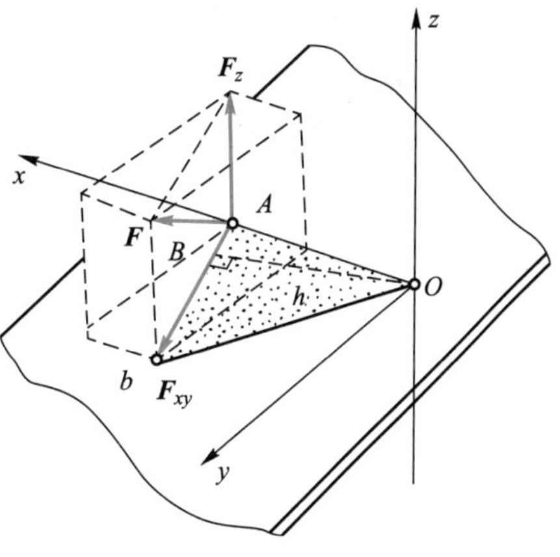

Fig.4-9

The moment of a space force about an axis is a scalar, which is equal to the moment of the projection of this force on a plane perpendicular to this axis about the intersecting point of this plane and this axis, denoted as $M_z(\boldsymbol{F})$, i.e.

$$M_z(\boldsymbol{F})=M_O(\boldsymbol{F}_{xy})=\pm F_{xy}h=\pm 2S_{\triangle OAB} \tag{4-16}$$

where h is the distance from point O to the force projection $\boldsymbol{F}_{xy}$. The positive and negative moments are determined by the right-handed screw rule, where the thumb pointing in the same direction as the positive direction of the axis is positive and the opposite is negative.

In particular, the moment of a force about an axis is equal to zero when the line of action of force passes through the axis ($h=0$) or when its line of action is parallel to the axis ($F_{xy}=0$). The unit of the moment of the force about the axis is N · m.

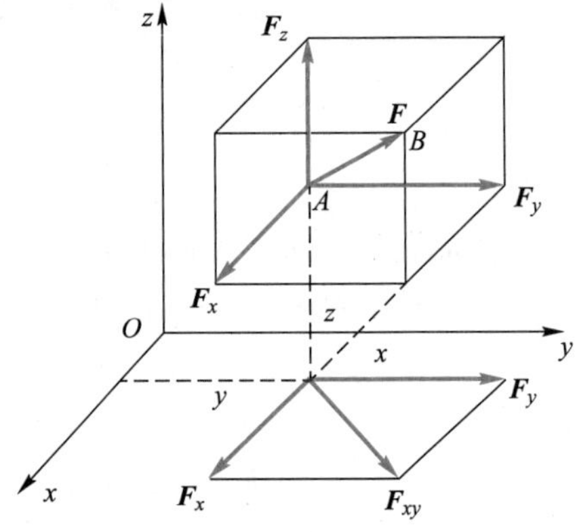

Fig.4-10

The moment of the force about the axis can also be expressed analytically. Make a rectangular coordinate system $Oxyz$, as shown in Fig.4-10. Let the coordinates of the point of action of the force $\boldsymbol{F}$ be (x, y, z), and the components and projections along the three axes be $\boldsymbol{F}_x$, $\boldsymbol{F}_y$, $\boldsymbol{F}_z$ and F_x, F_y, F_z, respectively. From equation (4-16), we have

$$\begin{aligned}M_z(\boldsymbol{F})&=M_O(\boldsymbol{F}_{xy})=M_O(\boldsymbol{F}_x)+M_O(\boldsymbol{F}_y)\\&=xF_y-yF_x\end{aligned}$$

The moments of the force $\boldsymbol{F}$ about the x and y axes can be derived similarly, therefore, the analytical expression for the moments of the force about the axes is obtained as

$$\left.\begin{aligned}M_x(\boldsymbol{F})&=yF_z-zF_y\\M_y(\boldsymbol{F})&=zF_x-xF_z\\M_z(\boldsymbol{F})&=xF_y-yF_x\end{aligned}\right\} \tag{4-17}$$

3. The relationship between the moment of the force about the point and the moment of the force about the axis passing through the point

Comparing equations (4-15) and (4-17), the projection of the force moment vector about a point on any axis passing through the point is equal to the moment of the force about that axis, that is

$$\left.\begin{aligned}M_x(\boldsymbol{F})&=[\boldsymbol{M}_O(\boldsymbol{F})]_x\\M_y(\boldsymbol{F})&=[\boldsymbol{M}_O(\boldsymbol{F})]_y\\M_z(\boldsymbol{F})&=[\boldsymbol{M}_O(\boldsymbol{F})]_z\end{aligned}\right\}$$

4.5 Reduction of space general force system

1. The translation theorem for space force

A force $\boldsymbol{F}$ acts on the object at point A, and there is another point B on the object, as shown in Fig.4-11a. Add two mutually equilibrium forces $\boldsymbol{F}'$ and $\boldsymbol{F}''$ at point B, and take $\boldsymbol{F}'=-\boldsymbol{F}''=\boldsymbol{F}$, as shown in Fig.4-11b. $\boldsymbol{F}$ and $\boldsymbol{F}''$ form a couple, and the couple moment vector is equal to the moment vector $\boldsymbol{M}=\boldsymbol{M}_B(\boldsymbol{F})$ of the force F about point B, that is, $\boldsymbol{M}=\boldsymbol{M}_B(\boldsymbol{F})$, as shown in Fig.4-11c. It can be seen that the original force $\boldsymbol{F}$ acting at point A is equivalent to the force $\boldsymbol{F}'$ and the couple $(\boldsymbol{F}, \boldsymbol{F}'')$.

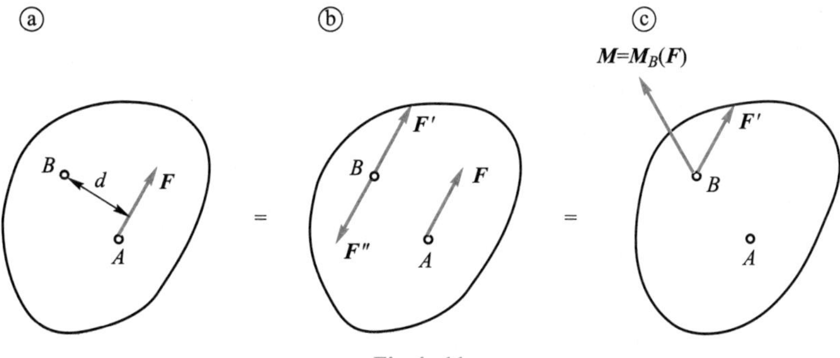

Fig.4–11

About the translation theorem of space force: a force acting on a rigid body can be moved parallel to any specified point in the rigid body, but a couple must be added at the same time, and the moment vector of the couple is equal to the moment vector of the original force about the specified point.

2. Reduction of space general force system

A space general force system $\boldsymbol{F}_1$, $\boldsymbol{F}_2$, $\cdots$, $\boldsymbol{F}_n$ act on each point A_1, A_2, $\cdots$, A_n of the rigid body (Fig.4–12a). The point O in the rigid body is chosen as the simplified center, and the forces are translated to point O by applying the force translation theorem, but the corresponding couples are introduced. The moment vector of couple is equal to the moment vector of the force about the simplified center O. In this way, the original force system is equivalently transformed into the space concurrent force system $\boldsymbol{F}_1'$, $\boldsymbol{F}_2'$, $\cdots$, $\boldsymbol{F}_n'$ acting at point O and the space couple system with the couple moment vectors $\boldsymbol{M}_1$, $\boldsymbol{M}_2$, $\cdots$, $\boldsymbol{M}_n$, as shown in Fig.4–12b, where

$$\boldsymbol{F}_1' = \boldsymbol{F}_1, \ \boldsymbol{F}_2' = \boldsymbol{F}_2, \ \cdots, \ \boldsymbol{F}_n' = \boldsymbol{F}_n$$

$$\boldsymbol{M}_1 = \boldsymbol{M}_O(\boldsymbol{F}_1), \ \boldsymbol{M}_2 = \boldsymbol{M}_O(\boldsymbol{F}_2), \ \cdots, \ \boldsymbol{M}_n = \boldsymbol{M}_O(\boldsymbol{F}_n)$$

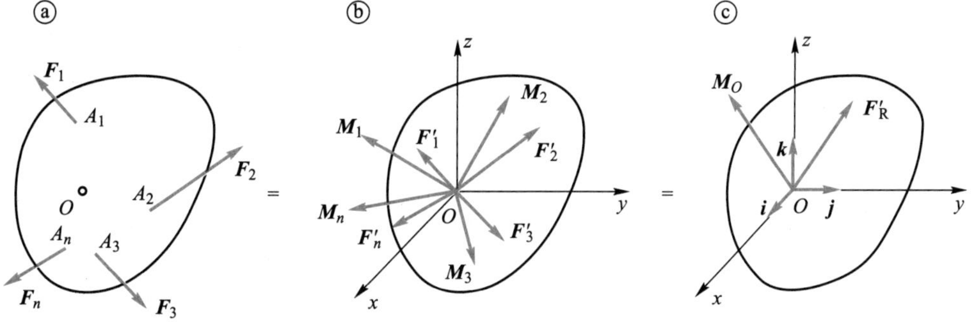

Fig.4–12

The space concurrent force system acting at point O can be reduced into a resultant force $\boldsymbol{F}_\mathrm{R}'$ acting at point O, and

$$\boldsymbol{F}_\mathrm{R}' = \sum \boldsymbol{F}_i' = \sum \boldsymbol{F}_i \tag{4-18}$$

That is, the resultant force vector $\boldsymbol{F}'_R$ is equal to the vector sum of each force in the original force system, which is called the principal vector of the original space force system.

The additional couple system in space can be reduced into a couple, whose resultant couple moment vector is

$$\boldsymbol{M}_O = \sum \boldsymbol{M}_i = \sum \boldsymbol{M}_O(\boldsymbol{F}_i) \tag{4-19}$$

where the vector $\boldsymbol{M}_O$ is equal to the vector sum of the moments of each force in the original force system about the simplified center O, which is called the principal moment of the force system about the point O.

Based on the above discussion, a space general force system can be reduced to a force and couple. The force acts on the simplified center, the force vector is equal to the principal vector of the force system; the moment vector of the couple is equal to the principal moment of the force system about the simplified center. In general, the principal vector is independent of the choice of the simplification center, while the principal moment is related to the choice of the simplification center.

In order to calculate the principal vector and principal moment, a rectangular coordinate system $Oxyz$ with the simplified center O as the origin can be taken, and equation (4-18) is projected onto the coordinate axes x, y and z, then we have

$$F'_{Rx} = \sum F_{ix}, \quad F'_{Ry} = \sum F_{iy}, \quad F'_{Rz} = \sum F_{iz} \tag{4-20}$$

Therefore, the magnitude and direction of the principal vector are

$$F'_R = \sqrt{R'^2_{Rx} + F'^2_{Ry} + F'^2_{Rz}} \tag{4-21}$$

$$\cos\alpha = \frac{F'_{Rx}}{F'_R}, \quad \cos\beta = \frac{F'_{Ry}}{F'_R}, \quad \cos\gamma = \frac{F'_{Rz}}{F'_R} \tag{4-22}$$

where α, β, γ denote the angle between the principal vector $\boldsymbol{F}'_R$ and the positive x, y, and z axes, respectively.

Similarly, projecting equation (4-19) onto the coordinate axes x, y, z, and applying the relationship between the moment of the force about the point and the moment of the force about the axis we obtain

$$\left.\begin{aligned} M_{Ox} &= \sum M_x(\boldsymbol{F}_i) \\ M_{Oy} &= \sum M_y(\boldsymbol{F}_i) \\ M_{Oz} &= \sum M_z(\boldsymbol{F}_i) \end{aligned}\right\} \tag{4-23}$$

Therefore, the magnitude and direction of the principal moment $\boldsymbol{M}_O$ are

$$M_O = \sqrt{M^2_{Ox} + M^2_{Oy} + M^2_{Oz}} = \sqrt{\left[\sum M_x(F_i)\right]^2 + \left[\sum M_y(F_i)\right]^2 + \left[\sum M_z(F_i)\right]^2} \tag{4-24}$$

$$\cos\alpha' = \frac{M_{Ox}}{M_O}, \quad \cos\beta' = \frac{M_{Oy}}{M_O}, \quad \cos\gamma' = \frac{M_{Oz}}{M_O} \tag{4-25}$$

where α', β', γ' denote the angles between the principal moments $\boldsymbol{M}_O$ and the axes x, y, and z positive direction, respectively.

The space general force system can be reduced to a force and a couple passing through the center of the simplification. The final results of the force system are further discussed in terms of the principal vector $\boldsymbol{F}'_R$ and the principal moment $\boldsymbol{M}_O$ about the simplified center as follows.

(1) $\boldsymbol{F}'_R=0, \boldsymbol{M}_O=0$

Both the principal vector and principal moment are equal to zero, which means that the original space general force system is an equilibrium force system.

(2) $\boldsymbol{F}'_R=0, \boldsymbol{M}_O\neq 0$

The principal vector is equal to zero, while the principal moment is not equal to zero, which means that the original force system is equivalent to a space couple system, that is, the original force system can be reduced into a resultant couple. The resultant couple moment vector is equal to the principal moment of the original force system about the simplified center. For this case the principal moment of the force system is independent of the position of the simplified center.

(3) $\boldsymbol{F}'_R\neq 0, \boldsymbol{M}_O=0$

The principal vector is equal to zero, while the principal moment is not equal to zero, which means that the original force system can be reduced into a resultant force that acts on the simplified center, and the force vector is equal to the principal vector $\boldsymbol{F}'_R$ of the force system. This case is related to the position of the simplification center.

(4) $\boldsymbol{F}'_R\neq 0, \boldsymbol{M}_O\neq 0$

Both the principal vector and principle moment are not equal to zero. According to the position relationship between the principal vector and the principal moment, they are discussed as follows.

① $\boldsymbol{M}_O \perp \boldsymbol{F}'_R$, and the force $\boldsymbol{F}'_R$ and the couple ($\boldsymbol{F}''_R$, $\boldsymbol{F}_R$) are in the same plane (Fig.4−13a). As in Fig.4−13b, let $\boldsymbol{F}_R=\boldsymbol{F}'_R=-\boldsymbol{F}''_R$, so $\boldsymbol{F}'_R$ and $\boldsymbol{F}''_R$ are a pair of equal, opposite and co-linear equilibrium forces, which can be removed. At this point the original force can be reduced to a resultant force $\boldsymbol{F}_R$, the magnitude and direction of the resultant force $\boldsymbol{F}_R$ is the same as $\boldsymbol{F}'_R$, but the resultant force does not act at the simplified center O point, but deviates from the distance d (Fig.4−13c), that is

$$d=\frac{M_O}{F'_R} \tag{4-26}$$

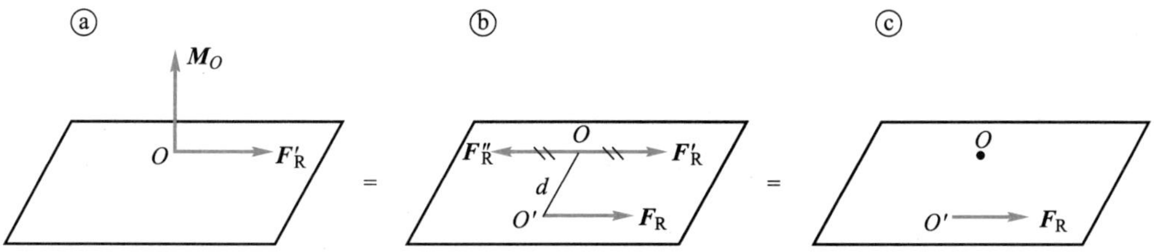

ⓐ $\boldsymbol{M}_O$ O $\boldsymbol{F}'_R$ = ⓑ $\boldsymbol{F}''_R$ O $\boldsymbol{F}'_R$ d O' $\boldsymbol{F}_R$ = ⓒ O O' $\boldsymbol{F}_R$

Fig.4−13

② $\boldsymbol{M}_O /\!/ \boldsymbol{F}'_R$, the force system can no longer be further reduced, as shown in Fig.4−14.Such a combination of a force and a couple in the plane perpendicular to them is called a force wrench. The force wrench is the simplest force system consisting of the two basic elements of statics (force and couple) and cannot be further reduced. The line of action of the force wrench is called the central axis of this force system. In the above case, the central axis passes through the center of simplification.

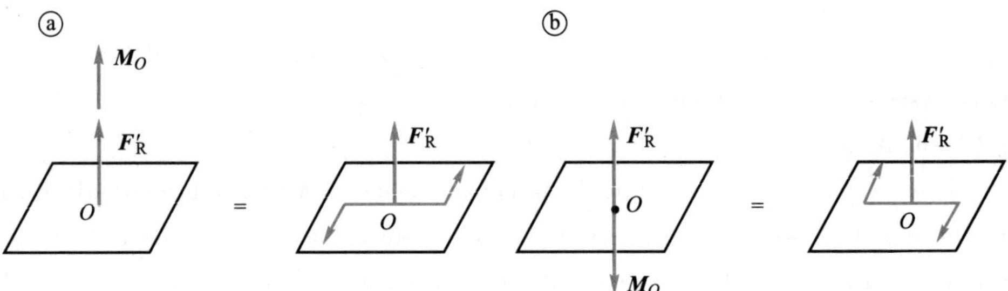

Fig.4-14

③ F'_R and M_O form an arbitrary angle α, as shown in Fig.4-15a, which is the most general case obtained by reducing the force system. M_O can be decomposed into two space couples M'_O and M''_O, which are perpendicular to F'_R and parallel to F'_R, respectively. As shown in Fig.4-15b, since $M''_O \perp F'_R$, they can be replaced by the force F''_R acting on point O'. Since the couple moment vector is a free vector, M'_O can be shifted in parallel so that it is co-linear with F''_R.

$$d=\frac{M''_O}{F'_R}=\frac{M_O \sin\alpha}{F''_R}$$

That is, the force system can also be reduced to a force wrench, but the central axis of the force wrench does not pass through the simplified center.

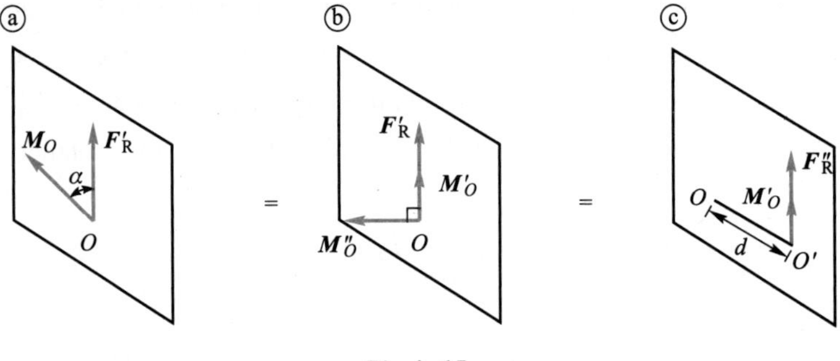

Fig.4-15

3. Resultant moment theorem for space general force system

When the space force system is reduced to point O with $F'_R \neq 0$, $M_O \neq 0$, and $M_O \perp F'_R$, the final result of the simplification is a resultant force F'_R acting at point O' (Fig.4-13), it is obviously

$$M_O = M_O(F_R)$$

and

$$M_O = \sum M_O(F_i)$$

So get

$$\boldsymbol{M}_O(\boldsymbol{F}_R) = \sum \boldsymbol{M}_O(\boldsymbol{F}_i) \tag{4-27}$$

Equation (4-27) shows that if a space general force system can be reduced into a resultant force, the moment of the resultant force about any point in space is equal to the vector sum of the moments of each force in the system about the same point, which is the theorem of the resultant moment of the space force about any point.

Projecting equation (4-27) onto the three axes through the point O and applying the relationship between the moment of the force about the point and the moment of the force about the axis, get

$$\left.\begin{aligned} M_x(\boldsymbol{F}_R) &= \sum M_x(\boldsymbol{F}_i) \\ M_y(\boldsymbol{F}_R) &= \sum M_y(\boldsymbol{F}_i) \\ M_z(\boldsymbol{F}_R) &= \sum M_z(\boldsymbol{F}_i) \end{aligned}\right\} \tag{4-28}$$

Equation (4-28) shows that if a space force system can be reduced into a resultant force, the moment of the resultant force about any axis is equal to the algebraic sum of the moments of each force in the system about the same axis, which is the theorem of the resultant moment of the space force system about the axis.

Example 4-4 The force system shown in Fig.4-16a consists of four forces, $F_1=60$ N, $F_2=400$ N, $F_3=500$ N, and $F_4=200$ N. Try to reduce the force system toward point A.

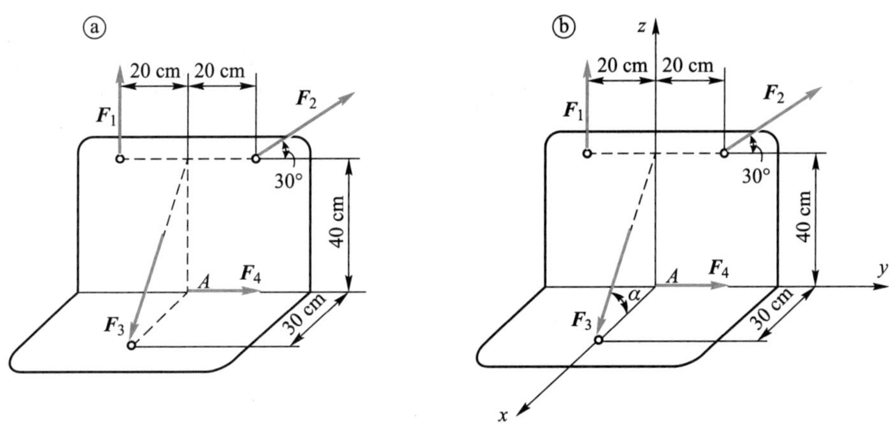

Fig.4-16

Solution From the geometric relationship in Fig.4-16b, we have

$$\sin\alpha = \frac{4}{5} = 0.8, \quad \cos\alpha = \frac{3}{5} = 0.6$$

Taking the $Axyz$ coordinate system as shown in Fig.4-16b, the principal vector $\boldsymbol{F}'_R$ of the force system is obtained by reducing the force system to point A:

$$F'_{Rx} = F_3\cos\alpha = 500\times0.6\ \text{N} = 300\ \text{N}$$

$$F'_{Ry} = F_2\cos 30° + F_4 = (400\times0.866+200)\ \text{N} = 546.4\ \text{N}$$

$$F'_{Rz} = F_1 + F_2\sin 30° - F_3\sin\alpha = (60+400\times0.5-500\times0.8)\ \text{N} = -140\ \text{N}$$

then

$$F'_R=\sqrt{F_{Rx}^2+F_{Ry}^2+F_{Rz}^2}=638.87\ \text{N}$$

$$\cos(\boldsymbol{F}'_R,\boldsymbol{i})=\frac{F'_{Rx}}{F'_R}=\frac{300\ \text{N}}{638.87\ \text{N}}=0.469\ 6$$

$$\cos(\boldsymbol{F}'_R,\boldsymbol{j})=\frac{F'_{Ry}}{F'_R}=\frac{546.4\ \text{N}}{638.87\ \text{N}}=0.855$$

$$\cos(\boldsymbol{F}'_R,\boldsymbol{k})=\frac{F'_{Rz}}{F'_R}=\frac{-140\ \text{N}}{638.87\ \text{N}}=-0.219$$

The principal moment of the force system $\boldsymbol{M}_A$:

$$M_x=\sum M_x(\boldsymbol{F}_i)=-F_1\cdot 0.2\ \text{m}+F_2\sin 30°\times 0.2\ \text{m}-F_2\cos 30°\times 0.4\ \text{m}=-110.56\ \text{N}\cdot\text{m}$$

$$M_y=\sum M_y(\boldsymbol{F}_i)=F_3\sin\alpha\cdot 0.3\ \text{m}=120\ \text{N}\cdot\text{m}$$

$$M_z=\sum M_z(\boldsymbol{F}_i)=0$$

then

$$M_A=\sqrt{M_x^2+M_y^2+M_z^2}=\sqrt{(-110.56)^2+120^2+0}\ \text{N}\cdot\text{m}=163.17\ \text{N}\cdot\text{m}$$

$$\cos(\boldsymbol{M}_A,\boldsymbol{i})=\frac{M_x}{M_A}=\frac{-110.56\ \text{N}\cdot\text{m}}{163.17\ \text{N}\cdot\text{m}}=-0.678$$

$$\cos(\boldsymbol{M}_A,\boldsymbol{j})=\frac{M_y}{M_A}=\frac{120\ \text{N}\cdot\text{m}}{163.17\ \text{N}\cdot\text{m}}=0.735$$

$$\cos(\boldsymbol{M}_A,\boldsymbol{k})=\frac{M_z}{M_A}=\frac{0\ \text{N}\cdot\text{m}}{163.17\ \text{N}\cdot\text{m}}=0$$

4.6 Equilibrium equations for space general force system

The space general force system can be reduced to obtain a force and a couple. The force vector is equal to the principal vector, and the moment of the couple is equal to the principal moment. A sufficient necessary condition for the equilibrium of space general force system is that both the principal vector and the principal moment are equal to zero, i.e.

$$\boldsymbol{F}'_R=0,\quad \boldsymbol{M}_O=0 \tag{4-29}$$

According to the formula for the principal vector and principal moment

$$F'_R=\sqrt{\left(\sum F_{ix}\right)^2}+\sqrt{\left(\sum F_{iy}\right)^2}+\sqrt{\left(\sum F_{iz}\right)^2}=0 \tag{4-30}$$

$$M_O=\sqrt{\left[\sum M_x(\boldsymbol{F}_i)\right]^2+\left[\sum M_y(\boldsymbol{F}_i)\right]^2+\left[\sum M_z(\boldsymbol{F}_i)\right]^2}=0 \tag{4-31}$$

get

$$\left.\begin{aligned}&\sum F_{ix}=0,\quad \sum F_{iy}=0,\quad \sum F_{iz}=0\\&\sum M_x(\boldsymbol{F}_i)=0,\quad \sum M_y(\boldsymbol{F}_i)=0,\quad \sum M_z(\boldsymbol{F}_i)=0\end{aligned}\right\} \tag{4-32}$$

Equation (4−32) is the equilibrium equation of space general force system. The sufficient and necessary condition for the equilibrium of space general force system is that the algebraic sum of the projections of all forces on each axis of the rectangular coordinate system is equal to zero, and the algebraic sum of the moments about each axis is also equal to zero. The space general force system has six independent equilibrium equations, which can solve six unknown forces.

A force system in which the lines of action are parallel to each other but not in the same plane, which is called a space parallel force system. The space parallel force system is a special case of the space general force system. $\boldsymbol{F}_1$, $\boldsymbol{F}_2$, $\cdots$, $\boldsymbol{F}_n$ is the space parallel force system acting on the rigid body (Fig.4−17), and establish the coordinate system as shown in Fig.4−17, and make the z−axis parallel to the lines of action of the forces. Clearly, we have $\sum F_{ix} = 0$, $\sum F_{iy} = 0$, and $\sum M_z(\boldsymbol{F}_i) = 0$ regardless of whether the force system is equilibrious or not. Equation (4−32) leaves only three independent equilibrium equations for the space parallel force system, namely

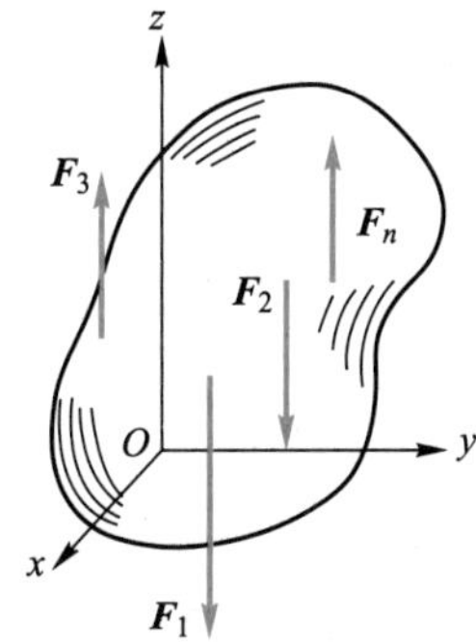

Fig.4−17

$$\sum F_{iz} = 0, \quad \sum M_x(\boldsymbol{F}_i) = 0, \quad \sum M_y(\boldsymbol{F}_i) = 0 \quad (4-33)$$

Equation (4−33) shows that a sufficient and necessary condition for the equilibrium of the space parallel force system is that the algebraic sum of the projections of each force in the system on the axes parallel to the lines of action of the forces is equal to zero, and the algebraic sum of the moments of the forces about the two axes perpendicular to the lines of action of the forces is equal to zero.

Example 4−5 A homogeneous equilateral triangular plate weighing G with its center of gravity at point O is suspended by three equal lengths of lead rope, which is horizontal, see Fig.4−18. At point D on the plate, a lead downward force $\boldsymbol{F}$ is applied, the height of the triangle is h, $CD = h/3$, try to calculate the tensile force of the rope.

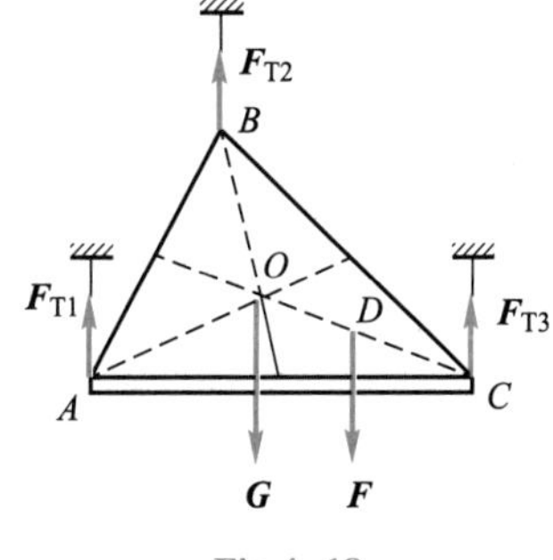

Fig.4−18

Solution The force analysis of the triangular plate is shown in Fig.4−18, three unknown forces $\boldsymbol{F}_{T1}$, $\boldsymbol{F}_{T2}$ and $\boldsymbol{F}_{T3}$, gravity $\boldsymbol{G}$ and applied force $\boldsymbol{F}$. All these forces form a space parallel force system, thus we can list three independent equilibrium equations. The three unknown forces can be determined by all the equilibrium equations, namely

$$\sum M_{AB}(\boldsymbol{F}_i) = 0, \quad F_{T3}h - G \cdot \frac{1}{3}h - F \cdot \frac{2}{3}h = 0$$

$$\sum M_{BC}(\boldsymbol{F}_i) = 0, \quad F_{T1}h - G \cdot \frac{1}{3}h - F \cdot \frac{1}{3}h\sin 30° = 0$$

$$\sum F_z = 0, \quad F_{T1} + F_{T2} + F_{T3} - G - F = 0$$

The solution can be obtained as

$$F_{T1}=\frac{1}{3}G+\frac{1}{6}F, \quad F_{T2}=\frac{1}{3}G+\frac{1}{6}F, \quad F_{T3}=\frac{1}{3}G+\frac{2}{3}F$$

Example 4-6 The object weighing G_1 is wound on a drum wheel of radius r by a rope, and the object weighing G_2 is wound on a wheel of radius R by a rope, and A and B are sliding bearings (Fig. 4-19). The structure is in equilibrium under the gravities $\boldsymbol{G}_2$, $\boldsymbol{G}_1$ and the reaction force of the bearing. $R=6r$ and $G_2=6$ N, try to determine the magnitude of gravity $\boldsymbol{G}_1$ and the reaction force at bearings A and B. The weights of drums, wheels and axle are neglected.

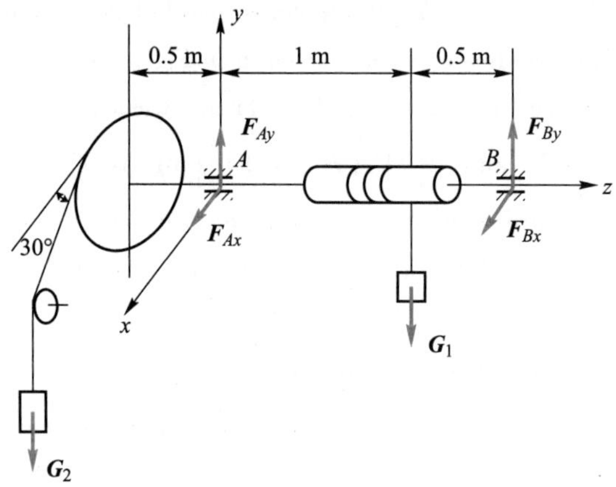

Fig.4-19

Solution The force analysis for the whole object is shown in Fig.4-19, gravity $\boldsymbol{G}_2$ and five unknown forces ($\boldsymbol{F}_{Ax}$, $\boldsymbol{F}_{Ay}$, $\boldsymbol{F}_{Bx}$, $\boldsymbol{F}_{By}$, $\boldsymbol{G}_1$), all these forces form a space general force system. There are six equilibrium equations for the space general force system, so the five unknown forces can be determined by all the equations.

Taking the coordinate system as shown in Fig.4-19, we first take the moments about the three axes and have

$$\sum M_z(\boldsymbol{F}_i)=0, \quad G_2\times R-G_1\times r=0$$

$$\sum M_x(\boldsymbol{F}_i)=0, \quad G_2\sin 30°\times 0.5\ \text{m}-G_1\times 1\ \text{m}+F_{By}\times 1.5\ \text{m}=0$$

$$\sum M_y(\boldsymbol{F}_i)=0, \quad G_2\cos 30°\times 0.5\ \text{m}-F_{Bx}\times 1.5\ \text{m}=0$$

can be solved for

$$G_1=36\ \text{N}, \quad F_{By}=23\ \text{N}, \quad F_{Bx}=\sqrt{3}\ \text{N}$$

and then projected onto the x and y axes, i.e.

$$\sum F_{ix}=0, \quad F_{Ax}+G_2\cos 30°+F_{Bx}=0$$

obtains

$$F_{Ax}=4\sqrt{3}\ \text{N}$$

$$\sum F_{iy}=0, \quad F_{Ay}-G_2\sin 30°-G_1+F_{By}=0$$

obtains

$$F_{Ay} = 16 \text{ N}$$

Problems

4-1 The top of the column OA is subjected to a rope tension $F_T = 10$ kN, which are shown in Fig.P4-1. Determine the projection of the force $\boldsymbol{F}_T$ on the coordinate axes and the moments about point O and the three coordinate axes.

Answer: $F_{Tx} = 3\sqrt{2}$ kN, $F_{Ty} = 4\sqrt{2}$ kN, $F_{Tz} = -5\sqrt{2}$ kN

4-2 As shown in Fig.P4-2, determine the moment of the force $\boldsymbol{F}$ about point A.

Answer: $M_A = F\sqrt{a^2+b^2}$, direction: $\boldsymbol{n} = a\boldsymbol{i} + b\boldsymbol{j}$

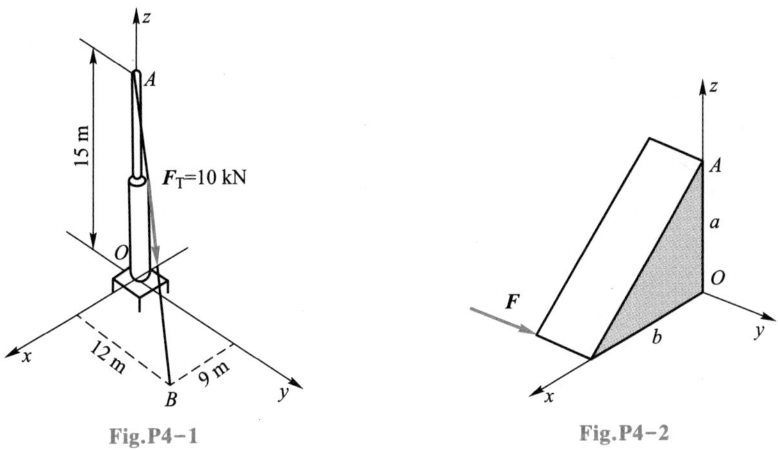

Fig.P4-1　　Fig.P4-2

4-3 As shown in Fig.P4-3, determine the moment of the force $\boldsymbol{F}$ about point A.

Answer: $\boldsymbol{M}_A = \frac{4}{5}Fb\boldsymbol{i} - \frac{7}{5}Fb\boldsymbol{j} - \frac{3}{5}Fb\boldsymbol{k}$

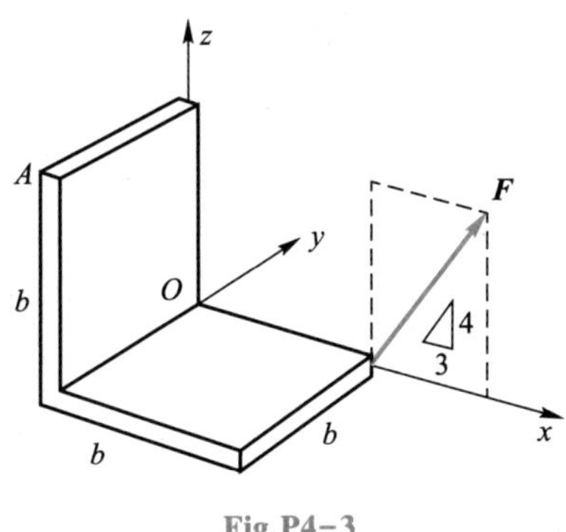

Fig.P4-3

4-4 As shown in Fig. P4-4, the two forces acting on the handle of the pipe wrench form a couple. Try to find the magnitude and direction of this couple moment vector.

Answer: $M=78.3$ N · m, the direction is negative along the x axis

4-5 The positions of the forces $\boldsymbol{F}_1$, $\boldsymbol{F}_2$, $\boldsymbol{F}_3$, and $\boldsymbol{F}_4$ in space are shown in Fig.P4-5, and we know that $F_1=F_2=F_3=F_4=10$ N, write the analytical expressions for each force and the moment of each force about point O.

Answer: $\boldsymbol{F}_1=10\boldsymbol{k}$, $\boldsymbol{F}_2=10\boldsymbol{i}$, $\boldsymbol{F}_3=5\boldsymbol{i}+5\sqrt{3}\boldsymbol{j}$, $\boldsymbol{F}_4=\frac{5}{2}\sqrt{3}\boldsymbol{i}+\frac{15}{2}\boldsymbol{j}-5\boldsymbol{k}$;

$$\boldsymbol{M}_O(\boldsymbol{F}_1)=15\boldsymbol{i}-5\sqrt{3}\boldsymbol{j},\ \boldsymbol{M}_O(\boldsymbol{F}_2)=10\boldsymbol{j}-15\boldsymbol{k},\ \boldsymbol{M}_O(\boldsymbol{F}_3)=\frac{15}{2}\boldsymbol{k};$$

$$\boldsymbol{M}_O(\boldsymbol{F}_4)=-15\boldsymbol{i}+\frac{5}{2}\sqrt{3}\boldsymbol{j}-\frac{15}{4}\sqrt{3}\boldsymbol{k}$$

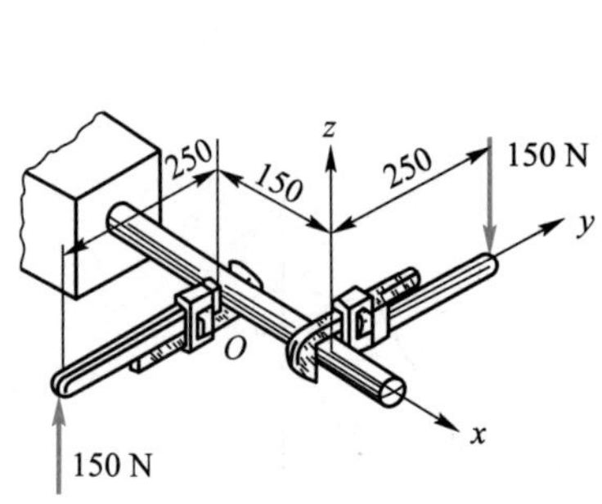

Fig.P4-4

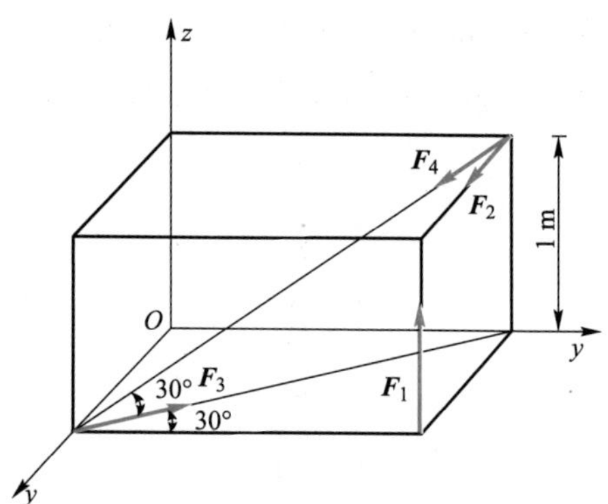

Fig.P4-5

4-6 As shown in Fig.P4-6, the five forces are known to be $F_1=2$ N, $F_2=2$ N, $F_3=1$ N, $F_4=4$ N, and $F_5=7$ N. Determine the resultant force of these five forces.

Answer: $F_x=6$ N, $F_y=4$ N, $F_z=4$ N

4-7 As shown in Fig.P4-7, determine the magnitude of the moment $\boldsymbol{M}_z$ of the force $F=1000$ N about the z-axis.

Answer: $M_z=-101.41$ N · m

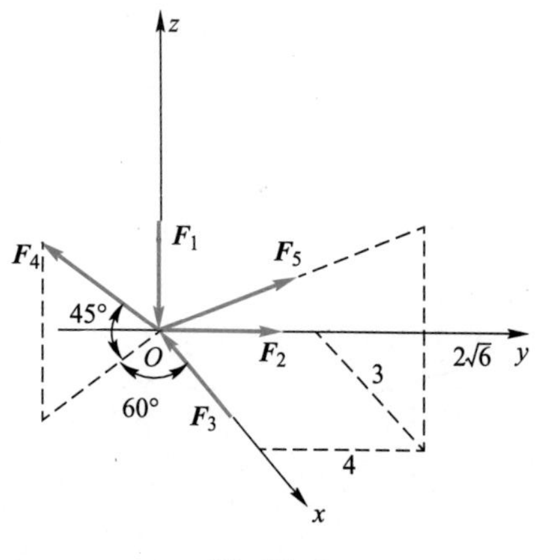

Fig.P4-6

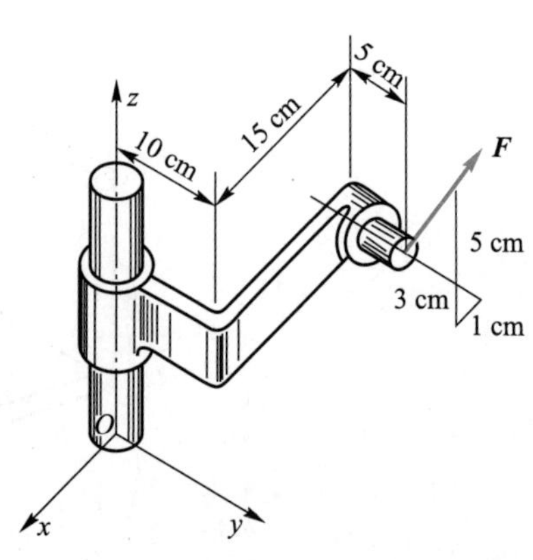

Fig.P4-7

4-8 As shown in Fig.P4-8, the couple moment vectors $\boldsymbol{M}_1$ and $\boldsymbol{M}_2$ represent the couples acting on the planes ABC and ACD, respectively, and it is known that $M_1 = M_2 = M$. Determine the resultant couple.

Answer: $\boldsymbol{M} = \frac{2}{\sqrt{13}} M_2 \boldsymbol{i} + \left(\frac{M_1}{\sqrt{5}} + \frac{3}{\sqrt{13} M_2} \right) \boldsymbol{j} + \frac{2M_1}{\sqrt{5}} \boldsymbol{k}$

4-9 The force system has $F_1 = 100$ N, $F_2 = 300$ N, and $F_3 = 200$ N. The positions of the lines of action of the forces are shown in Fig.P4-9.Try to reduce the force system to point O.

Answer: Principal vector: $\boldsymbol{F}'_R = (-345.4\boldsymbol{i} + 249.6\boldsymbol{j} + 10.56\boldsymbol{k})$ N;

Principal moment: $\boldsymbol{M}_O = (-51.78\boldsymbol{i} - 36.05\boldsymbol{j} + 103.6\boldsymbol{k})$ N · m

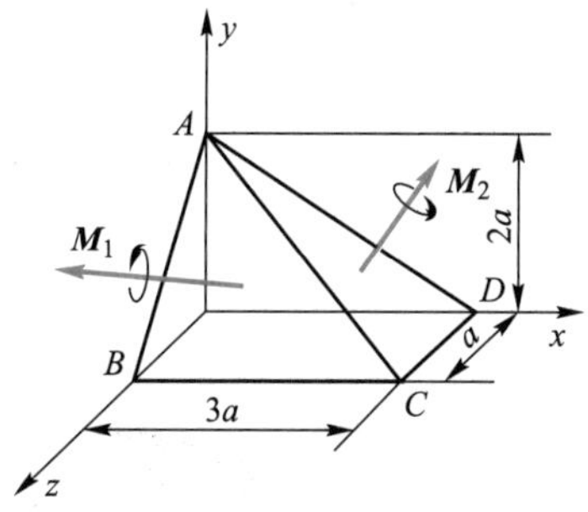

Fig.P4-8

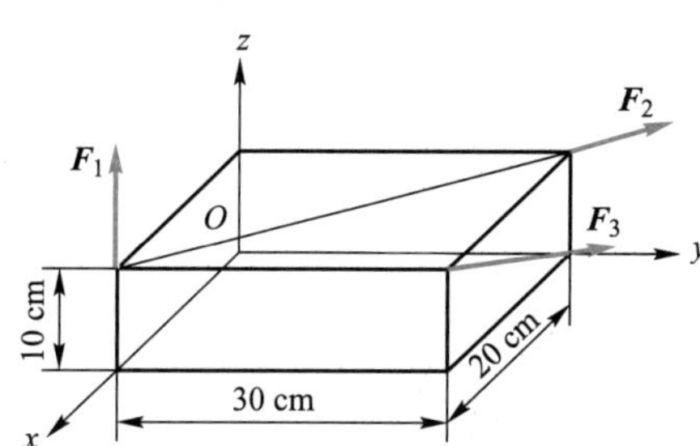

Fig.P4-9

4-10 As shown in Fig.P4-10, the forces are distributed along the edges, face diagonals, and body diagonals of the square. It is known that $F_1 = 10$ N, $F_2 = 10$ N, $F_3 = 10$ N, $F_4 = 10$ N, and the length of the square $a = 1$ m. Determine the reduction of the force system to point O.

Answer: Principal vector: $\boldsymbol{F}'_R = (-10\boldsymbol{i} + 10\boldsymbol{k})$ N;

Principal moment: $\boldsymbol{M}_O = (-10\boldsymbol{i} - 10\boldsymbol{j})$ N · m

4-11 As shown in Fig.P4-11, a square homogeneous plate weighing 18 kN has its center of gravity at point G. The plate is suspended by three ropes and kept horizontal. Try to determine the tension of each rope.

Answer: $F_{AD} = \frac{9}{4}\sqrt{6}$ kN, $F_{BD} = \frac{9}{4}\sqrt{6}$ kN, $F_{CD} = \frac{9}{2}\sqrt{5}$ kN

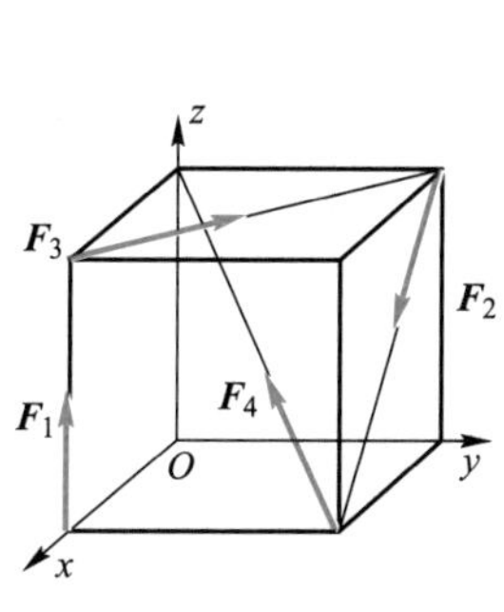

Fig.P4-10

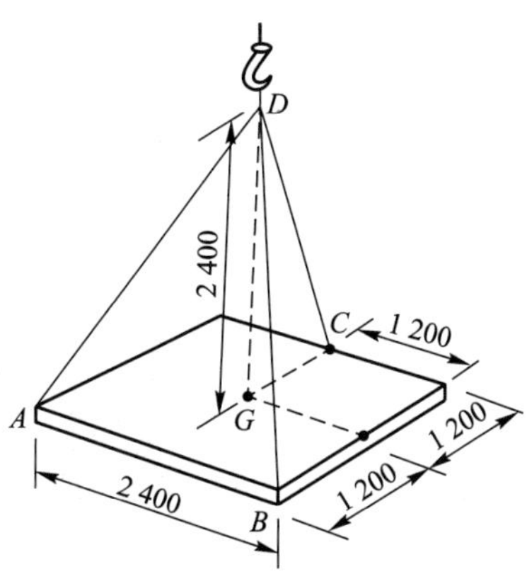

Fig.P4-11

4-12 As shown in Fig.P4-12, the three poles of the tripod are connected by ball hinges D and supported by ball hinge supports A, B, and C. There is a horizontal force of 1 kN acting at the point D of the tripod, and the weight of each pole is neglected. Try to determine the force on each pole.

Answer: $F_{AD}=-\frac{5}{6}\sqrt{6}$ kN, $F_{BD}=\frac{2}{9}\sqrt{15}$ kN, $F_{CD}=\frac{\sqrt{58}}{6}$ kN

4-13 The space frame consists of three straight bars connected by ball hinges at the end D. The ends A, B and C are fixed to the horizontal floor with ball hinges as shown in Fig.P4-13.If the weight of the object hanging from the end D is $G=10$ kN, determine the reaction forces of hinges A, B, and C (regardless of the weight of the bars).

Answer: $F_A=F_B=-26.4$ kN(press), $F_C=33.5$ kN(pull)

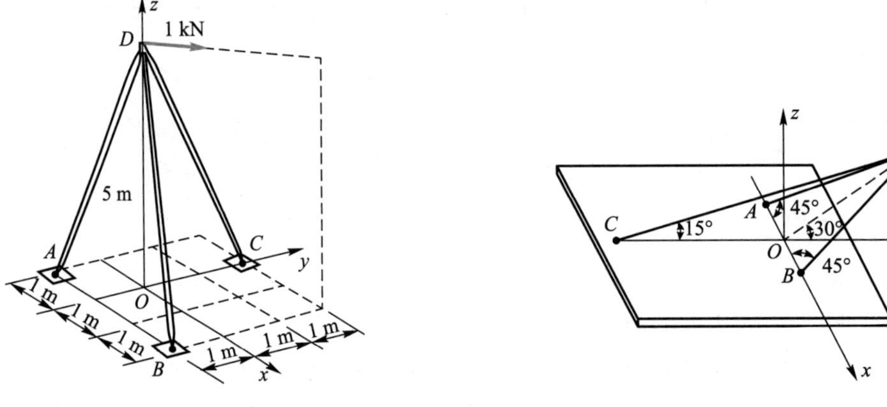

Fig.P4-12   Fig.P4-13

4-14 As shown in Fig.P4-14, the radii of three disks A, B and C are 15 cm, 10 cm and 5 cm, respectively. The three axes OA, OB and OC are in the same plane and $\angle AOB$ is a right angle. The forces acting on the three disks are equal to 10 N, 20 N and F. If the system formed by the three disks is free to move, determine the magnitude of the force $\boldsymbol{F}$ and angle α that make the system equilibrium.

Answer: $F=50$ N, $\alpha=143°8'$

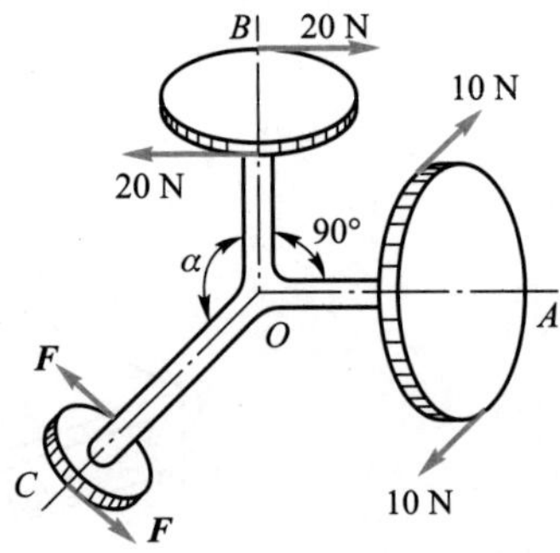

Fig.P4-14

4-15 As shown in Fig.P4-15, a rectangular homogeneous plate $ABCD$ weighs 800 N and is supported in a horizontal position by butterfly hinges H and K and a spreader bar EC. Moreover, AB = 1.5 m, BC = 0.6 m, AH = BK = 0.25 m, CE = 0.75 m. Determine the internal force F_{CE} of the bar CE and the reaction force of the butterfly hinge K and H.

Answer: F_{CE} = 666.7 N, F_{Hx} = −133.34 N, F_{Hz} = −133.3 N, F_{Kx} = −666.7 N, F_{Kz} = 226.7 N

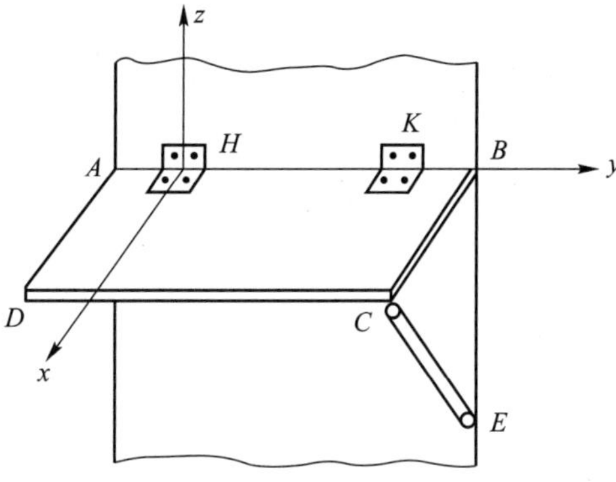

Fig.P4-15

4-16 The bar system is connected by hinges, located on the sides and diagonals of the cube, as shown in Fig.P4-16. The force $\boldsymbol{F}_{LD}$ acting at node D is along the diagonal LD. The force $\boldsymbol{F}_{CH}$ acting at node C is along the CH side. If the hinges B, L and H are fixed, determine the internal force of each bar (without the weight of the bar).

Answer: $F_1 = F_{LD}$ (pull), $F_2 = -\sqrt{2}F_{LD}$ (press), $F_3 = -\sqrt{2}F_{LD}$ (press), $F_4 = \sqrt{6}F_{LD}$ (pull), $F_5 = -P-\sqrt{2}F_{LD}$ (press), $F_6 = F_{LD}$ (pull)

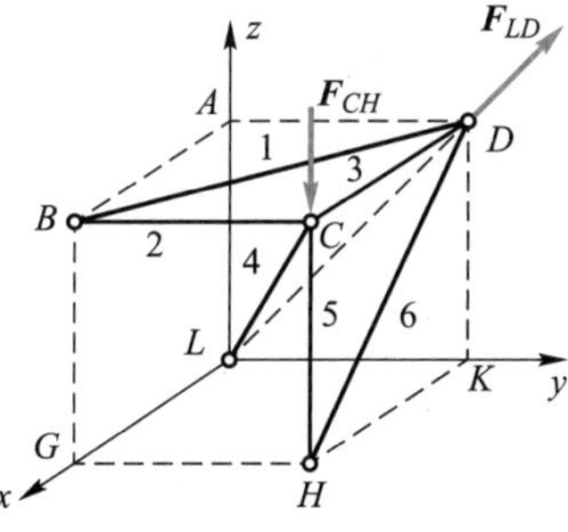

Fig.P4-16

Chapter 5 Friction

Teaching Scheme of Chapter 5

5.1 Sliding friction

Two objects with rough surfaces contact each other, when there is a relative sliding or relative sliding trend between their contact surfaces, a force along the common tangent of the contact surface is generated to prevent this relative sliding or relative sliding trend, this phenomenon is called sliding friction. Before these two objects begin to slide relative to each other, the friction force is called static sliding friction. After sliding, the friction force is called as kinetic friction force.

Since friction is a force that prevents the relative sliding between two objects, the direction of friction force on the object is always opposite to the relative sliding or relative sliding tendency of the object. The friction is analyzed according to the different cases of applied force, which can be divided into three cases, namely, static friction $\boldsymbol{F}_s$, maximum static friction $\boldsymbol{F}_{smax}$ and kinetic friction $\boldsymbol{F}_d$.

An object with gravity $\boldsymbol{W}$ is placed on a rough horizontal surface, then the object is in equilibrium under the action of gravity $\boldsymbol{W}$ and the reaction force $\boldsymbol{F}_N$ on the support surface (Fig.5-1a). When a force $\boldsymbol{F}$ is applied on the horizontal direction of the object, there is a tendency of relative sliding between the object and the horizontal surface, and the object can remain equilibrium as long as the force $\boldsymbol{F}$ is not very large (Fig.5-1b). At this point, the frictional force generated between the contact surfaces is called static friction, noted as $\boldsymbol{F}_s$. The object is in equilibrium and the magnitude of the static friction can be determined by the equilibrium equation, i.e.

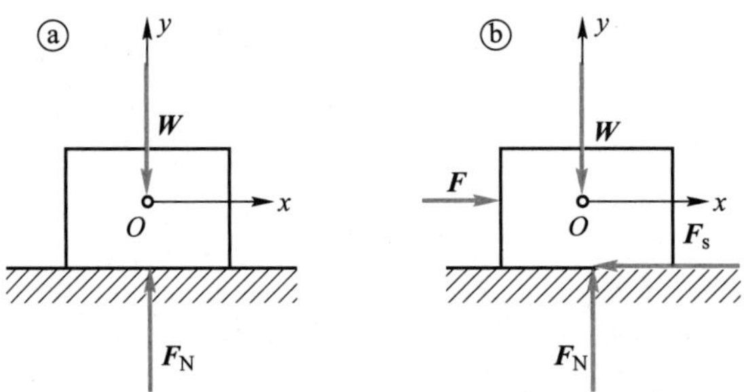

Fig.5-1

$$\sum F_{ix} = 0, \quad F - F_s = 0, \quad F_s = F$$

$$\sum F_{iy} = 0, \quad F_N - W = 0, \quad F_N = W$$

It can be seen that, as long as the block is equilibrious, the static friction is equal to the horizontal force $\boldsymbol{F}$. When the horizontal force $\boldsymbol{F}$ increases, the static friction $\boldsymbol{F}_s$ also increases, which is the common nature of static friction and reaction force.

However, the static friction force is different from the reaction force, and it does not increase indefinitely with the increase of the applied force $\boldsymbol{F}$. When the force $\boldsymbol{F}$ increases to a certain value, the object is in the critical state of sliding, but not yet sliding. When the static friction reaches its maximum value, that is the maximum static friction $\boldsymbol{F}_{smax}$, as shown in Fig. 5 – 2a. Thereafter, if the force $\boldsymbol{F}$ continues to increase, the static friction can no longer increase, the object will lose equilibrium and begin to slide (Fig.5-2b), the friction force at this time is called the kinetic friction, recorded as $\boldsymbol{F}_d$.

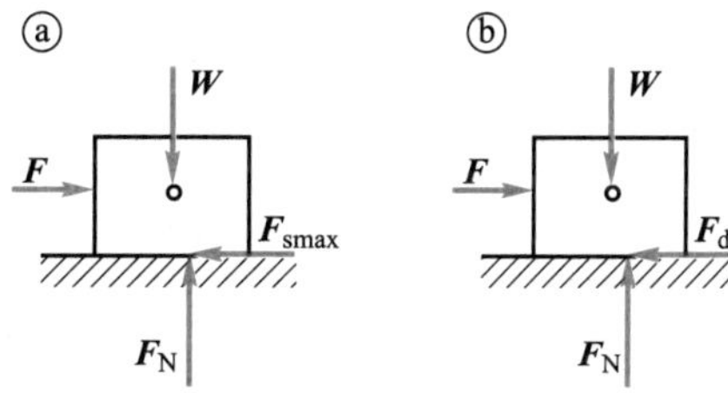

Fig.5-2

It can be seen that the magnitude of the static friction force changes with the magnitude of the applied force, and its magnitude can be determined by the equilibrium equation, and its direction is opposite to the tendency of the relative sliding of the two contacting objects. The range of the static friction is

$$0 \leqslant F_s \leqslant F_{smax}$$

According to a large number of experiments: the magnitude of the maximum static friction is proportional to the normal force F_N between the two objects, that is

$$F_{smax} = f_s F_N \tag{5-1}$$

where f_s is the coefficient of static friction, it is dimensionless. The equation (5-1) is called the law of static friction (also known as Coulomb law).

The coefficient of static friction f_s is mainly related to the characteristics of the two surfaces in contact (e.g. roughness, temperature, humidity and lubrication) and can be determined by experimental method.

As above mentioned, once the tension force just exceeds the maximum static friction, the friction on the contact surface of the object cannot allow the object to continue to maintain equilibrium, and then the object will slide. The friction force at this point is called kinetic friction, and its direction is opposite to the direction of the relative sliding between the objects. Experiments show that the magnitude of the kinetic friction force is proportional to the normal force F_N between the contacting objects, i.e.

$$F_d = f_d F_N \tag{5-2}$$

where f_d is the coefficient of kinetic friction, it is dimensionless. The equation (5-2) is called the law of kinetic friction.

The kinetic friction is different from the static friction. Generally, the coefficient of kinetic friction is smaller than the coefficient of static friction, i.e.

$$f_d < f_s \tag{5-3}$$

In addition to the material and surface conditions of the contacting objects, the coefficient of kinetic friction is also related to the magnitude of the relative sliding speed between the contacting objects. In general, the coefficient of kinetic friction decreases with increasing relative velocity. When the relative velocity is not large, f_d can be approximated as a constant.

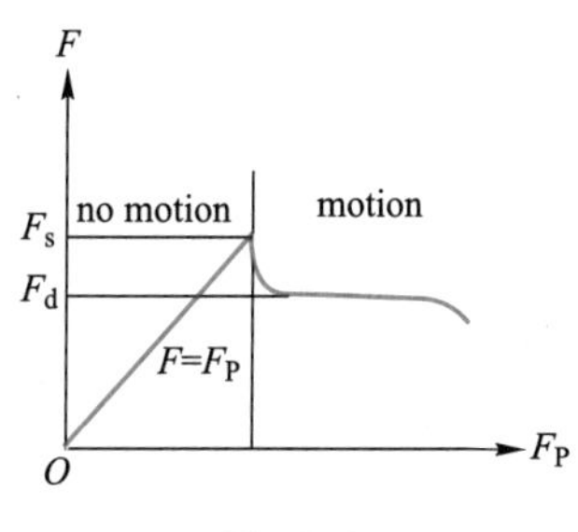

Fig.5-3

The above discussion about friction can be summarized by a graph in Fig.5 – 3, which shows the variation of the friction force $\boldsymbol{F}$

versus the applied force $\boldsymbol{F}_P$.

5.2 Friction angle and self-locking phenomenon

Static friction is also a reaction force, which can be called tangential reaction force.

When considering friction, the reaction force on the contact surface of the object includes the normal reaction force $\boldsymbol{F}_N$ and the tangential reaction force $\boldsymbol{F}_s$.

The resultant force $\boldsymbol{F}_R$ of the two reaction forces $\boldsymbol{F}_N$ and $\boldsymbol{F}_s$ is called the full reaction force (Fig.5-4a). There is a declination φ between the common normal of the contact surface and the full reaction force. The parallelogram law of force shows that the angle φ increases with the increase of the static friction $\boldsymbol{F}_s$. When the block is in the critical state of sliding, the static friction $\boldsymbol{F}_s$ reaches a maximum value $\boldsymbol{F}_{smax}$, and the angle φ also reaches a maximum value. The angle between the full reaction force in the critical state and the normal to the contact surface is called as friction angle, denoted as φ_m (Fig.5-4b).

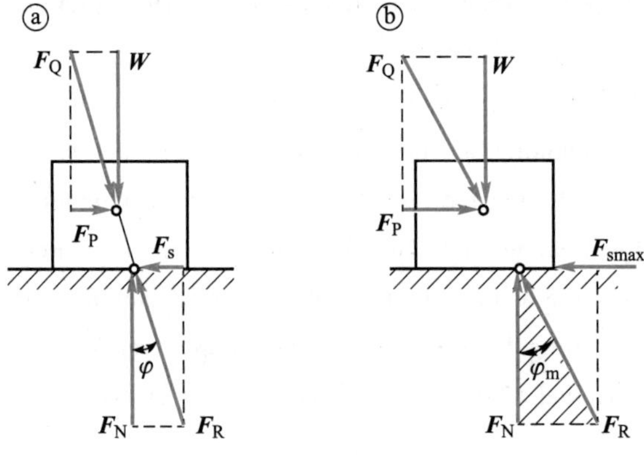

Fig.5-4

From the definition of friction angle, we have

$$\tan \varphi_m = \frac{F_{smax}}{F_N} = \frac{f_s F_N}{F_N} = f_s \tag{5-4}$$

It follows that the tangent of the friction angle is equal to the coefficient of static friction.

Based on the equation (5-4), the coefficient of static friction between two objects can be easily determined experimentally.

The two materials to be measured are attached to the surface of the block and the inclined plane, respectively. The block is placed on the inclined plane, which is initially inclined at an angle φ(Fig.5-5a), and the block is at rest. At this time, the gravity $\boldsymbol{W}$ and the full reaction force $\boldsymbol{F}_R$ should be equal in magnitude and opposite in direction. The angle φ between the full reaction force $\boldsymbol{F}_R$ and the normal of the inclined plane is equal to the inclination angle α of the inclined plane (Fig.5-5b), i.e. $\varphi = \alpha$.

Therefore, to determine the extreme value of φ angle, gradually increase the angle of inclination α, until the object will reach the critical equilibrium state of motion. When the angle of inclination α is equal to φ_m, we can measure the angle of inclination α, and then using the equation (5-4) to determine its tangent value, which is the coefficient of static friction (Fig.5-5c), that is

$$f_s = \tan \varphi_m = \tan \alpha_{max} \tag{5-5}$$

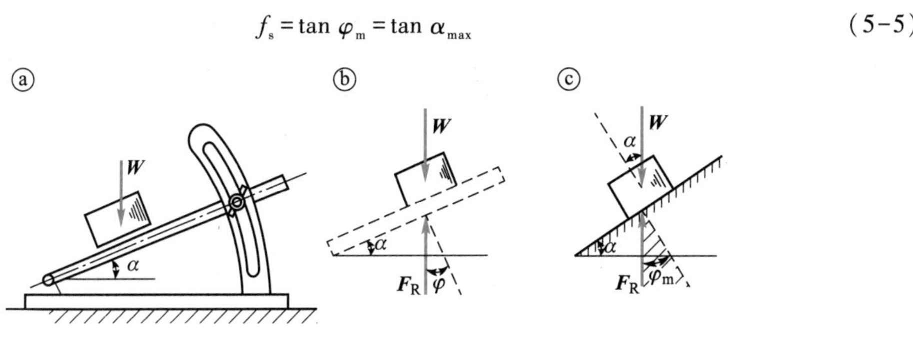

Fig.5-5

From the above experiment, it can be seen that when the block reaches the critical equilibrium state, the inclination angle α of the inclined plane is irrelevant to the mass of the block.

When the block is at rest, the static friction is always less than or equal to the maximum static friction, and thus the angle φ between the full reaction force $\boldsymbol{F}_R$ and the normal to the contact surface is always less than or equal to the friction angle φ_m. In addition, according to the two-force equilibrium condition, when the block is equilibrious, the resultant force $\boldsymbol{F}_Q$ of the applied force $\boldsymbol{F}_P$ and $\boldsymbol{W}$ acting on the object must be equal, opposite and colinear to the full reaction force $\boldsymbol{F}_R$ (Fig. 5-6). As a result, the angle between the resultant force $\boldsymbol{F}_Q$ of the applied force acting on the object and the normal to the contact surface is φ, and $\varphi \leqslant \varphi_m$. If the line of action of the resultant applied force lies in the shaded area of Fig.5-6, the full reaction force will always be equilibrious with the resultant applied force no matter how much it increases.

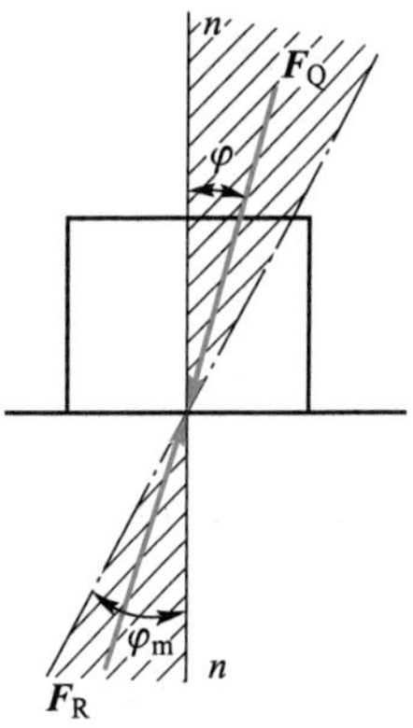

Fig.5-6

In other words, the resultant force $\boldsymbol{F}_Q$ of the applied force acting on the object, regardless of its magnitude, as long as $\varphi \leqslant \varphi_m$, the block can always be in equilibrium, this phenomenon is called self-locking phenomenon. This equilibrium is not related to the magnitude of the applied force, but only to the friction angle. The condition $\varphi \leqslant \varphi_m$ is called the self-locking condition.

5.3 Equilibrium problems with friction

The friction force is usually unknown, while the direction of friction force is opposite to the direction of the relative sliding tendency. To solve the equilibrium problem with the friction force, it is necessary to list the supplementary equations, i.e., $F_s \leqslant f_s F_N$, and the number of supplementary equations is the same with the number of friction forces.

Example 5-1 Fig.5-7a shows a block weighing 400 N that is placed on a rough horizontal surface and subjected to a force $\boldsymbol{F}$, $F=80$ N and $\alpha=45°$, with a coefficient of static friction $f_s=0.2$ between the block and the ground. (1) Determine whether the block is at rest under the force $\boldsymbol{F}$. (2) Determine the minimum value of the force F that causes the block to move toward the right.

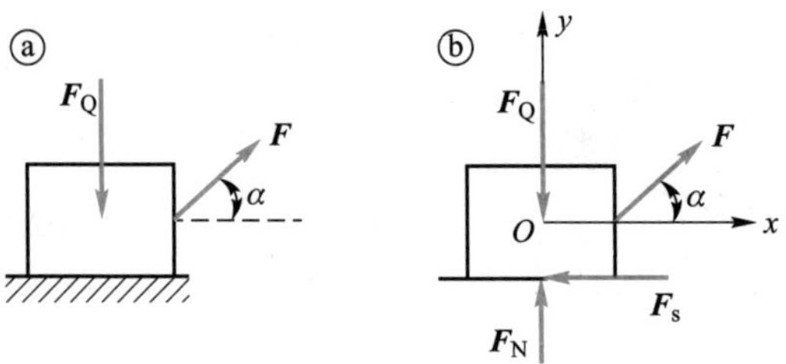

Fig.5-7

Solution (1) We do not know whether the block is in equilibrium, so firstly assume that the block is at rest, the force analysis of the block is shown in Fig.5-7b, $\boldsymbol{F}_N$ and $\boldsymbol{F}_s$ are unknown forces, the force system acting on the block can be regarded as a coplanar concurrent force system, the equilibrium equation is

$$\sum F_{ix}=0,\quad F\cos\alpha - F_s = 0$$

$$F_s = F\times\sqrt{2}/2 = 56.56 \text{ N}$$

$$\sum F_{iy}=0,\quad F_N + F\sin\alpha - F_Q = 0$$

$$F_N = F_Q - F\times\sqrt{2}/2 = 343.44 \text{ N}$$

The maximum static friction force F_{smax} is

$$F_{smax} = f_s F_N = 0.2\times 343.44 \text{ N} = 68.69 \text{ N}$$

Since $F<F_{smax}$, the block is at rest, and the friction between the block and the ground is calculated by the equilibrium equation, which is 56.56 N.

(2) From the known condition, the block is in a critical state. Let the minimum value of $\boldsymbol{F}$ be F', the block is in a critical state, so it can be listed in three independent equations (two equilibrium equations and a Coulomb law supplementary equation), that is

$$\sum F_{ix}=0,\quad F'\cos\alpha - F_{smax} = 0$$

$$\sum F_{iy}=0,\quad F_N + F'\sin\alpha - F_Q = 0$$

$$F_{smax} = f_s F_N$$

Solving the above three equations

$$F' = \frac{f_s F_Q}{\cos\alpha + f_s\sin\alpha} = \frac{0.2\times 400}{0.707+0.2\times 0.707}\text{ N} = 94.3 \text{ N}$$

Example 5-2 As shown in Fig.5-8a, a block of weight G is placed on a rough inclined surface with an angle of α. The block is at rest under the action of a horizontal force $\boldsymbol{F}_Q$. It is known that the angle of inclination α is greater than the angle of friction, if there is no horizontal force $\boldsymbol{F}_Q$, the block can not be equilibrious on the inclined surface. The static friction coefficient between the block and inclined surface is f_s. Try to determine the maximum and minimum values of the horizontal force $\boldsymbol{F}_Q$.

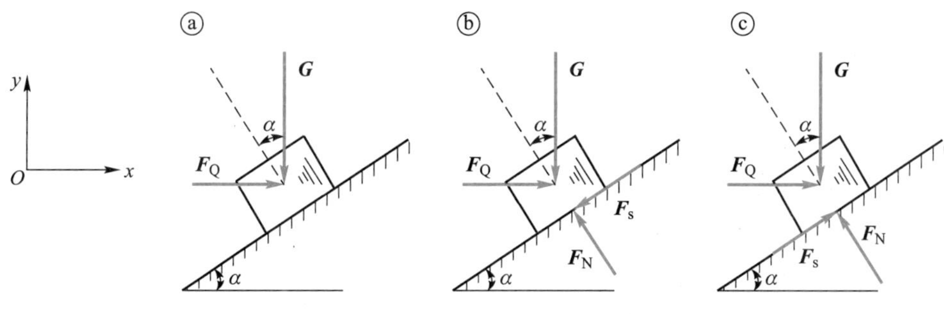

Fig.5-8

Solution This is a problem about critical state equilibrium, the maximum value of the horizontal force $\boldsymbol{F}_Q$ occurs when the block has a tendency to slide upward, and the minimum value of the horizontal force $\boldsymbol{F}_Q$ occurs when the block has a tendency to slide downward. The direction of friction force acting on the block is different in these two cases.

Considering that the block has a tendency to slide upward, force analysis as shown in Fig.5-8b, there are three unknown forces $\boldsymbol{F}_N$, $\boldsymbol{F}_Q$ and $\boldsymbol{F}_s$, and the direction of friction $\boldsymbol{F}_s$ is downward along the slope, the two equilibrium equations and a Coulomb law supplementary equation, there are

$$\sum F_x = 0,\quad F_Q\cos\alpha - G\sin\alpha - F_s = 0$$

$$\sum F_y = 0,\quad F_N - G\cos\alpha - F_Q\sin\alpha = 0$$

$$F_s \leqslant f_s F_N$$

Solving the above three equations together

$$F_Q \leqslant G\frac{\tan\alpha + f_s}{1 - f_s\tan\alpha} = G\tan(\alpha + \varphi)$$

Considering the block tends to slide downward, force analysis as shown in Fig.5-8c, there are three unknown forces $\boldsymbol{F}_N$, $\boldsymbol{F}_Q$ and $\boldsymbol{F}_s$, and the direction of the friction force $\boldsymbol{F}_s$ is upward along the slope, two equilibrium equations and a Coulomb law supplementary equation, there are

$$\sum F_{ix} = 0,\quad F_Q\cos\alpha - G\sin\alpha + F_s = 0$$

$$\sum F_{iy} = 0,\quad F_N - G\cos\alpha - F_Q\sin\alpha = 0$$

$$F_s \leqslant f_s F_N$$

Solving the above three equations together

$$F_Q \geqslant G\tan(\alpha - \varphi)$$

Therefore, when the object is equilibrious, the range of horizontal force F_Q is

$$G\tan(\alpha - \varphi) \leqslant F_Q \leqslant G\tan(\alpha + \varphi)$$

Example 5-3 A homogeneous ladder AB with length l and weight P_1 is placed against a wall, as shown in Fig.5-9a. Knowing that $\theta = \arctan(4/3)$ and the coefficient of friction between the ladder and the wall $f_B = 1/3$, a person weighing $P_2 = 3P_1$ goes up the ladder. What should be the coefficient of friction f_A between the ladder and the ground when people can safely reach to the top of the ladder?

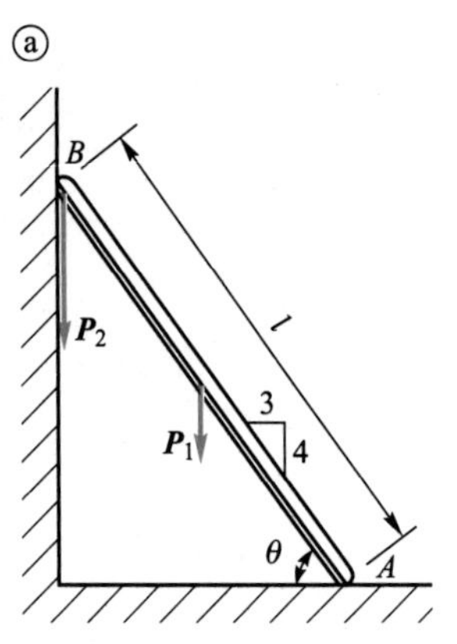

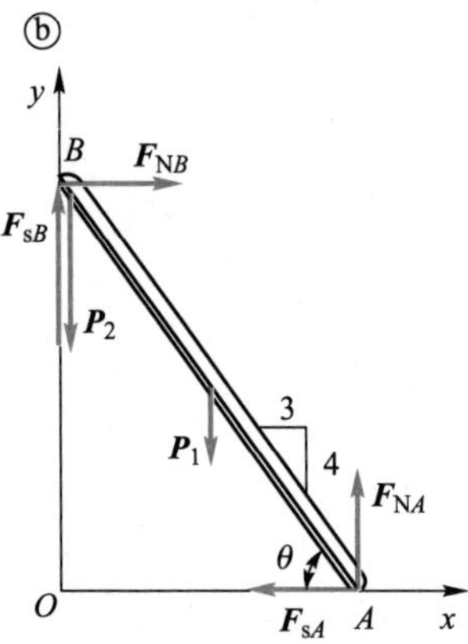

Fig.5-9

Solution Take the ladder AB as the object of study, the force analysis is shown in Fig.5-9b. Let a person reach to the top of the ladder, the ladder is in a critical equilibrium state. Choose the coordinate system shown in the figure, and list the equilibrium equation as

$$\sum M_A(\boldsymbol{F}_i) = 0, \quad P_2 l\cos\theta - F_{sB} l\cos\theta - F_{NB} l\sin\theta + P_1 \frac{l}{2}\cos\theta = 0 \tag{a}$$

$$\sum F_x = 0, \quad F_{NB} - F_{sA} = 0 \tag{b}$$

$$\sum F_y = 0, \quad F_{NA} + F_{sB} - P_1 - P_2 = 0 \tag{c}$$

Supplementary equations

$$F_{sA} = f_A F_{NA} \tag{d}$$

$$F_{sB} = f_B F_{NB} \tag{e}$$

From equation (a), equation (b) and equation (e), we get

$$F_{sA} = F_{NB} = \frac{3.5P_1}{f_B + \tan\theta}$$

Substituting into equation (c), we get

$$F_{NA} = \frac{0.5f_B + 4\tan\theta}{f_B + \tan\theta} P_1$$

Then from equation (d), we get

$$f_A = \frac{F_{sA}}{F_{NA}} = \frac{3.5}{0.5f_B + 4\tan\theta} = 0.636$$

5.4 Rolling resistance

There is a wheel on the horizontal plane with weight P and radius r. A horizontal force $\boldsymbol{F}$ acts on its center O, as shown in Fig.5-10. If the wheel and the horizontal surface are seen as a rigid body, then the contact is only a point. As a result, as long as a small horizontal force $\boldsymbol{F}$ is applied to the wheel center, the wheel will be out of balance and roll up due to force couple. But in practice, to pull or push a wheel, often add a certain force. If the force is less than a certain value, the wheel is not rolling, why?

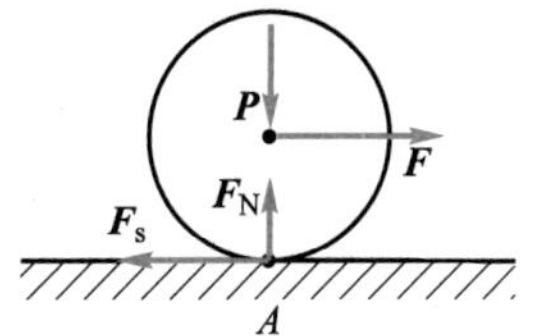

Fig.5-10

In fact, both wheel and horizontal surface are not rigid bodies. Due to the deformation, the contact surface is not a point but an arc, as shown in Fig.5-11a. The horizontal force on the wheel is also distributed in this section of the arc to form a coplanar general force system. The force system can be simplified to point A, and obtain a force $\boldsymbol{F}_{\mathrm{R}}$ and a couple. The couple moment is M_{s}, as shown in Fig.5-11b. The force $\boldsymbol{F}_{\mathrm{R}}$ can be decomposed to obtain the mutually perpendicular normal force $\boldsymbol{F}_{\mathrm{N}}$ and the frictional force $\boldsymbol{F}_{\mathrm{s}}$, as shown in Fig.5-11c. When the wheel is equilibrious, $\boldsymbol{F}_{\mathrm{N}}$, $\boldsymbol{F}_{\mathrm{s}}$ and M_{s} can be found by the equilibrium equation.

$$\sum F_x = 0, \quad F = F_{\mathrm{s}}$$

$$\sum F_y = 0, \quad F_{\mathrm{N}} = P$$

$$\sum M_A = 0, \quad M_{\mathrm{s}} = Fr$$

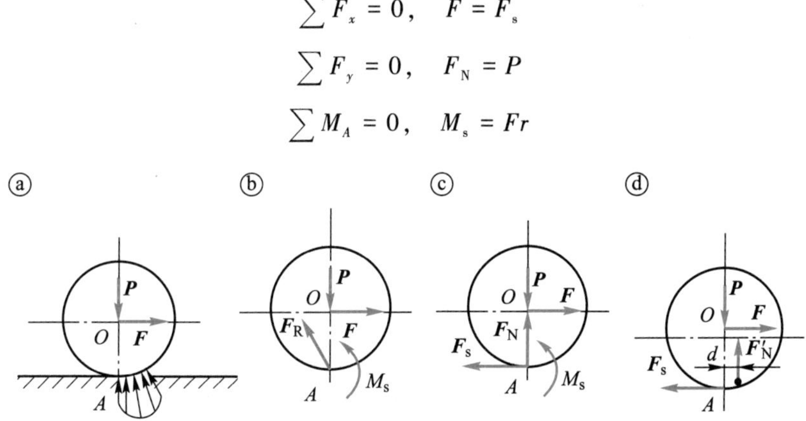

Fig.5-11

The couple moment M_{s} is called rolling resistance couple moment. From the above derivation, we can see that similar to the static friction, the rolling resistance couple moment M_{s} increases with the increase of the applied force. When the force $\boldsymbol{F}$ increases to a certain value, the roller is in the critical equilibrium, and the rolling resistance couple moment reaches its maximum value, called the maximum rolling resistant couple moment M_{smax}. If $\boldsymbol{F}$ becomes larger, the wheel will roll, and the rolling resistant couple moment is approximately equal to M_{smax} during the rolling process, and there is $0 \leqslant M_{\mathrm{s}} \leqslant M_{\mathrm{smax}}$.

Rolling resistance law: The maximum rolling resistance couple moment M_{smax} is independent of the roller radius and proportional to the magnitude of the normal force $\boldsymbol{F}_N$ on the support surface, i.e.

$$M_{smax}=\delta F_N \tag{5-6}$$

where δ is a proportionality constant, called rolling resistance coefficient, can be determined by the experiment.

Although the rolling resistance coefficient can be similar to the friction coefficient, f is a dimensionless constant, while δ has a length dimension and physical meaning. The result of reducing the reaction force system toward point A is further reduced as shown in Fig.5-11d. The normal force $\boldsymbol{F}_N$ on the support surface and the maximum rolling resistance couple moment M_{smax} can be reduced a force F'_N with $F'_N=F_N$, and the distance between the line of action of the force $\boldsymbol{F}'_N$ and the centerline is $d=M_{smax}/F'_N$. Comparing with equation (5-6), we get $\delta=d$. Thus, the rolling resistance coefficient is the maximum distance that the normal reaction force $\boldsymbol{F}_N$ deviates from point A in the direction of rolling when the wheel starts rolling.

When the wheel is subjected to horizontal pull, there is both sliding friction and rolling resistance couple moment, if the sliding frictional force reaches firstly then there is

$$F_s=F_{smax}=fF_N=fP$$

If the rolling resistance couple moment is reached firstly, there is

$$M_{smax}=\delta F_N=F_r r \Rightarrow F_r=\frac{\delta}{r}P$$

In general, there are

$$\frac{\delta}{r} \ll f$$

It means that the rolling needs much less effort than sliding.

Problems

5-1 As shown in Fig.P5-1, a rope is used to pull an object with a gravity of 500 N and a tension $\boldsymbol{F}_T$ of 150 N. (1) If the friction coefficient f is 0.45, determine whether the object is in equilibrium, and the magnitude and direction of friction force between the object and ground; (2) If the friction coefficient f is 0.577, what is the required tension to pull the object?

Answer: (1) Equilibrium, $F_f=F_x=130$ N, The direction is horizontal to the right;

(2) $F_T=250$ N

5-2 As shown in Fig.P5-2, a bar placed in a V-groove can be turned with a couple moment $M=15$ N·m. Determine the coefficient of static friction between the bar and the V-groove, knowing that the bar has a gravity of 400 N and a diameter of 25 cm.

Answer: $f=0.223$

5-3 As shown in Fig.P5-3, a ladder AB rests against a wall and weighs 200 N. The length of the ladder is l and the angle θ is 60°. The coefficient of static friction between the ladder and the contact surface is 0.25. A person weighing 650 N climbs up the ladder, what should be the distance s from the highest point C to point A that the person can reach?

Answer: $s = 0.456l$

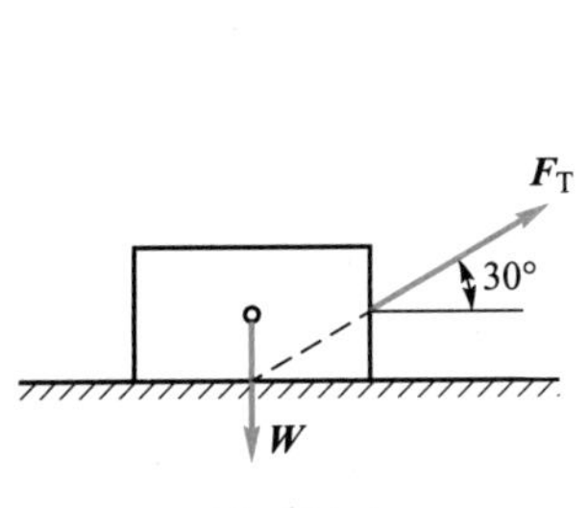

Fig.P5-1

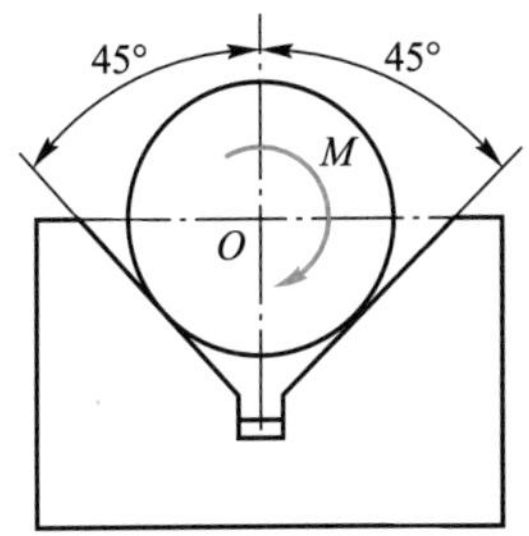

Fig.P5-2

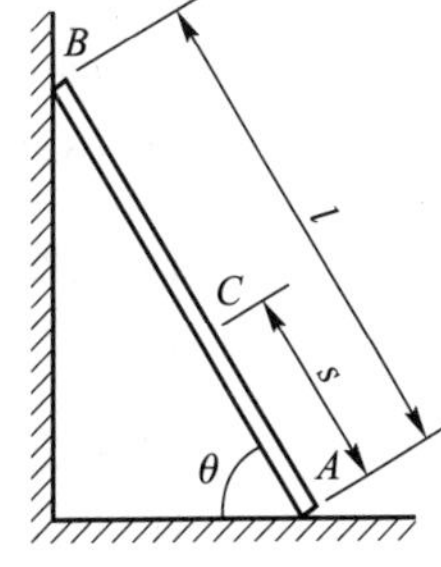

Fig.P5-3

5-4 A steel rolling mill consists of two wheels, both of which have a diameter of $d = 500$ mm and a gap of $a = 5$ mm between them. The wheels rotate in opposite directions, as shown in Fig.P5-4. It is known that the friction coefficient f between the heated steel plate and the cast iron wheel is 0.1, determine the thickness b of the iron plate that can be rolled and pressed. (Hint: the resultant force acting on the steel plate must be horizontal toward the right.)

Answer: $b \leqslant 7.5$ mm

5-5 The top angle of the split A is θ, and the weight P acts on the block B, as shown in Fig.P5-5. The friction coefficient between A and B is f (friction is not considered for elsewhere). Determine the magnitude of the force $\boldsymbol{F}$ required to lift the weight.

Answer: $\dfrac{\sin\theta - f\cos\theta}{\cos\theta + f\sin\theta}P \leqslant F \leqslant \dfrac{\sin\theta + f\cos\theta}{\cos\theta - f\sin\theta}P$

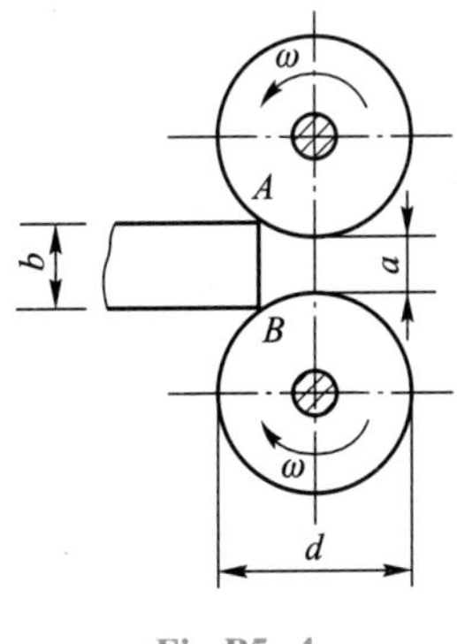

Fig.P5-4

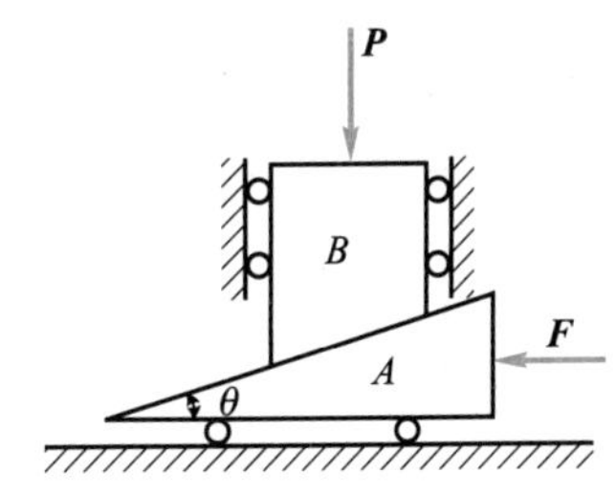

Fig.P5-5

5-6 The width of a brick clamp is 0.25 m. The curved bar AGB is articulated with $GCED$ at point G. The dimensions are shown in Fig.P5-6. Let the brick weigh $P = 120$ N, and the force $\boldsymbol{F}$ acting on the center line of the brick clamp to lift the brick, and the friction coefficient between the brick clamp and the brick is 0.5. Determine how long b is to hold the brick.

Answer: $b \leqslant 110$ mm

5-7 As shown in Fig.P5-7, the length of plate AB is l, and the ends of A and B are placed on two inclined planes at angles $\alpha_1 = 50°$ and $\alpha_2 = 30°$, respectively. The angle of friction between the end of the plate and the inclined plane is known to be $\varphi_m = 25°$. If the object M is to be placed on the plate while the plate remains horizontal, determine the range of placement of the object.

Answer: $0.183l < x < 0.545l$

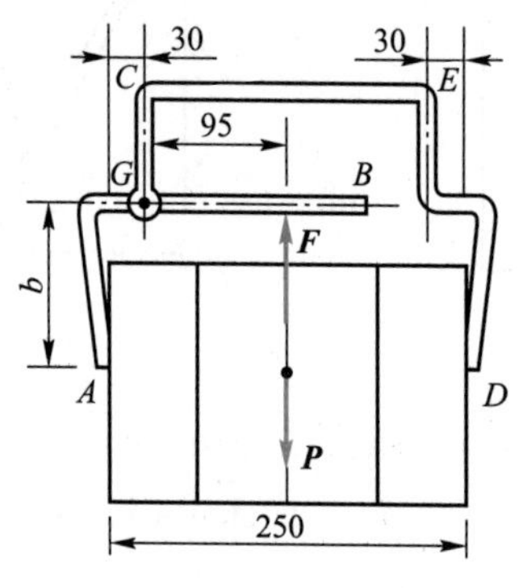

Fig.P5-6

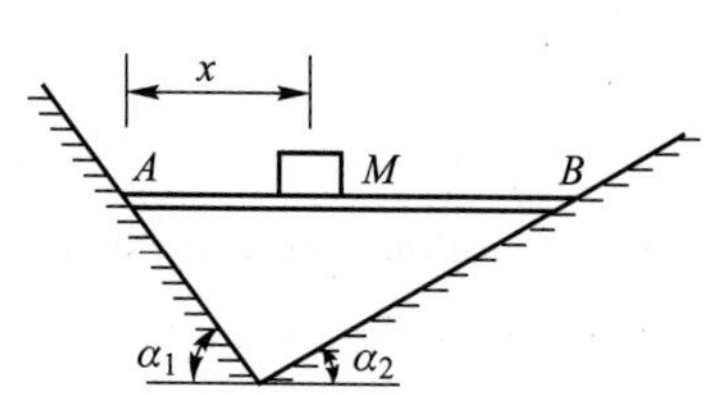

Fig.P5-7

5-8 As shown in Fig.P5-8, the object having a weight $P = 100$ kN is pressed against a straight wall by a horizontal force $\boldsymbol{F}$. The coefficient of static friction between the object and the wall is $f_s = 0.4$. How much $\boldsymbol{F}$ is required to prevent the object from sliding downward?

Answer: $F \geqslant 250$ kN

5-9 As shown in Fig.P5-9, a cylindrical roller weighs 3 kN, which has a radius of 0.3 m and is placed on a horizontal surface. If the rolling resistant coefficient $\delta = 0.5$ cm, determine the magnitude and direction of the force $\boldsymbol{F}$ required to pull the roller in the two cases of $\alpha = 0$ and $\alpha = 30°$.

Answer: $\alpha = 0$: $F = 5$ kN; $\alpha = 30°$: $F = 2.94$ kN

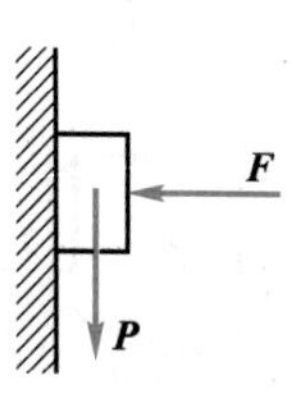

Fig.P5-8

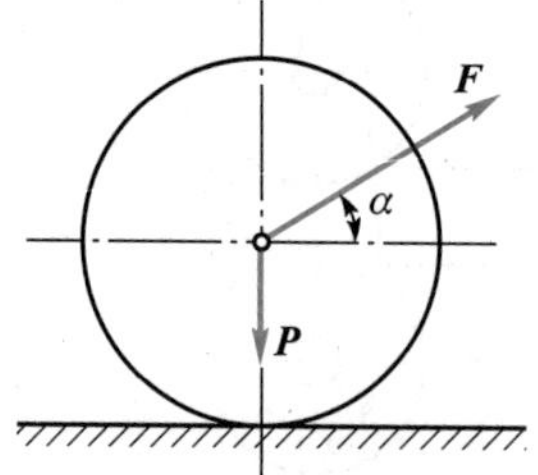

Fig.P5-9

Chapter 6 Fundamental Concepts of Bar Deformation

Teaching Scheme of Chapter 6

6.1 Tasks of deformation statics mechanics

Any structure or mechanical device is made up of a combination of parts. These are called structural member. In rigid statics, the calculation of external forces on members has been solved according to the equilibrium relationship of forces. However, how to ensure that a member works properly under external forces is still a problem that needs to be solved further.

When an engineering structure or machine works, the members are subjected to loads. Members are generally made of solids, which will undergo a change in shape and size under the action of external forces. This is known as deformation. When the load exceeds a certain extent, the member will be over-deformed or fractured. To ensure the structure or mechanical equipment can be used properly, the members should have sufficient capacity to withstand the load. Thus, each member must meet several basic requirements.

6.1.1 Members should be of sufficient strength

Strength refers to the ability of a member to resist damage under load. Usually, the damage of a member means that the member breaks or plastic deformation occurs under load. If the deformation can disappear after the forces are unloaded, this deformation is called elastic deformation. If the deformation cannot disappear after the forces are unloaded, it is called plastic deformation. Any member is not allowed to be damaged during use. For example, crane slings are not allowed to break; tooth surface of gears is not allowed to appear pressure pits.

6.1.2 Members should have sufficient rigidity

Rigidity refers to the ability of a member to resist elastic deformation under load. When elastic deformation is beyond the requirements of normal work allowed, is also not allowed. As shown in Fig.6-1a, when the gear is overload, the deformation of Fig.6-1b will occur. As a result, the gears do not mesh properly. Therefore, the deformation of the members under load can not exceed the permissible value, i.e. it must have sufficient rigidity.

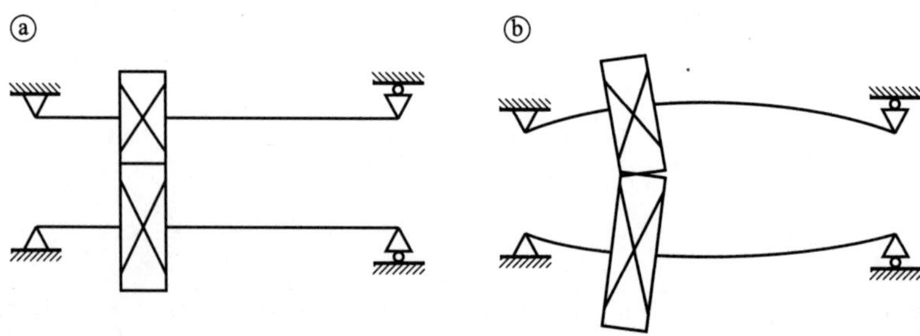

Fig.6-1

6.1.3 Members should have adequate stability

The so-called stability refers to the ability of the member to maintain its original form of balance. When the pressure is small, they can maintain their original linear equilibrium form. However, when the pressure exceeds a certain value, these members may suddenly become bent under the action of disturbing forces. This sudden change in the original form of equilibrium is called loss of stability, or instability. Thus, it is necessary to keep its original equilibrium form intact, i.e. the members are required to have sufficient stability.

Different members have different degrees of strength, rigidity and stability requirements. For example, a gas storage tank should not be broken and the key factor is to ensure strength. A lathe spindle is mainly to ensure rigidity while a piston bar under pressure should maintain stability. The ability of a member to meet the requirements of strength, rigidity and stability is called the load-bearing capacity of the member.

In addition, the economic factors should also be considered when a complex structure is designed. In general, the safety requires the use of more and better materials, while the latter requires the use of less materials. These two factors are often contradictory. The task of deformation statics mechanics is to provide a theoretical basis for the calculation of the strength, rigidity and stability of load-bearing members, to select the appropriate materials for the members and to determine a reasonable shape and size. Finally, the designed load-bearing members can meet the requirements of both safety and economy.

Compared with other fundamental sciences, this book is closely related to engineering practice. Its research method includes the whole process of experimental, theoretical and practical cycle development. The strength, rigidity and stability of a member are related to the mechanical properties of the materials, which must be determined through experiments. On the other hand, the theoretical results need to be verified by experiments. There are some complex problems that cannot be solved by existing theories alone, which need to be solved by experiments.

6.2 Simplification of engineering members

In the theoretical analysis of engineering members, some unimportant factors must be ignored and a simplified model may be established by appropriate assumptions and simplifications. In this book, the construction of simplified models is usually considered from the following aspects.

6.2.1 Deformable bodies and assumptions

Engineering members are generally made of solid materials. All solid materials will deform under the action of external forces. They are called deformable bodies. In rigid statics, solids can be regarded as rigid bodies. This because the small deformations which have little effect on the study of equilibrium problems of solids can be ignored. However, in the following part of the book, the main study focusses on the relationship between force and deformation. Thus, all members must be regarded as deformable bodies. The microstructure and mechanical properties of deformable bodies are complex, and there are differences between different materials or different parts of the same material. However, since this book is a study of the load-bearing capacity of a member from a macroscopic point of view, the influence of some microscopic factors is ignored to simplify the analysis and apply mathematical tools. Several basic assumptions are introduced as follows.

1. The assumption of continuity

It is considered that the member is continuously and void-free throughout its geometric volume. Based on this assumption, physical quantities within the object, such as deformation and displacement, can be considered as continuously varying and can be expressed as a continuous function of coordinates.

2. The assumption of homogeneity

It is assumed that the mechanical properties are identical at all points in the member. Under this assumption, any small segment of the object can be removed for analysis and the conclusions obtained can be applied to the whole object. The mechanical properties of the material measured in tests with large specimens can also be applied to any small segment of the object.

3. The assumption of isotropic

It is considered that materials have the same mechanical properties along any direction. Commonly used engineering materials, such as metals, glass and well-cast concrete, can be considered isotropic. In the study of this book, materials are generally assumed to be isotropic.

6.2.2 Basic forms of members

The members in engineering are diverse. In geometry, they can be simplified into three categories: bars, plates (shells) and blocks.

1. Bar

A bar is a member whose length is much greater than its transverse dimensions (height and width). The cross section and the axis are the two main geometric features of a bar. The cross section is the section perpendicular to the length of the bar. The axis is the line connecting the form centroids of each cross section. Bars with a straight axis are called straight bars and those with a curved axis are called curved bars, as shown in Fig.6-2. All straight bars with equal cross section areas are called equal straight bars. Bars with different cross section sizes are called variable section bars.

2. Plate and shell

A plate (shell) is a member whose thickness is much smaller than the dimensions of other two directions. The two main geometric features of a plate and shell are the mid-surface (the face that bisects the thickness) and the thickness perpendicular to the mid-surface. Those with a flat mid-surface are called plates and those with a curved mid-surface are called shells. As shown in Fig.6-3.

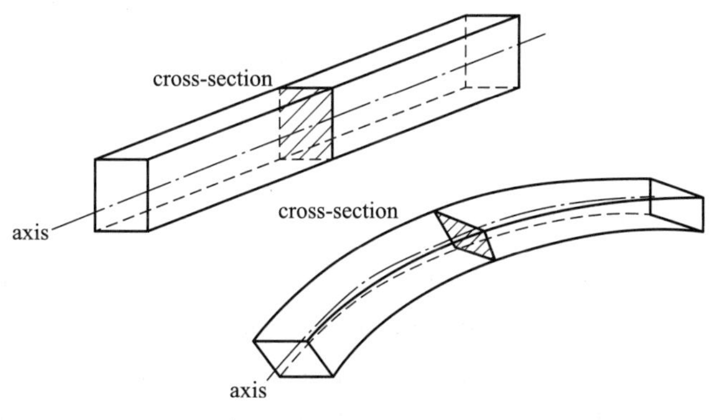

Fig.6-2

3. Block

Members with equivalent dimensions in the length, width and height directions are called blocks. As shown as Fig.6-4.

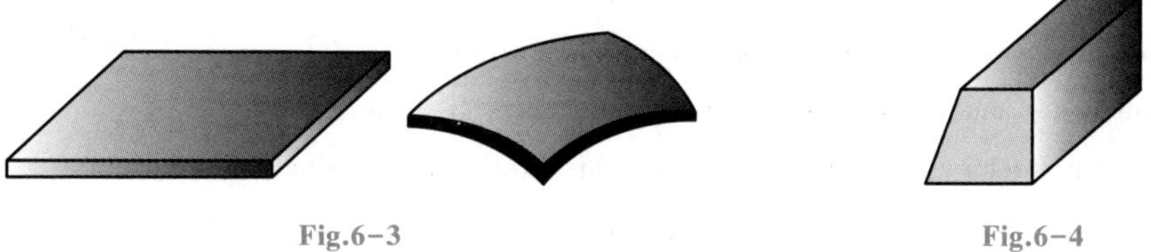

Fig.6-3 Fig.6-4

This book focuses mainly on the straight bars with uniform cross section.

6.2.3 Small deformation constraints

In engineering, the deformation of a member under the action of an external force is usually very small compared to the original dimensions of the member. Therefore, when establishing the equilibrium equations for a member or analysing some other problem, the deformation can be ignored. Some calculations can carry out by using its original dimensions. It is called a small deformation problem. For example, when calculating the reaction moment of the fixed end in Fig.6-5, Δ can be ignored and use $M = Fl$. To simply calculations, quadratic power and their products in the calculations of deformation process can be ignored since the errors are very small. When the elastic deformation of the member is very large, its impact can't be ignored. This type of problem is called large deformation problem. This book focuses mainly on small deformation problems.

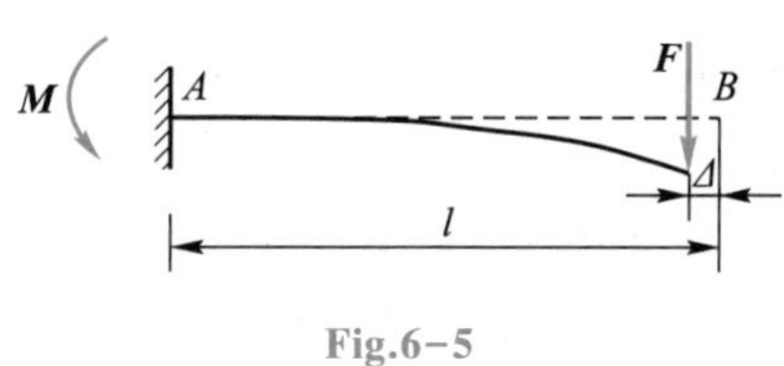

Fig.6-5

In summary, the studies of bars are made of homogeneous, continuous, isotropic materials (mainly straight bars of equal cross section) and is limited to the study of its small elastic deformation.

6.3 Internal force and stress

6.3.1 The concept of internal force

It is well known that internal forces exist within an unloaded member, such as the force $\boldsymbol{F}$ between points A and B in Fig. 6 – 6. This internal force is an intermolecular interaction force. It keeps the particles in a certain relative position to each other and keeps the members in a certain shape. Since this internal force is determined by the nature of the substance itself, it is also known as the intrinsic internal force. When an external force is applied to a member, the member is deformed and the relative positions of the adjacent segments are changed. The equilibrium position of neighboring particles is destroyed and the intrinsic internal force is readjusted. As a result, the internal force between the adjacent segments will increase, as shown in Fig.6 – 6. The additional internal force is caused by the external force and is a resistance to deformation. It increases as the external force increases. But for members made of various materials, there is a limit to the amount of additional internal force. If beyond the critical value, the member will be damaged. This additional internal force is closely related to the load-bearing capacity of the member and is the internal force studied in this book.

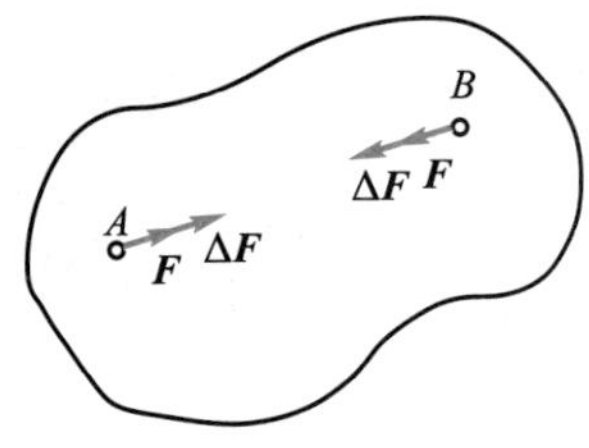

Fig.6–6

6.3.2 Section method

The following is an introduction to the basic method of calculating internal forces—the section method. The section method is based on the idea that an object in equilibrium should have all its segments in equilibrium. The object shown in Fig.6–7a is in equilibrium by the forces $\boldsymbol{F}_1$, $\boldsymbol{F}_2$, $\cdots$, $\boldsymbol{F}_n$. To find the internal forces on its arbitrary section m—m, imagine cutting the object from the section m—m and taking a part of it. For example, part Ⅰ is the selected object and part Ⅱ is removed. These two segments are interacting at the cut. After cutting, this action is replaced by the corresponding forces. In general, these forces are spatial general force system on the cross section m—m. As shown in Fig.6 – 7b, the spatial rectangular coordinate system over the section centroid is established. The forces at each point on the section are simplified towards the centroid O. The result of this simplification is three forces $\boldsymbol{X}$, $\boldsymbol{Y}$ and $\boldsymbol{Z}$ along the x, y and z axes and three couple moments M_x, M_y and M_z against the three axes. Because the object is balanced, these six internal force components can be solved by establishing the equilibrium equations. The above method of finding the internal forces is known as the section method. In general, some of the six internal force components are zero and the calculation can be simplified.

The steps in the sectional method for finding internal forces:

(1) **Cutting** Imagine cutting the member along that section and dividing it into two segments. Any segment of the member is selected and the other is removed.

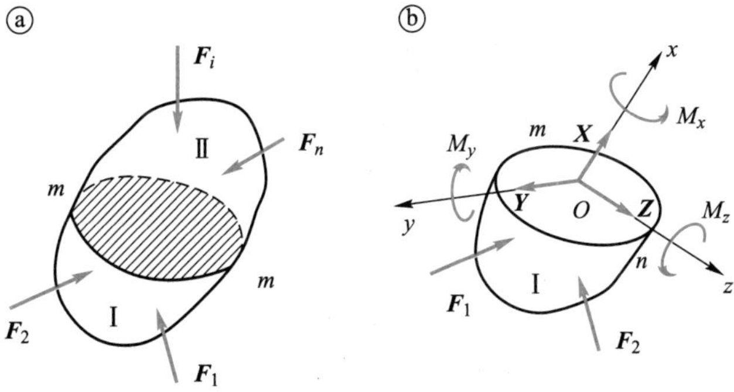

Fig.6-7

(2) **Substitute force** The action of the removed segment on the selected segment is replaced by the corresponding internal forces. The nature of the corresponding forces can be determined from the equilibrium analysis.

(3) **Equilibrium** Use the equilibrium equations and solve for the internal forces.

6.3.3 The concept of stress

Finding the internal forces alone does not solve the problem of the strength of the member. The same internal force, acting on cross sections of different sizes, will produce different results. For example, if two straight bars of the same material and different cross section area apply the same axial tensile load (the internal force in the cross section is also the same), the thin bar will break first as the tensile force increases. It is indicated that the degree of danger of a member depends on the aggregation of the distributed internal forces in the cross section, rather than on the sum of the distributed internal forces. In the above example, the same axial load is more dangerous when it is applied in a smaller cross section. The stress, which is the internal forces distributed over a unit area, can be used to describe the strength.

In general, the stresses at various points of cross section are different. In order to obtain the stress at point C on a specified section, a tiny area ΔA can be taken around point C (Fig.6-8). If the internal force acting on this tiny area is $\Delta \boldsymbol{P}$, the average stress at each point on this micro-area is

Fig.6-8

$$p_{m} = \frac{\Delta P}{\Delta A} \tag{6-1}$$

In order to obtain an accurate value of the stress at this point, ΔA can be infinitely small. When ΔA approaches zero, the limit of the average stress p_m is

$$p=\lim_{\Delta A\to 0}\frac{\Delta P}{\Delta A}=\frac{\mathrm{d}P}{\mathrm{d}A} \tag{6-2}$$

p is called the total stress of point C. In stress analysis, the total stress is usually decomposed into a normal stress component σ perpendicular to the section and a tangential stress component τ parallel to the section. The former is called the normal stress; the latter is called the shear stress. From the parallelogram law of force, we get

$$\left.\begin{aligned}\sigma&=p\cos\alpha\\ \tau&=p\sin\alpha\end{aligned}\right\} \tag{6-3}$$

The unit of stress is the Pascal, symbolized as Pa. 1 Pa is equal to 1 Newtons per square meter (1 Pa = 1 N/m^2). This unit is very small and for convenience, megapascals (MPa) and gigapascals (GPa) are often used as units in engineering, and the relationship between them is

$$1\ \mathrm{MPa}=10^6\ \mathrm{Pa}$$

$$1\ \mathrm{GPa}=10^9\ \mathrm{Pa}$$

6.4 Displacement and strain

6.4.1 Displacement

When a member is deformed, each point, line or surface within it may undergo a change in spatial position, which is called displacement. The vector represented by the line between the original position of a point and its new position is called the line displacement of point. The angle at which a line or plane is rotated during deformation is called the angular displacement. For example, the straight bar shown in Fig.6-9 is deformed under the force **F**. The total displacement of point A on the end face is AA_1, the axial line displacement is u, the transverse line displacement is v and the angular displacement of the end face is θ.

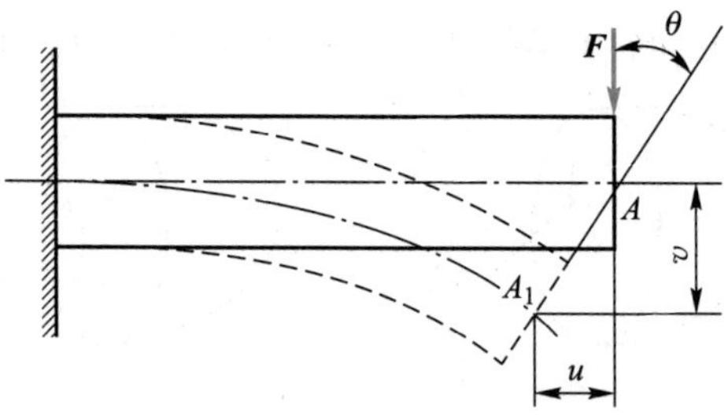

Fig.6-9

6.4.2 Strain

To study the deformation of a point within a member, it is envisaged to take a tiny square hexahedron around that point (Fig.6-10a), which is called an element. The deformation of the element has the form of a change in the length of the sides and a change in the angle of each side. In Fig.6-10b, the original side length of the element in the horizontal direction is dx, and the deformed side length is dx+ Δdx. Δdx is called the absolute line deformation of the side length dx, or line deformation. In order to reflect the degree of deformation in the dx direction, the average line deformation per unit length ε_m is introduced, that is

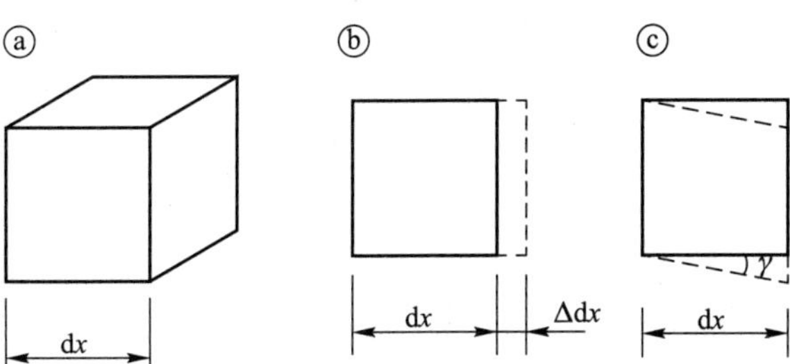

Fig.6-10

$$\varepsilon_m = \frac{\Delta dx}{dx} \tag{6-4}$$

ε_m is called the avenge normal strain. When Δdx approaches to zero, we have

$$\varepsilon = \lim_{\Delta dx \to 0} \varepsilon_m = \lim_{\Delta dx \to 0} \frac{\Delta dx}{dx} \tag{6-5}$$

ε is called the normal strain of a point. The increase in length after deformation is the tensile strain and the reverse is the compressive strain.

The sides of the original element are perpendicular to each other. Change in right angle after deformation, γ, is called angular strain or shear strain (Fig.6-10c).

Both line and shear strains are dimensionless. There is a close relationship between the normal stress σ and the normal strain ε, the shear stress τ and the shear strain γ, which will be discussed in detail in the future.

6.5 Basic forms of bar deformation

The forms of deformation under different external forces are different. For the bars, there are four basic forms.

1. Axial tension and compression

The deformation of a bar under the action of a pair axial forces at the two ends. Deformation as extension or shortening along the axis (Fig.6-11a, b).

2. Shear

The deformation of a bar under the action of a pair of transverse forces, which are perpendicular to the axis and close to each other. Deformation as the relative misalignment of the two segments of the bar along the load direction (Fig.6-11c).

3. Torsion

Under the action of a pair of opposite couples at the two ends. Deformation as the relative rotation of any two cross sections about the axis (Fig.6-11d).

4. bending

Under the action of a transverse force perpendicular to the axis or a couple whose plane is parallel to the axis. The deformation is expressed as the axis of the bar changing from a straight line to a plane curve (Fig.6-11e, f).

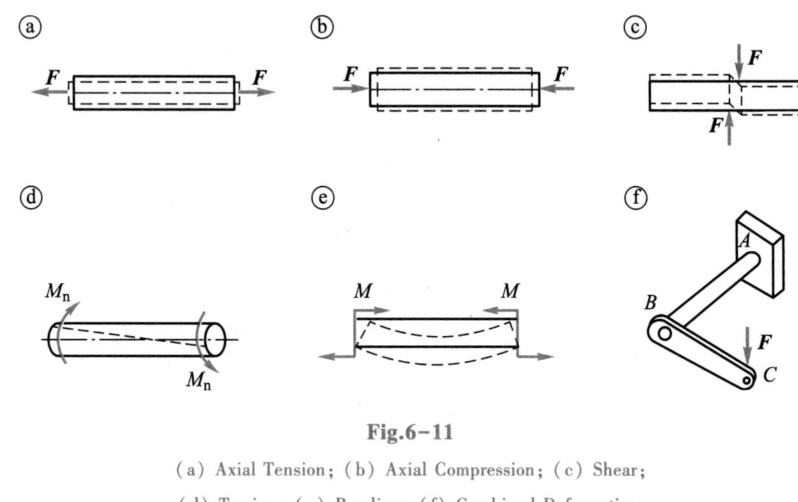

Fig.6-11

(a) Axial Tension; (b) Axial Compression; (c) Shear;
(d) Torsion; (e) Bending; (f) Combined Deformation.

Members in actual engineering may be subjected to many forms of external forces at the same time and undergo complex deformation. But any complex deformation can be seen as a combination of several of the above basic deformations.

Chapter 7 Axial Tension and Compression

Teaching Scheme of Chapter 7

Axial tensile and compressive deformation is one of the basic forms of deformation of the bars. Through the study of tensile and compressive deformation, the reader will develop a preliminary concept of the relationship between deformation and internal forces, the basic mechanical properties of materials and the steps involved in strength calculations. Some of the basic concepts and research methods presented in this chapter are the basis for the other chapters.

7.1 Introduction of axial tension and compression

The axial tensile and compressive deformations are common in engineering. For example, the turbine cover fastening bolt shown in Fig.7-1a is subjected to the reverse force of the cylinder block and the cylinder head. As shown in Fig.7-1b, a pair of tensile forces are applied along the axis. Similarly, the bar *AB* of a cantilever crane (Fig.7-2), the tension bar in a truss (Fig.7-3), etc., can be simplified to this situation. Such members are called axial tension bars.

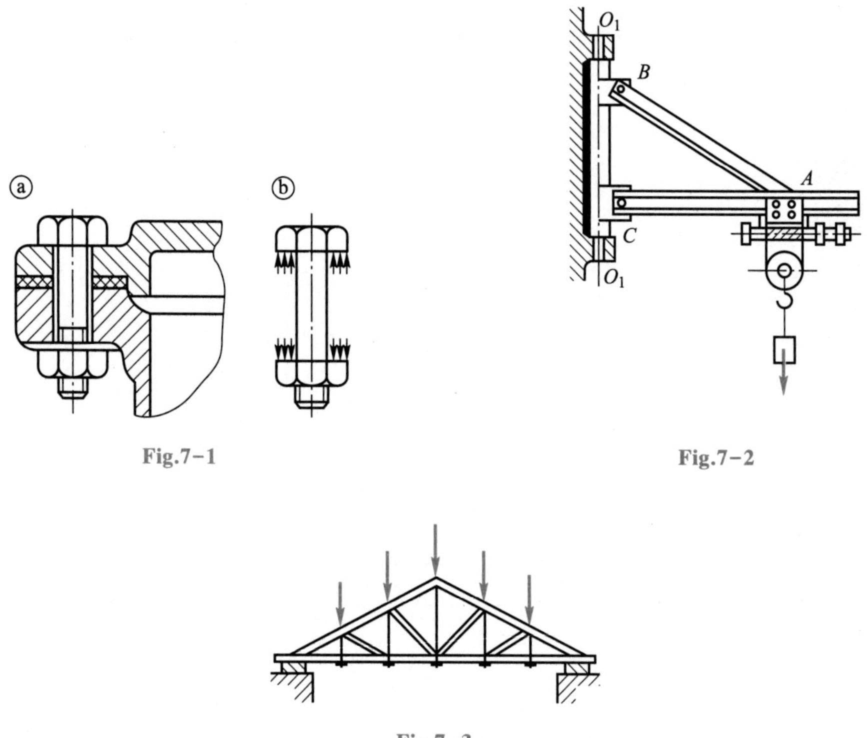

Fig.7-1

Fig.7-2

Fig.7-3

The legs of a truck-mounted crane, as shown in Fig.7-4a, are subjected to forces that can be simplified to the case shown in Fig.7-4b. They are subjected to a pair of compression forces along the axis. This type of member is called an axial compression bars. Some bars in trusses (Fig.7-3) are also axial compression bars.

In summary, the axial tension and compression of the bar is characterised by: the line of external forces acting on the bar coincides with the axis of the bar. The deformation is characterised by: the extension or shortening of the bar in the direction of the axis. The calculation sketch of the axial tension and compression bar is shown in Fig.7-5.

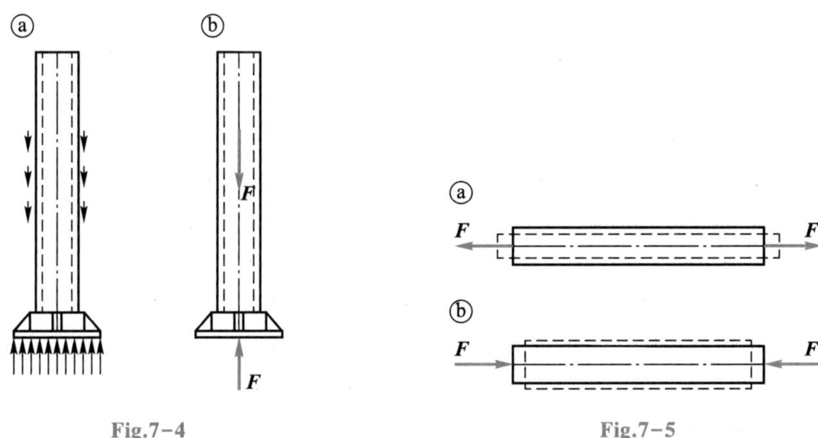

Fig.7-4

Fig.7-5

7.2 Internal forces in axial tension or compression

Applying the section method, an imaginary plane is used to divide the bar into two segments along the cross section m—m, as shown in Fig.7-6a. The internal forces interacting between the left and right segments of the bar at cross section m—m are a distributed forces system as shown in Fig.7-6b and Fig.7-6c. Let its combined force be $\boldsymbol{F}_{\mathrm{N}}$, and by the equilibrium condition

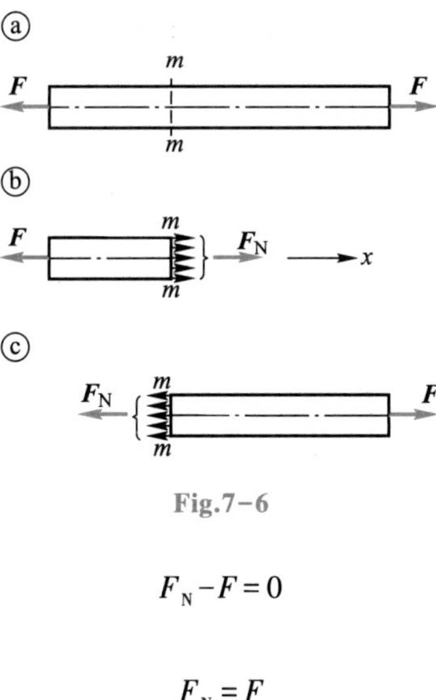

Fig.7-6

$$F_{\mathrm{N}}-F=0$$

We have

$$F_{\mathrm{N}}=F$$

Since the line of the external force $\boldsymbol{F}$ coincides with the axis of the bar, the line of the combined internal force $\boldsymbol{F}_N$ must also coincide with the axis of the bar. This internal force is called the axial force and denoted as $\boldsymbol{F}_N$. Its signs are specified that: the axial force is positive when it departs from the section and is called tensile force; the axial force is negative when it points to the section. Before establishing the equilibrium equations, let the axial force be positive (tensile force). If the result obtained is negative, it means that it is compressive force. The axial forces $\boldsymbol{F}_N$ in Fig.7-6b and Fig.7-6c are both positive.

As previously noted, the internal forces in a section are a distributed forces system. Only the resultant forces of these distributed forces can be found using the section method. In future, when studying various problems, the internal forces referred to are the resultant forces of the distributed internal forces in the section.

When there are more than two external forces acting along the axis of the bar, the axial force can still be calculated by the section method. The bar ABC shown in Fig.7-7a is in equilibrium under the action of the three forces $\boldsymbol{F}_1$, $\boldsymbol{F}_2$, $\boldsymbol{F}_3$. When calculating the axial force in the AB section, the bar is cut off along the 1—1 section and the left section is selected. As shown in Fig.7-7b, $\boldsymbol{F}_{N1}$ is the axial force in the 1—1 section. From the equilibrium condition of the left segment

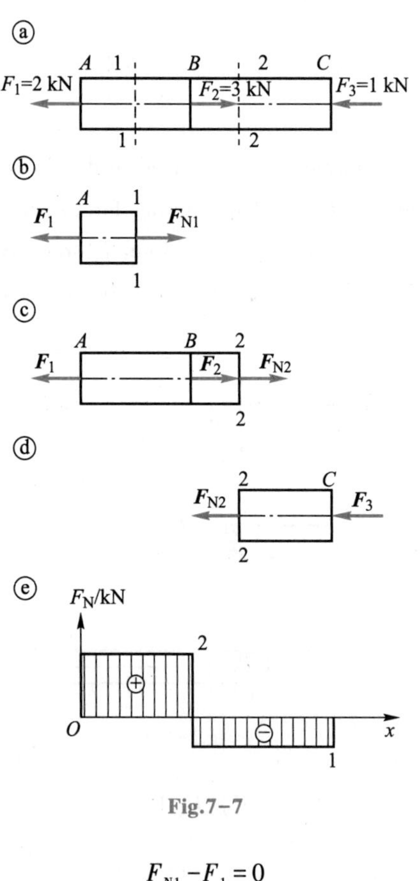

Fig.7-7

$$F_{N1}-F_1=0$$

We have

$$F_{N1}=F_1=2\ \text{kN}$$

Similarly, to calculate the axial force in the *BC* section, the bar is cut off along the 2-2 section in the *BC* section, and the left section is selected (Fig.7-7c). The axial force in the 2-2 section is expressed in terms of $\boldsymbol{F}_{N2}$. F_{N2} is first set to a positive value, and the equilibrium condition is given by

$$F_{N2}-F_1+F_2=0$$

and

$$F_{N2}=F_1-F_2=-1\ \text{kN}$$

This means that the axial force in the *BC* section is $F_{N2}=-1$ kN. The negative sign means that the assumed direction of the axial force is opposite to the actual direction. This segment is compressed. When calculating the axial force in the *BC* section, the right section can also be selected (Fig.7-7d), and F_{N2} is also first set as the tensile force. The same equilibrium conditions can be obtained from the right section

$$F_{N2}=-F_3=-1\ \text{kN}$$

In general, after the bar has been cut off by the section method, the segment of the bar with the smallest number of external forces should be selected for simplicity of calculation.

As can be seen from the above discussion, the values of the axial forces along the axia may differ from each other under the action of a number of external forces. In strength and rigidity calculations, graphs are often used to show how the values of internal forces vary with the position of the section. The position of each cross section is generally indicated by coordinates parallel to the bar axis. The corresponding internal force is indicated by coordinates perpendicular to the bar axis. The internal forces in each cross section are drawn to scale in this coordinate system. The graphs are called internal force diagrams. The graph representing axial forces are called axial force diagrams. Fig.7-7e shows the axial force diagram for bar *ABC*.

Example 7-1 A straight bar of equal section is loaded as shown in Fig.7-8a with $F_1=120$ kN, $F_2=90$ kN and $F_3=60$ kN. Try to draw the axial force diagram.

Solution (1) Find the support reaction force

Before calculating the internal force of a bar, its support reaction force is generally calculated. Let the support reaction force of the bar be $\boldsymbol{F}_A$, as shown in Fig. 7-8b, according to the equilibrium condition of the whole bar

$$-F_A+F_1-F_2+F_3=0$$

obtain

$$F_A=F_1-F_2+F_3=120\ \text{kN}-90\ \text{kN}+60\ \text{kN}=90\ \text{kN}$$

(2) Calculate the axial force of each section of the bar

Section *AB*: The bar is cut within section *AB* using an imaginary plane, taking the left section as the object of study, as shown in Fig.7-8c. The axial force in the section is assumed to be the tensile force F_{N1}. From the equilibrium conditions

$$F_{N1}-F_A=0$$

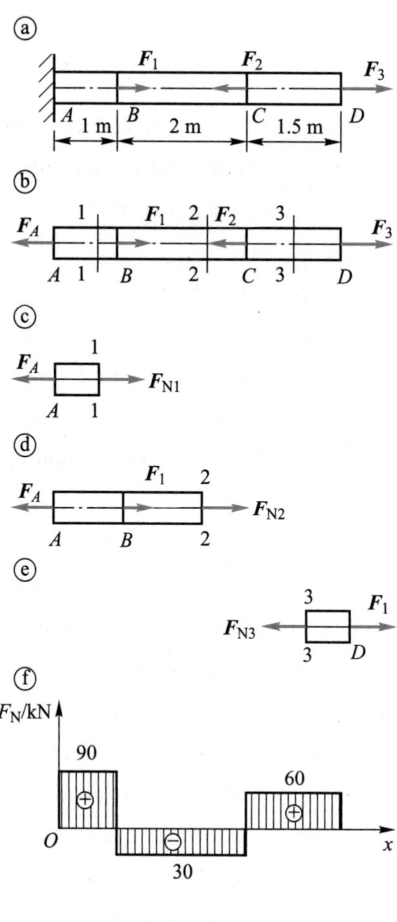

Fig.7-8

We have

$$F_{N1}=F_A=90\ \text{kN}$$

BC section: use the imaginary plane in *BC* section to cut the bar, still take the left section as the object of study, as shown in Fig.7-8d.From the equilibrium conditions, we get

$$F_{N2}=F_A-F_1=90\ \text{kN}-120\ \text{kN}=-30\ \text{kN}$$

The minus sign indicates that F_{N2} is actually compressive force.

CD section: Using an imaginary plane to cut the bar within the *CD* segment and select the right segment, as shown in Fig.7-8e. From the equilibrium conditions we can get

$$F_{N3}=F_3=60\ \text{kN}$$

(3) Draw axial force diagrams

The axial force diagram for the bar is shown in Fig.7-8f. Note that this diagram should correspond to the top and bottom of Fig.7-8a. From the axial force diagram we can get that the maximum value of the axial force in section *AB* is $F_{N\max}=F_{N1}=90$ kN.

The internal force in each section of the BC section cannot be considered to be the external force acting in the section, i.e. $\boldsymbol{F}_{N2} = -\boldsymbol{F}_2$. The axial force is the internal force, which is related to, but different from the external force.

7.3 Stress in axial tension or compression

The internal forces in the cross section of an axially tensioned (compressed) bar are only axial. Therefore, the corresponding stresses are only σ. Since the bar is assumed to be a uniformly continuous deformable body, the internal forces are continuously distributed in the cross section. The area of the cross section is denoted as A. On the differential area $\mathrm{d}A$, the internal force elements $\sigma \mathrm{d}A$ form a parallel force system perpendicular to the cross section. The resultant of $\sigma \mathrm{d}A$ is the axial force F_N. Thus the static force relationship is obtained

$$F_N = \int_A \sigma \mathrm{d}A$$

Because the law of variation of stress σ in the cross section is unknown, the specific relationship between σ and F_N cannot be determined from the above static equation alone. Before deformation, draw a series of longitudinal and transverse lines of a straight bar with equal section, which is shown in the Fig.7-9. After deformation, the following phenomenon can be observed.

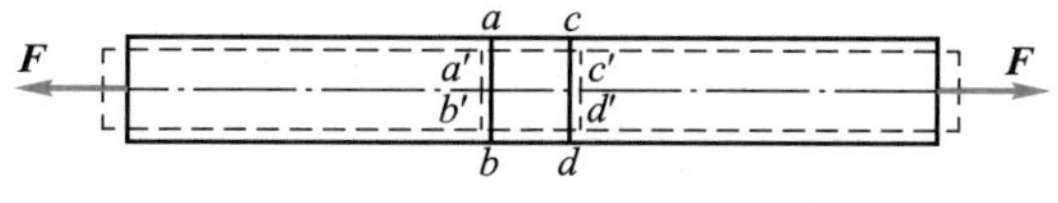

Fig.7-9

(1) The bar is stretched, but the transverse lines remain straight and any two adjacent transverse lines are shifted by a distance parallel to each other along the axis.

(2) The transverse line remains perpendicular to the axis after deformation.

The plane section hypothesis: a cross section remains plane after deformation and is still perpendicular to the axis. It can be deduced that the elongation of all longitudinal lines between any two cross sections is the same and the deformation of all points on the cross section is the same. Because the material is assumed to be homogeneous and continuous, we can get that the internal forces are uniformly distributed in the cross section. It means that the distributed internal forces (i.e. the normal stress σ) at each point of the cross section are equal, and so we have

$$F_N = \int_A \sigma \mathrm{d}A = \sigma A$$

The normal stress in the cross section of the tension (compression) bar is therefore

$$\sigma = \frac{F_N}{A} \tag{7-1}$$

The positive and negative signs of σ are specified to be the same as F_N, with tensile stresses being positive and compressive stresses negative.

Equation(7 − 1) is the calculation of normal stresses in the cross section of the axial tension (compression) bar. The conditions for its application are: external forces acting must coincide with the axis of the bar. In addition, the formula is not applicable in the small area around the point of the external force. In this area, the stress distribution is very complex and not uniform. The Saint−Venant principle shows that this small area is no larger than the transverse dimension of the bar, so that the equation (7−1) is applicable at a distance from the point of action.

Example 7−2 The dumpling bracket is shown in Fig.7−10. AB is a bar of circular section with a diameter of $d=16$ mm and BC is a bar of square section with a side length of $a=14$ mm. If the load is $P=15$ kN, calculate the stress in the cross section of each bar.

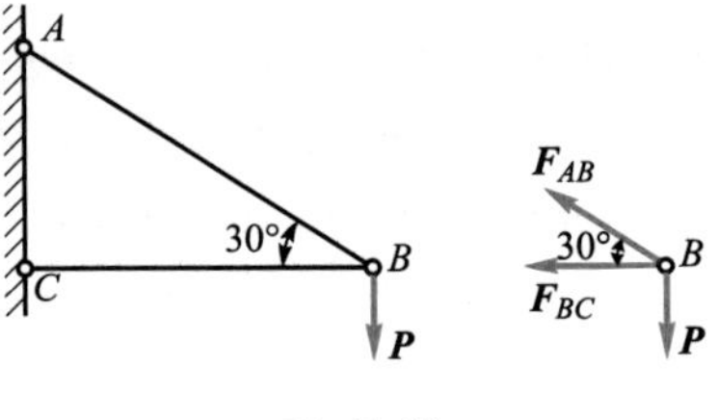

Fig.7−10

Solution (1) Calculate the axial force of each bar

Using the section method, node B is select as the object and each bar axial force is assumed to be in tension. From the equilibrium equation

$$\sum F_x = 0,\quad F_{AB}\cos 30^\circ + F_{BC} = 0$$

$$\sum F_y = 0,\quad F_{AB}\sin 30^\circ - P = 0$$

We get

$$F_{AB} = \frac{P}{\sin 30^\circ} = 30\ \text{kN}$$

$$F_{BC} = -F_{AB}\cos 30^\circ = -26\ \text{kN}$$

(2) Calculation of stresses in each bar

From the equation(7−1) we get

$$\sigma_{AB} = \frac{F_{AB}}{A_{AB}} = \frac{30\times10^3\ \text{N}}{\frac{\pi}{4}\times16^2\times10^{-6}\ \text{m}^2} = 149\times10^6\ \text{Pa} = 149\ \text{MPa}$$

$$\sigma_{BC} = \frac{F_{BC}}{A_{BC}} = \frac{-26\times10^3\ \text{N}}{14^2\times10^{-6}\ \text{m}^2} = -133\times10^6\ \text{Pa} = -133\ \text{MPa}$$

7.4 Elastic deformation in axial tension or compression

The analysis of the deformation of a bar under load is a fundamental element of the solid mechanics. The purpose of studying deformation: (1) to determine the law of distribution of stress from the law of deformation according to the relationship between deformation and force; (2) to carry out rigidity calculations. When axial tension or compression occurs in a bar, the deformation is happened in both axial direction and in transverse dimensions. The former is longitudinal deformation and the latter is transverse deformation.

7.4.1 Hooke law

An equally straight bar is shown in the Fig.7-11. Let the original length of the bar be l and the cross section area be A. Under the action of an axial tension $\boldsymbol{F}$, the length of the bar changes from l to l_1. The absolute deformation of the bar in the axial direction is

$$\Delta l = l_1 - l \tag{7-2}$$

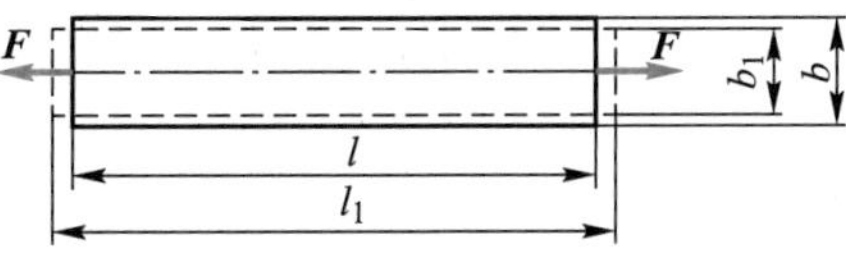

Fig.7-11

Experiments show the bar in the axial tension or compression, if the external force does not exceed a certain range, the relation of axial deformation Δl, external force $\boldsymbol{F}$, bar length l and the cross section area A is

$$\Delta l \propto \frac{Fl}{A}$$

Introducing a scale factor E, we have

$$\Delta l = \frac{Fl}{EA} \tag{7-3}$$

For a equally straight bar subjected to an external axial force at both ends (as shown in the Fig.7-11), the equation (7-3) can be rewritten as $F_N = F$

$$\Delta l = \frac{F_N l}{EA} \tag{7-4}$$

The above equation also applies to the situation in axial compression. Bar stretching, Δl is positive; bar compression, Δl is negative. The equation (7-4) is the formula for calculating the axial deformation of a straight bar in axial tension and compression, commonly known as Hooke law. The coefficient E in the formula is related to the properties of the material, which is called the modulus of elasticity of the material, and its value can be determined by experiment. The modulus of elasticity E reflects the ability of the material to resist elastic deformation: the larger of E, the stronger of the material to resist elastic deformation.

From the equation(7-4) we can see that, EA is larger, the deformation of the member is smaller. EA reflects the ability of the bar to resist tensile (compression) deformation, known as the tensile (compression) rigidity.

If we substitute $\frac{F_N}{A} = \sigma$ and $\frac{\Delta l}{l} = \varepsilon$ into equation (7-4), we can get

$$\sigma = E\varepsilon \tag{7-5}$$

This is another form of Hooke law. Hooke law can be expressed as follows: stress is proportional to strain when the stress does not exceed a certain limit value. Since the strain ε is dimensionless, the modulus of elasticity E has the same unit as the stress and is a material constant.

Finally, the equation(7-4) can only be applied when the axial force F_N, the cross section area A, the modulus of elasticity E in the bar length l is constant. For step bars or axial force changes in different parts of the bar, the equation (7-4) should be applied in every different segments. The generated value shall be superimposed, that is

$$\Delta l = \sum \frac{F_{Ni} l_i}{EA_i} \tag{7-6}$$

7.4.2 Poisson ratio

The transverse dimension of the bar before deformation is b and after deformation is b_1(see Fig.7-11), the transverse normal strain of the bar is

$$\varepsilon' = \frac{\Delta b}{b} = \frac{b_1 - b}{b}$$

Experiments have shown that, if the stress in a tension (compression) bar does not exceed a certain limit, the relationship between its transverse strain $\varepsilon'$and its longitudinal strain ε is satisfied as follows

$$\left|\frac{\varepsilon'}{\varepsilon}\right| = \mu \tag{7-7}$$

where μ is called Poisson ratio or transverse deformation factor. It is a dimensionless quantity whose value varies with the material and can be determined experimentally.

Considering that the transverse strain ε' is of opposite sign to the longitudinal strain ε, we have

$$\varepsilon' = -\mu\varepsilon \tag{7-8}$$

The modulus of elasticity E and the Poisson ratio μ are both elastic constants inherent to the material itself. They are parameters that reflect the material's ability to deform elastically. The values of E and μ for some materials are given in table 7-1.

Table 7-1 E and μ values for some materials

Name of material	E/GPa	μ
Carbon steel	196~216	0.24~0.28
Alloy steel	186~206	0.25~0.30
Grey cast iron	80~157	0.23~0.27
Copper and its alloys (brass, bronze)	74~128	0.31~0.42
Concrete	14~35	0.16~0.18
Rubber	0.007 8	0.47

Example 7-3 A stepped steel bar is shown in Fig.7-12a. The cross section area of section AC

is known to be $A_1 = 500\ \text{mm}^2$. The section CD is $A_2 = 200\ \text{mm}^2$ and that the modulus of elasticity of the bar is $E = 200$ GPa. Find: (1) the internal forces and stresses in the cross section of each section of the bar; (2) the total elongation of the bar.

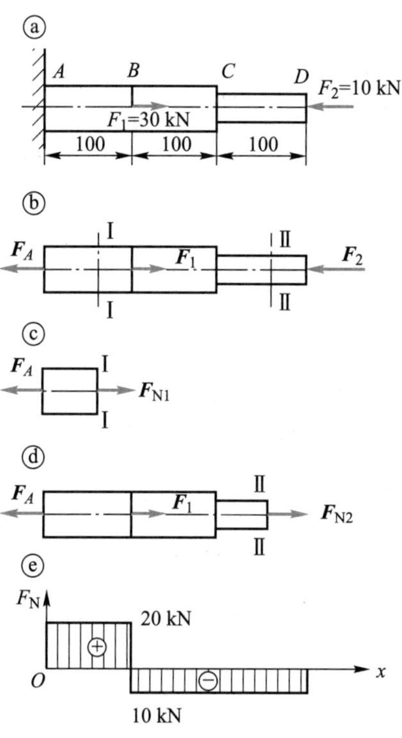

Fig.7-12

Solution (1) Calculate the internal forces

The fixed end restraint is removed and replaced by the restrained reaction force $\boldsymbol{F}_A$, as shown in Fig.7-12b. From the equilibrium equation of the whole bar

$$\sum F_x = 0, \quad -F_A + F_1 - F_2 = 0$$

This gives $F_A = 20$ kN.

The bar is cut off at sections I — I , II — II , respectively, as shown in Fig.7-12c, d. From the equilibrium conditions, the axial force in either section within section AB is

$$F_{N1} = F_A = 20\ \text{kN}$$

The axial forces in sections BC and CD are the same

$$F_{N2} = F_A - F_1 = -10\ \text{kN}$$

The resulting axial force diagram is plotted in Fig.7-12e.

(2) Calculation of stress

The normal stresses in any cross section can be calculated according to equation(7-1):

$$\sigma_{AB}=\frac{F_{N1}}{A_1}=\frac{20\times10^3\ \text{N}}{500\times10^{-6}\ \text{m}^2}=40\times10^6\ \text{Pa}=40\ \text{MPa}$$

$$\sigma_{BC}=\frac{F_{N2}}{A_1}=\frac{-10\times10^3\ \text{N}}{500\times10^{-6}\ \text{m}^2}=-20\times10^6\ \text{Pa}=-20\ \text{MPa}$$

$$\sigma_{CD}=\frac{F_{N2}}{A_2}=\frac{-10\times10^3\ \text{N}}{200\times10^{-6}\ \text{m}^2}=-50\times10^6\ \text{Pa}=-50\ \text{MPa}$$

(3) Calculate the total tension of the bar

$$\begin{aligned}\Delta l_{AD} &= \sum_{i=1}^{3}\frac{F_{Ni}l_i}{EA_i}\\ &=\frac{1}{200\times10^9}\times\left(\frac{20\times10^3\times100\times10^{-3}}{500\times10^{-6}}-\frac{10\times10^3\times100\times10^{-3}}{500\times10^{-6}}-\frac{10\times10^3\times100\times10^{-3}}{200\times10^{-6}}\right)\ \text{m}\\ &=-0.015\times10^{-3}\ \text{m}=-0.015\ \text{mm}\end{aligned}$$

A negative calculation indicates that the entire bar is shortened.

Example 7-4 A steel plate of dimensions $l\times h\times b=250\ \text{mm}\times50\ \text{mm}\times10\ \text{mm}$ is shown in the Fig.7-13. The modulus of elasticity $E=200$ GPa and Poisson ratio $\mu=0.25$. Caculate the change in thickness of the plate.

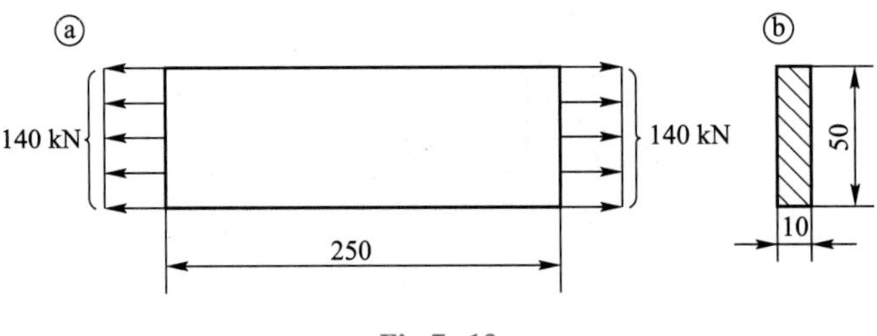

Fig.7-13

Solution Under the action of a uniformly distributed load at both ends, the steel plate undergoes axial tensile deformation. The normal stress in the cross section can be calculated according to equation (7-1), i.e.

$$\sigma=\frac{F}{A}\tag{7-9}$$

By Hooke law, we have

$$\varepsilon=\frac{\sigma}{E}\tag{7-10}$$

Lateral strain is

$$\varepsilon'=\frac{\Delta b}{b}=-\mu\varepsilon\tag{7-11}$$

So

$$\Delta b=-\mu\varepsilon b\tag{7-12}$$

Substituting equation(7-10) into equation (7-12), and considering equation (7-9), we obtain

$$\Delta b=-\mu \cdot \frac{F}{EA} \cdot b=-0.25\times\frac{140\times10^3}{200\times10^3\times50\times10}\times10\ \text{mm}=-0.003\ 5\ \text{mm}$$

This means that the thickness of the steel plate is reduced by 0.003 5 mm.

7.4.3 Strain energy in axial tension (compression)

The elastic system is deformed by an external force, which does work on the corresponding displacement. This work will be transformed into energy stored in the linear elastic body. When the external force decreases and the deformation gradually disappears, the linear elastic body will again release the stored energy and do work. The energy stored in the linear elastic body as a result of deformation is called deformation energy or strain energy.

Fig.7-14a shows a straight bar under axial tension. The tensile force acting on the lower end slowly increases from zero to the final value F_1. As the force increases, the displacement also gradually increases to its final value Δl_1. When the stress is less than the proportional limit, F is proportional to the Δl, as shown in Fig.7-14b. On the base of F (the elongation is Δl), the increasement $\mathrm{d}F$ leads to the corresponding deformation increment is $\mathrm{d}(\Delta l)$. Thus the force F on the bar does work due to the displacement $\mathrm{d}(\Delta l)$, and the work done is

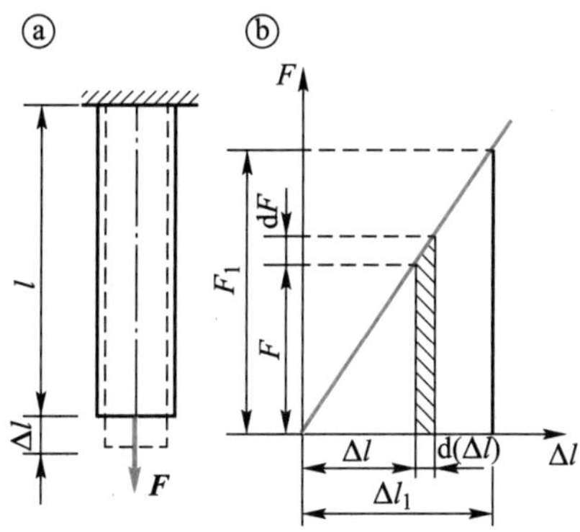

Fig.7-14

$$\mathrm{d}W=F\mathrm{d}(\Delta l)$$

Clearly, $\mathrm{d}W$ is equal to the differential area of the shaded part of the line drawn in Fig.7-14b. Considering the force as an accumulation of a series of $\mathrm{d}F$, the total work W should be the sum of the above differential areas, i.e. W is equal to the area under the $F-\Delta l$ curve:

$$W=\frac{F\Delta l}{2}$$

Because the external force starts from zero and gradually increases to the final value F, it is a linearly varying force doing work, unlike the case of a constant force doing work. Therefore, the work is equal to half of the final value of the force F and displacement Δl, i.e. the expression for the work of the external force has a factor of 1/2.

If no energy is lost, the strain energy U stored in the linear elastic body should be equal to the work W.

$$U=W=\frac{F\Delta l}{2} \tag{7-13}$$

Consider the axial force $F_N=F$, and apply Hooke law $\Delta l=\frac{F_N l}{EA}$, we have

$$U=\frac{F_N^2 l}{2EA} \tag{7-14}$$

The unit of strain energy is the joule(J), 1 J=1 N · m.

For bars with non-uniform stresses, the forces and deformations are not equal at all points. Therefore, the strain energy stored at each point is not the same. Thus, the concept of strain energy per unit volume is introduced and is known as the deformation specific energy or strain energy density (denote as u).

For a linearly elastic material, the force and deformation at all points are the same, so the strain energy density is

$$u=\frac{U}{Al}=\frac{F\Delta l}{2Al}=\frac{1}{2}\sigma\varepsilon \tag{7-15}$$

Hooke law $\sigma=E\varepsilon$, which can be written as

$$u=\frac{1}{2}\sigma\varepsilon=\frac{\sigma^2}{2E}=\frac{E\varepsilon^2}{2} \tag{7-16}$$

The unit of strain energy density is joules per metre^3(J/m^3).

7.5 Mechanical properties of materials in tension and compression

The mechanical properties of materials refer to the deformation, damage and other characteristics of materials during deformation by forces. The strength and rigidity of a member are not only related to its dimensions and the loads, but are also closely related to the mechanical properties of the materials. The modulus of elasticity E and Poisson ratio μ, are constants that reflect the mechanical properties of the material. The mechanical properties of a material are determined experimentally. There are many experiments to test the mechanical properties of materials. The tensile and compressive tests of materials at room temperature and under static loading conditions described in this section are the most basic experiments.

In order to facilitate comparison of the experimental results of different materials, the national standards have uniform provisions of the specimen shape, processing accuracy, experimental environment, etc. A standard specimen with a circular cross section in tension is shown in the Fig.7-15. The thicker ends of the specimen are the clamping part. In the middle segment of the specimen is drawn a length l_0 of the experimental section, which is called the gage length. The diameter of the circular specimen is indicated by d_0. There are two ratios between the pitch and the diameter in the national standard.

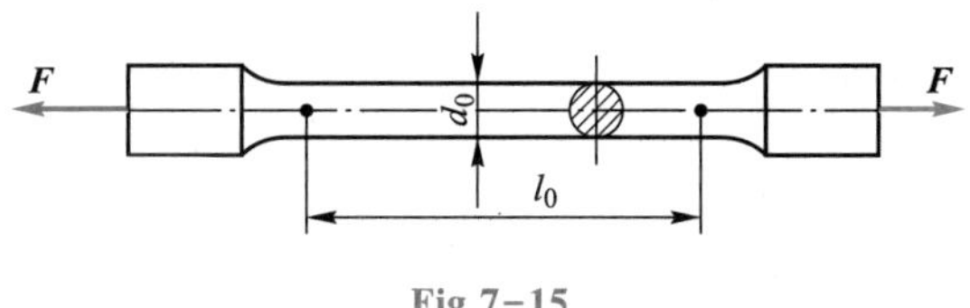

Fig.7-15

$$l_0 = 5d_0 \text{ and } l_0 = 10d_0$$

Axial tension (compression) tests are usually carried out on a universal materials testing machine. The test specimen is mounted on the machine and slowly loaded to deform the test specimen. The size of the force $\boldsymbol{F}$ can be read out from the force plate of the test machine, and the deformation Δl can be measured by the corresponding measuring instrument. Until the test specimen is destroyed, the load $\boldsymbol{F}$ and its corresponding deformation Δl is recorded step by step. At the same time, the $F-\Delta l$ curve is automatically plotted. The $F-\Delta l$ curve for standard specimens varies according to the material. Materials in engineering can be divided into two categories according to their mechanical properties: one category is called plastic materials, such as mild steel; the other category is called brittle materials, such as cast iron. The mechanical properties of these two types of materials in tensile and compressive deformation are described below.

7.5.1 Mechanical properties of low carbon steel in tension

Low-carbon steel is a variety of carbon steels with a carbon content of 0.25% or less, and is a widely used material in engineering. It is commonly used to reveal some of the mechanical properties of plastic materials.

The Fig.7-16 shows the $F-\Delta l$ curve drawn for the stretching of Low-carbon steel, which is also known as a tensile diagram. In order to eliminate the influence of the size of the specimen and to obtain a curve that reflects the inherent properties of the material, we use the normal stress $\sigma = F/A$. The Δl is divided by the original length l_0 of the standard distance to obtain the normal strain $\varepsilon = \Delta l / l_0$. With σ as the vertical coordinate and ε as the horizontal coordinate, the relationship between σ and ε can be made, called the stress-strain curve or $\sigma-\varepsilon$ curve. The stress-strain curve for low-carbon steel is shown in the Fig.7-17 and is similar in shape to its tensile diagram.

1. Four stages of the $\sigma-\varepsilon$ curve

The $\sigma-\varepsilon$ curve for low-carbon steel throughout the tensile test can be divided into four stages as follows.

(1) Elastic range

The elastic range is the initial range Ob in Fig.7-17. During this range the deformation of the specimen is completely elastic. If the load is removed at this stage, the deformation can disappear completely. The stress corresponding to point b is called the elastic limit of the material and is expressed as σ_e. Except for a very small section near point b, which is a slightly curved curve, the other segments (Oa section) are straight lines. Thus, the stress is basically proportional to the strain. The stress corresponding to point a is called the proportional limit of the material (σ_p). It can be seen that the Hookes law introduced earlier applies to $\sigma<\sigma_p$. From Hookes law it can be seen that

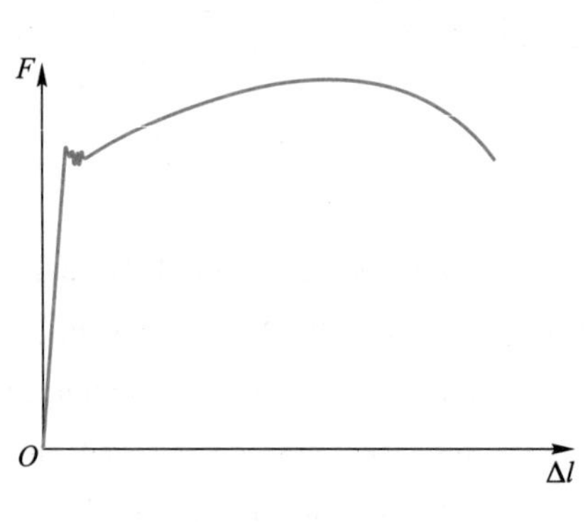

Fig.7-16

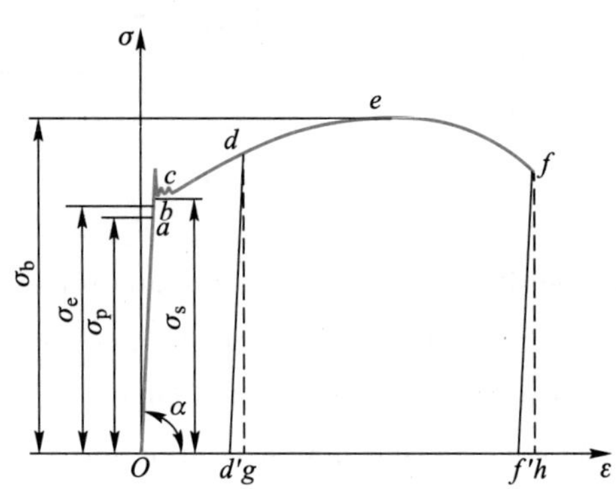

Fig.7-17

$$E=\sigma/\varepsilon$$

The slope of the straight line Oa in the Fig.7-17 is the modulus of elasticity E,

$$E=\frac{\sigma}{\varepsilon}=\tan\alpha$$

The greater the slope of the Oa stage, the greater the modulus of elasticity E of the material.

For metallic materials in general, points a and b on the $\sigma-\varepsilon$ curve are so close together that it is difficult to distinguish them experimentally. The proportional and elastic limits are often approximated as equal in engineering. However, for non-metallic materials such as rubber, their elastic properties can continue well beyond the proportional limit, so the above approximation is not reasonable.

(2) Yielding range

The yielding range is the section bc in Fig.7-17. As the stress exceeds the elastic limit of the material, the slope of the curve becomes smaller. It reachs a point where the stress suddenly drops and then fluctuates within a very small range. At this point, the strain increases rapidly while the stress remains almost constant. The $\sigma-\varepsilon$ curve is a small sawtooth shape close to horizontal. This phenomenon is called yielding or flow of the material. The stress corresponding to the lowest point of the yielding range is called the yield stress or yield limit (σ_s).

When the material yields, a stripe at 45° to the axis will appear on the surface of specimen, as shown in Fig.7-18. This stripe is known as a slip line. Slip lines are formed by the relative slippage between the lattices within a metallic material. The lattice slip causes irrecoverable plastic deformation of the material, so the yield stress σ_s is an important index of material

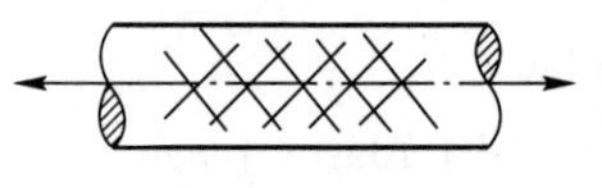
Fig.7-18

strength.

(3) Strain hardening range

The strain hardening range is the *ce* section in Fig.7-17. After the yielding range, the curve rises again gradually and the material regains its ability to resist deformation. The tensile force must be increased in order for the specimen to continue to deform. This phenomenon is known as strain hardening. The stress corresponding to the highest point *e* is called the ultimate stress and ultimate strength, expressed by σ_b, which is another important index of the strength of the material. During the strengthening phase, there is a significant reduction in the transverse dimension of the specimen, but the deformation remains uniform throughout the experimental section.

(4) Necking range

The necking range is the *ef* section in Fig.7-17. When the stress reaches the ultimate strength, there is a sudden and sharp reduction in the transverse dimension, forming the necking phenomenon shown in Fig. 7-19. Due to the reduction of the cross section area, the tensile force becomes smaller and smaller. In the $\sigma-\varepsilon$ diagram, the force $\sigma = P/A_0$ calculated using the original cross section area A_0 decreases. At point *f*, the specimen fails and the fracture is cup conical.

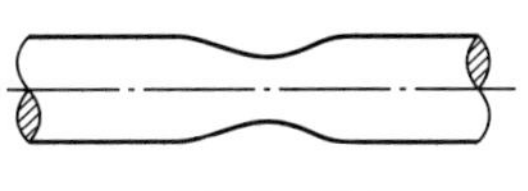

Fig.7-19

2. Percentage elongation and percentage reduction in Area

After the test specimen is break, the elastic deformation disappears and the plastic deformation retained within the standard distance is

$$\Delta l = l_1 - l_0$$

where l_0 is the original gage length and l_1 is the distance between the gage marks at fracture.

Dividing Δl by l_0 and expressing it as a percentage, this is Of' in the Fig.7-17 and is called percentage elongation, defined as

$$\delta = \frac{\Delta l}{l_0} \times 100\% = \frac{l_1 - l_0}{l_0} \times 100\%$$

Elongation is the main index of material plasticity. In engineering, materials with a δ value of more than 5% are called plastic materials, such as carbon steel, alloy steel, copper, aluminium, etc.; materials with a δ of less than 5% are called brittle materials, such as cast iron, stone, concrete, glass, etc.

The percentage reduction in area measures the amount of necking that occurs and is defined as

$$Z = \frac{A_0 - A_1}{A_0} \times 100\%$$

where A_0 is the original cross section area and A_1 is the final area at the fracture section. Z is also an index of the plasticity of the material.

3. Unloading law and cold hardening

If a specimen is loaded to strain hardening range, such as point *d* in the Fig.7-17, and then the load is unloaded. The unloading path is along a line *dd'* almost parallel to *Oa* back to point *d'* on the horizontal axis, During the process of unloading, the stress and strain vary in a linear relationship. This known as the unloading law. After all the stress has been unloaded, in the Fig.7−17, $d'g$ represents the elastic strain while Od' represents the plastic strain retained. The sum of them Og being the total strain generated by loading to point d. Thus, the total strain corresponding to any point d beyond the elastic range contains both elastic and plastic strains.

If unloading is followed by reloading within a short period of time, the stress−strain relationship rises roughly along the straight line $d'd$ and then def. Comparing the curves $Oabcdef$ and $d'def$ in the Fig.7 − 17, it can be seen that the proportional limit is increased, but the plastic deformation is reduced. This phenomenon is known as cold hardening.

For certain elements that do not require high plasticity, cold hardening can be used to improve its strength. For the production of large plastic deformation process, should try to eliminate the impact of cold hardening. The annealing process used in engineering is to play this role.

4. Mechanical properties of other plastic materials in tension

In addition to low carbon steel, there are also medium carbon steels, high carbon steels, alloy steels, brasses, aluminium alloys and other plastic materials commonly used in engineering. The Fig.7−20 gives the $\sigma-\varepsilon$ curves for several plastic materials in tension. Some of these curves do not have a distinct yielding range. Some do not have a necking range, but the common feature is that they all have an elastic range and a large percentage elongation. For plastic materials with no obvious yielding range, the nominal yield stress $\sigma_{0.2}$ that the stress corresponding to a plastic strain of 0.2% is used as the yield stress(Fig. 7−21).

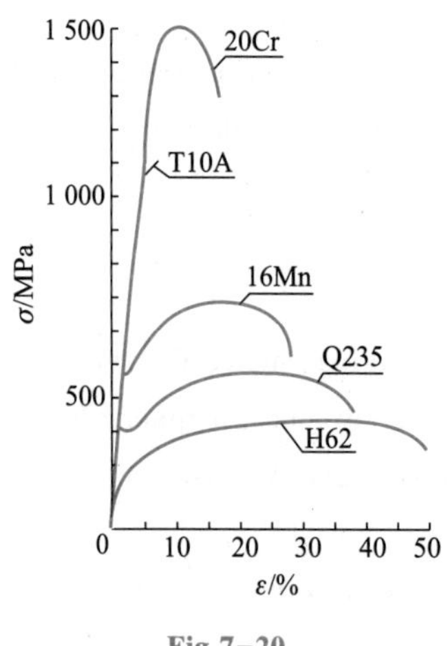

Fig.7−20

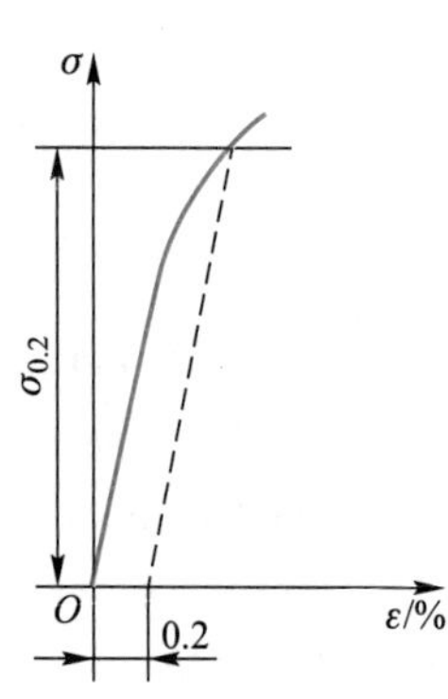

Fig.7−21

7.5.2 Mechanical properties of cast iron in tension

Grey cast iron is a typical brittle material and its $\sigma-\varepsilon$ diagram in tension is a slightly curved (Fig.7−22). It can be seen that the deformation of the specimen before failure is very insignificant and the elongation is $\delta=0.4\%\sim0.5\%$. The specimen is break along the cross section and the fracture is rough. There is no yielding or necking range in the tensile process. The maximum stress is the ultimate strength σ_b. It is the only index of the strength of cast iron. Cast iron has a very low tensile ultimate strength σ_b of about 140 MPa. $\sigma - \varepsilon$ curve does not have a distinct linear part. To determine the modulus of elasticity E, the beginning of the curve is often replaced by a cut line (Fig.7-22). The slope of the cut line is taken as the modulus of elasticity, called the cut line modulus of elasticity.

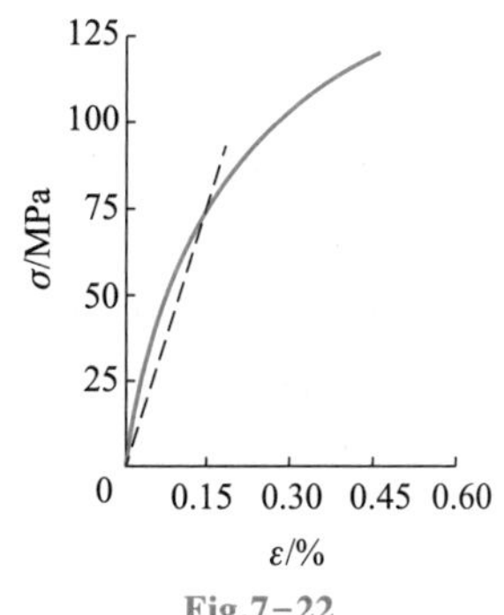

Fig.7-22

Cast iron becomes ductile iron after spheroidization. Its mechanical properties have changed significantly, not only has a better strength, there are better plastic properties.

7.5.3 Mechanical properties of low carbon steel in compression

Compression test of metal materials in accordance with *Metallic materials—Compression test method at room temperature* (GB/T 7314—2017), usually using cylindrical specimens. In order to avoid being bent during the experiment, the cylinder can not be too high, usually take $h = (1.5 \sim 3)\ d$, as shown in the Fig.7-23.

The $\sigma - \varepsilon$ curve for low-carbon steel in compression are shown as solid lines in Fig.7-24 (For comparison, the $\sigma - \varepsilon$ curve for low-carbon steel in tension are plotted as dashed lines). Prior to the yielding range, the compression and tension curves almost coincide. It is indicate that the modulus of elasticity E, the proportional limit σ_p and the yield stress σ_s are all approximately the same for low-carbon steel in compression as in tension. However, in the strain hardening range, the specimen becomes flatter and flatter. The compressive capacity becomes stronger and stronger, which no longer reflects the compressive capacity of the original specimen. Thus, the compressive ultimate strength of the material cannot be measured. Sometimes, it can be assumed that the tensile and compressive properties of low-carbon steel are the same.

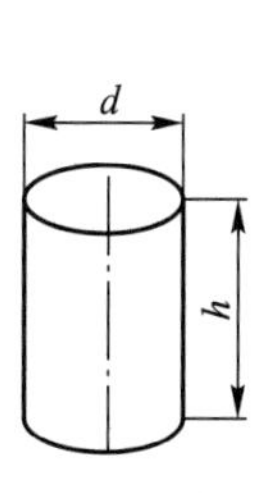

Fig.7-23

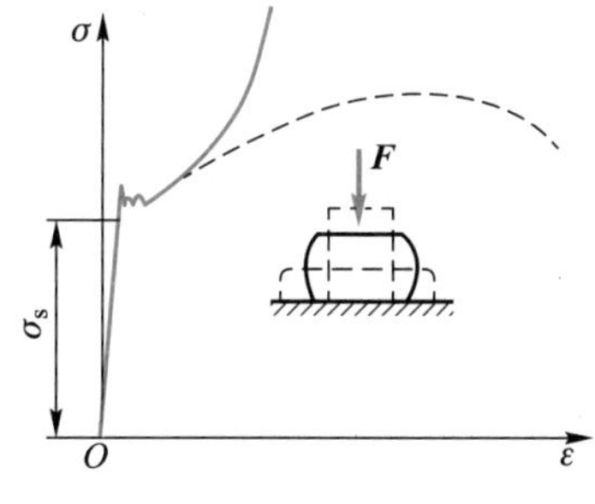

Fig.7-24

7.5.4 Mechanical properties brittle materials in compression

The $\sigma - \varepsilon$ curve for cast iron in compression is shown in Fig.7-25. When compared with the $\sigma - \varepsilon$ curve for cast iron in tension, there is also no obvious straight part and no yielding. However, the elongation of cast iron in compression is much greater than that in tension. The compressive ultimate strength σ_c is much greater than the tensile ultimate strength σ_b about 4 ~ 5 times. Usually, the fracture of a cast iron specimen in compression occurs in a diagonal direction of 45° to 55° from the axis. The damage process of brittle material specimens is a complex mechanical process. The end friction of the specimen is an important factor affecting this process. For cast iron specimens treated with good polishing, lubrication and other measures to reduce friction, the fracture is approximately in the axial direction. Other brittle materials, such as concrete and stone, have a much higher compressive strength than tensile strength.

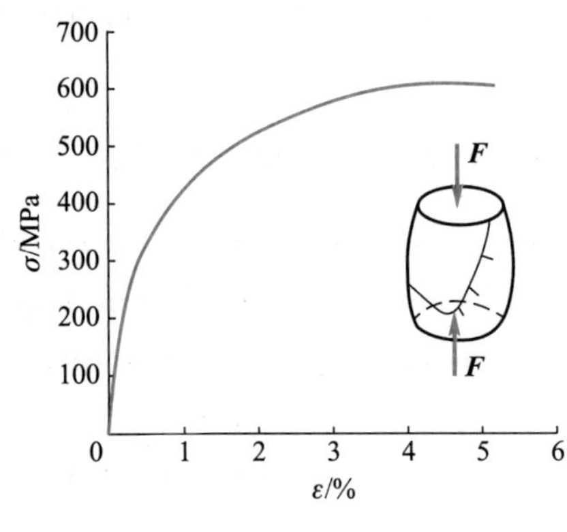

Fig.7-25

The brittle material has low tensile strength and poor plasticity, but its high compressive strength and low price make it suitable for making pressure-bearing members. It should note that the mechanical properties of the material are related to the temperature and the nature of the load. The above conclusions are limited to room temperature and static loading conditions. When the temperature and loading conditions change, the mechanical properties of the material will vary. Generally, at low temperatures and high rates of force, the material tends to become brittle; at high temperatures and low rates of force, the material tends to have increased plastic properties.

7.6 Strength calculation in axial tension or compression

7.6.1 Factor of Safety and Allowable Stress

It is known from experiments that when the stress reaches a certain limit value, the material will fail. The stress that causes the material to break is called the ultimate stress, expressed as σ_u. The ultimate stress of a material is determined by its nature. For plastic materials, the ultimate stress is their yield stress σ_s(or $\sigma_{0.2}$). For brittle materials, the ultimate stress is their ultimate strength σ_b or compression ultimate strength σ_c.

To ensure that the member has sufficient strength, the actual working stress σ should obviously be lower than its ultimate stress. A certain safety reserve must be left for the strength of the member. In strength calculations, the ultimate stress is usually divided by a factor greater than 1 as the upper limit of the working stress. It is called the allowable stress $[\sigma]$.

$$[\sigma]=\frac{\sigma_u}{n} \tag{7-17}$$

where the factor n, which is greater than 1, is called the factor of safety.

The determination of the factor of safety n is quite important and complex. It is a design factor that is supplemented for the possibility of inaccuracy of the actual member dimensions and other conditions of use. If n becomes larger, it becomes uneconomical because the mass increases more than necessary. A reasonable selection of n should take into account many safety aspects as well as economic aspects. The selection of a reasonable factor of safety is closely related to the level of technological development.

The factor of safety or the value of the stress used for various materials under different working conditions can be found in the relevant design codes or materials manuals. For plastic materials, $n_s=1.2\sim2.5$; for brittle materials, $n_b=2\sim3.5$, up to $3\sim9$, specific data can be found in the relevant design codes.

7.6.2 Strength calculation

To ensure that the axial tension (compression) bar can have sufficient strength, the maximum working stress shall not exceed the allowable stress of the material in tension (compression), i.e. the requirement

$$\sigma_{max}=\left(\frac{F_N}{A}\right)_{max}\leqslant[\sigma] \tag{7-18}$$

Maximum working stress requirements in bars for equal section bars

$$\sigma_{max}=\frac{F_{Nmax}}{A}\leqslant[\sigma] \tag{7-19}$$

The equation (7-18) and (7-19) are the strength conditions of the axial tension (compression) bar. The section where the maximum working stress is generated is called the dangerous section. Using strength conditions, three aspects of strength calculations in engineering can be solved.

1. Strength check

The material, cross section dimensions and the load to which the bar is subjected are known. The bar is checked to see if it meets the strength conditions. If so, the bar is sufficiently strong; if not, the bar is unsafe.

2. Section design

The load and the permissible stresses are known and the minimum cross section area are required. The strength conditions can be transformed into the following form

$$A\geqslant\frac{F_{Nmax}}{[\sigma]} \tag{7-20}$$

From this equation the required cross section area is calculated and then the section size is determined.

3. Allowable loads

Knowing the dimensions of the bar and the permissible stresses in the material, it is necessary to determine the maximum load that the bar can withstand. This is done by calculating the maximum permissible axial force according to the following formula

$$F_{\text{Nmax}} \leqslant A[\sigma] \tag{7-21}$$

Example 7-5 Check the strength of the steel cable of the hoist shown in Fig.7-26a. The cross section area $A=2.8\ \text{cm}^2$, the ultimate strength $\sigma_b = 400$ MPa and the factor of safety $n_b = 4.5$ are known.

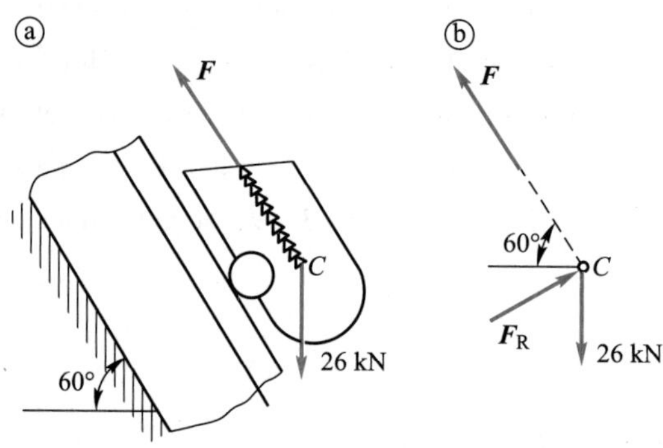

Fig.7-26

Solution (1) Calculate the tension of the steel cable

The centre of gravity C is the centre of the cart and the force diagram is shown in Fig.7-26b. The support reaction force of the rail is F_R, the tension force of the cable is $\boldsymbol{F}$, and the weight of the car is 26 kN. They are in the same plane and intersect at point C. From the equilibrium condition, we can get

$$F = 26\ \text{kN} \times \cos 30° = 22.5\ \text{kN}$$

(2) Calibrate the strength of the steel cable

The failure stress of a steel cable is generally taken as the ultimate strength σ_b, so the allowable stress of a steel cable is

$$[\sigma] = \frac{\sigma_b}{n_b} = \frac{400}{4.5}\ \text{MPa} = 88.9\ \text{MPa}$$

The working stress of the steel cable is

$$\sigma = \frac{F}{A} = \frac{22.5\times10^3}{2.8\times10^{-4}}\ \text{Pa} = 80.4\ \text{MPa} < [\sigma]$$

The steel cable is sufficiently strong.

Example 7-6 A straight bar of equal circular cross section is shown in Fig.7-27a. It is made of cast iron with a tensile allowable stress $[\sigma_t] = 60$ MPa, a compressive allowable stress $[\sigma_c] =$ 120 MPa and a modulus of elasticity $E = 80$ GPa. (1) Draw the axial force diagram; (2) Design cross section diameter; (3) Calculate the total elongation of the bar.

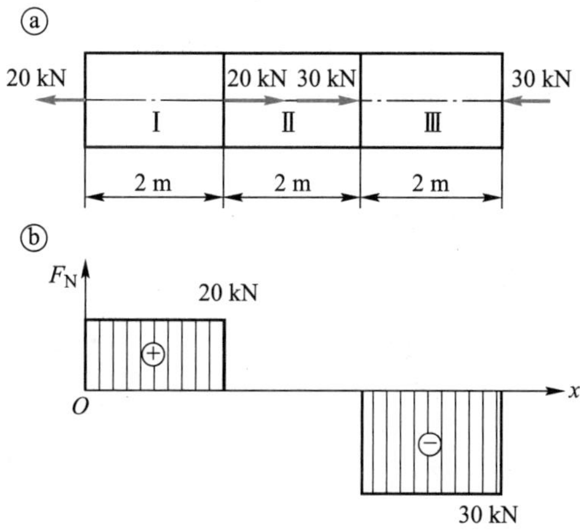

Fig.7-27

Solution (1) Draw the axial force diagram

The internal forces in the cross section of Ⅰ, Ⅱ and Ⅲ sections of the bar are $F_{N1}=20$ kN, $F_{N2}=0$ and $F_{N3}=-30$ kN, respectively, which can be drawn from the axial force diagram as shown in Fig.7-27b.

(2) Design cross section diameter

The negative axial force in section Ⅲ is greater in absolute value than the positive axial force in section Ⅰ. However, the cross sections in sections Ⅰ and Ⅲ are both dangerous sections. The design by tensile strength has

$$d'\geqslant\sqrt{\frac{4F_{N1}}{\pi[\sigma_t]}}=\sqrt{\frac{4\times20\times10^3}{3.14\times60\times10^6}}\ \text{m}=20.6\ \text{mm}$$

Designed for compression strength with

$$d''\geqslant\sqrt{\frac{4F_{N2}}{\pi[\sigma_c]}}=\sqrt{\frac{4\times30\times10^3}{3.14\times120\times10^6}}\ \text{m}=17.8\ \text{mm}$$

The diameter of the bar should be taken as $d=d'=20.6$ mm. The results show that the cross section dimensions of the bar are determined by the axial tension, even though the axial tension of the bar is smaller than the axial pressure. Because the tensile capacity of cast iron is lower than the compressive capacity.

(3) Calculate the total elongation of the bar

According to equation(7-6), the total elongation of the bar is

$$\Delta l=\Delta l_1+\Delta l_2+\Delta l_3=\frac{20\times10^3\times2\times10^3}{80\times10^3\times\frac{3.14\times(20.6)^2}{4}}\ \text{mm}+0\ \text{mm}-\frac{30\times10^3\times2\times10^3}{80\times10^3\times\frac{3.14\times(20.6)^2}{4}}\ \text{mm}=-0.75\ \text{mm}$$

Where the "−" sign indicates that the bar is actually shortened.

Example 7-7 The structure shown in Fig.7-28a has an allowable stress of $[\sigma]_1 = 140$ MPa, $[\sigma]_2 = 100$ MPa for the bar 1 and bar 2, respectively. The cross section area is $A_1 = 4\ \text{cm}^2$ and $A_2 = 3\ \text{cm}^2$. Find the allowable load for the structure.

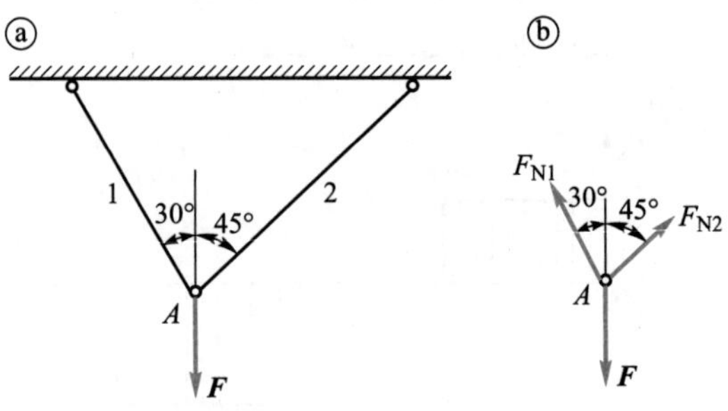

Fig.7-28

Solution (1) Calculate the internal force of each bar

Intercepting node A as the object of study, as shown in Fig.7-28b, the equilibrium equation of node A is given by

$$F_{N2}\sin 45° - F_{N1}\sin 30° = 0$$

$$F_{N1}\cos 30° + F_{N2}\cos 45° - F = 0$$

The solution is

$$F_{N1} = \frac{2F}{\sqrt{3}+1} = 0.732F$$

$$F_{N2} = \frac{\sqrt{2}F}{\sqrt{3}+1} = 0.518F$$

(2) Determination of allowable loads

Strength conditions by bar 1:

$$\sigma_1 = \frac{F_{N1}}{A_1} = \frac{0.732F}{A_1} \leqslant [\sigma]_1$$

It is possible to obtain

$$F \leqslant \frac{A_1[\sigma]_1}{0.732} = \frac{4\times10^{-4}\times140\times10^6}{0.732}\ \text{N} = 76.5\ \text{kN}$$

Strength conditions by bar 2:

$$\sigma_2 = \frac{F_{N2}}{A_2} = \frac{0.518F}{A_2} \leqslant [\sigma]_2$$

It is possible to obtain

$$F \leqslant \frac{A_2[\sigma]_2}{0.518} = \frac{3\times10^{-4}\times100\times10^6}{0.518}\ \text{N} = 57.9\ \text{kN}$$

Taking the smaller of the two, the permissible structural load $[F] = 52.9$ kN.

Example 7-8 The structure is subjected to the forces shown in Fig.7-29a. q is a uniform load distributed over the horizontal length. The bar AC is a rigid bar and the bar BD is a circular section. The allowable tensile stress is $[\sigma] = 150$ MPa and the modulus of elasticity is $E = 200$ GPa. Try to calculate the diameter of the bar BD and the plumb displacement at point C.

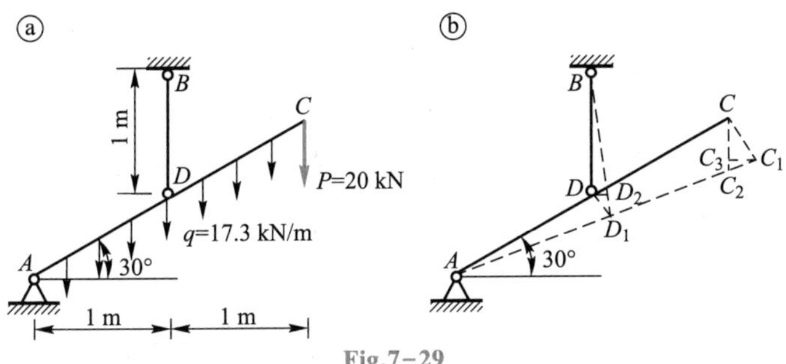

Fig.7-29

Solution (1) Calculate the diameter of the bar BD

Let the tension on the bar BD be F_N. From the equilibrium equation of the bar AC

$$\sum M_A = 0, \quad F_N \times 1\ \text{m} - 17.3\ \text{kN/m} \times 2\ \text{m} \times 1\ \text{m} - 20\ \text{kN} \times 2\ \text{m} = 0$$

We get

$$F_N = 74.6\ \text{kN}$$

Then from the strength condition we get

$$A \geqslant \frac{F_N}{[\sigma]} = \frac{74.6\times10^3}{150\times10^6}\ \text{m}^2 = 497\ \text{mm}^2$$

$$d \geqslant \sqrt{\frac{4A}{\pi}} = \sqrt{\frac{4\times497}{\pi}}\ \text{mm} = 25.2\ \text{mm}$$

Take a diameter of $d = 26$ mm.

(2) Calculate the plumb displacement at point C

Due to the elongation of the bar BD, the rigid bar AC is turned to the new position AC_1 as shown in Fig.7-29b. Point D is moved to point D_1 and point C is moved to point C_1. Making a plumb line CC_2 and a horizontal line C_1C_3. CC_2 is intersecting C_1C_3 at point C_3. CC_3 is the plumb displacement of point C. If DD_2 is made perpendicular to BD_1, then D_2D_1 is the elongation of the BD bar.

$$D_2D_1 = \Delta l = \frac{F_N l}{EA} = \frac{74.6\times10^3\times1}{200\times10^9\times497\times10^{-6}}\ \text{m} = 0.75\ \text{mm}$$

Then by geometric relations

$$CC_1 = 2DD_1, \quad D_2D_1 = DD_1\cos 30°, \quad CC_3 = CC_1\cos 30°$$

So

$$CC_3 = 2DD_1\cos 30° = 2D_2D_1 = 2 \times 0.75\ \text{mm} = 1.5\ \text{mm}$$

Discussion: For this question, if the lead displacement at point C is not more than $[\Delta]$, that is, the whole structure is required to have a certain rigidity. In this case, we can calculate the lead displacement Δ at point C and compare it with the allowable displacement $[\Delta]$. If $\Delta \leqslant [\Delta]$, the rigidity is sufficient and we call this condition as rigidity condition. For certain structures or systems, such as trusses and valve machinery, the rigidity condition should be considered. The displacement at certain points should not be too large. For most engineering components subjected to tension and compression, only the strength condition is required to be satisfied, without discussing its rigidity.

Example 7–9 A joint structure is shown in Fig.7–30a, b, where the main plate with the width $b = 80$ mm and the thickness $t = 12$ mm. The axial tension is $F = 110$ kN and the allowable material stress is $[\sigma] = 160$ MPa. The rivets made of the same material have uniform diameter $d = 16$ mm. Try to check the strength of the plate.

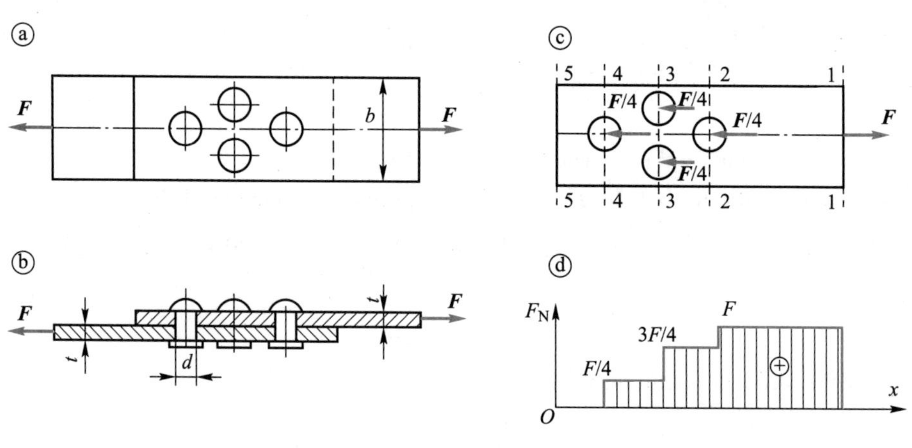

Fig.7–30

Solution (1) Analyse the internal forces

The rivets are of the same material and diameter and the distribution of the rivet group is symmetrical to the axial external force. It is usually assumed that each rivet is subjected to the same magnitude of force. In this case, each rivet is subjected to a force of $F/4$. Imagine that the upper main plate is taken out separately and is subjected to the forces shown in Fig.7–30c. The axial force diagram for the upper plate can then be made as shown in Fig.7–30d.

(2) Determine the dangerous section

The main board is divided into four sections. Sections 1—2, 3—4 have the same minimum cross section area. The internal force in section 3—4 is 1/4 of the internal force in section 1—2. Thus, the minimum cross section in section 3—4 cannot be a dangerous section. A comparison of sections 2—3 and 1—2 shows that the former has a smaller axial force but also a smaller minimum cross section area, while the latter has a larger axial force and a larger minimum cross section area. Therefore, the minimum cross section in both sections is a dangerous section.

section 1—2:

$$\sigma_{\max} = \frac{110 \times 10^3}{(80-16) \times 12} \text{ MPa} = 143 \text{ MPa} < [\sigma]$$

section 2—3:

$$\sigma_{max} = \frac{3\times110\times10^3}{4\times(80-16\times2)\times12}\ \text{MPa} = 143\ \text{MPa} < [\sigma]$$

The board is safe because all sections of the board meet the strength requirements.

Problems

7-1 Try to draw the axial forces diagram for each of the bars shown in Fig.P7-1.

Answer: (a) $F_{N\max}=F$; (b) $F_{N\max}=F$; (c) $F_{N\max}=50$ kN;

(d) $F_{N\max}=F$; (e) $F_{N\max}=4F$; (f) $F_{N\max}=ql$

7-2 A lifting ring screw (Fig.P7-2) with a diameter $d=48$ mm, an internal diameter $d_1=42.6$ mm and a lifting weight $P=50$ kN. Find the stress in the cross section of the screw.

Answer: $\sigma=35$ MPa

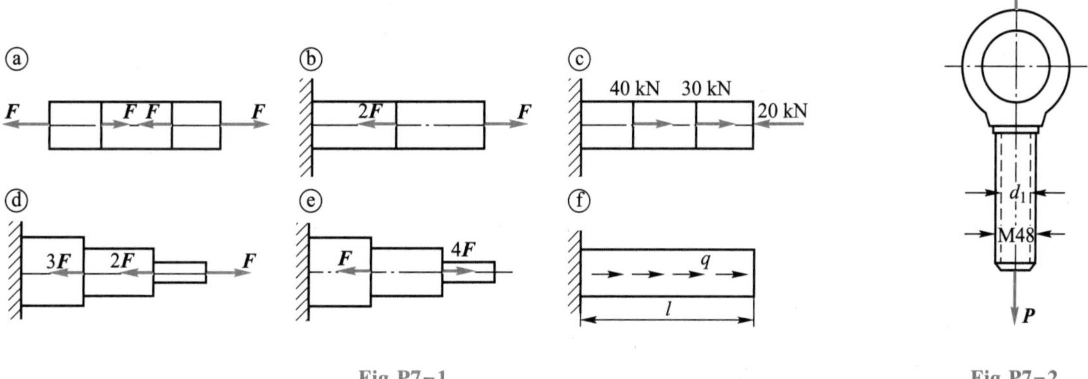

Fig.P7-1 Fig.P7-2

7-3 Fig.P7-3 shows a riveted joint structure with $F=7$ kN, $t=1.5$ mm, $b_1=4$ mm, $b_2=5$ mm and $b_3=6$ mm. (1) Draw the axial force diagram for the cover plate; (2) calculate the maximum normal stress in the plate.

Answer: (1) $F_{N\max}=F$; (2) $\sigma_{\max}=389$ MPa

7-4 A stepped straight bar is subjected to the forces shown in Fig.P7-4. Try to find: (1) the axial forces on cross sections 1—1, 2—2 and 3—3 of the bar and make an axial force diagram; (2) if the areas of cross sections 1—1, 2—2 and 3—3 are $A_1=200\ \text{mm}^2$, $A_2=300\ \text{mm}^2$, $A_3=400\ \text{mm}^2$, respectively, find the stresses in each cross section.

Answer: (1) $F_{N1}=20$ kN, $F_{N2}=10$ kN, $F_{N3}=10$ kN; (2) $\sigma_1=100$ MPa, $\sigma_2=33.3$ MPa, $\sigma_3=$ 25 MPa

7-5 A stepped straight bar is subjected to the forces shown in Fig.P7-5. The cross section areas of the two sections are A and $2A$. The modulus of elasticity of the material is known to be E. Try to

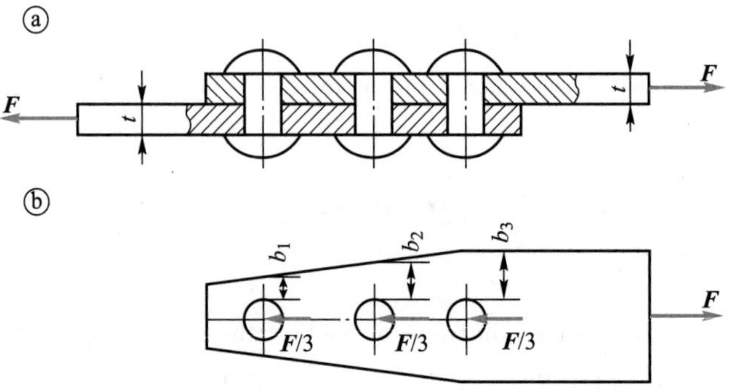

Fig.P7-3

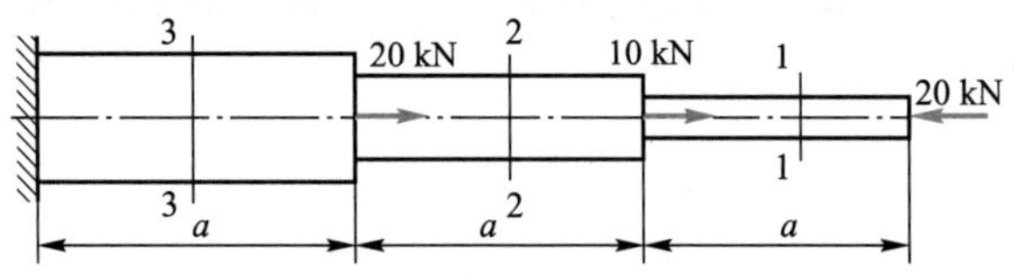

Fig.P7-4

find: (1) the maximum tensile and compressive stresses in the bar; (2) the total elongation of the bar.

Answer: (1) $\sigma^{+}_{\max}=\dfrac{Fl}{EA}$, $\sigma^{-}_{\max}=-\dfrac{Fl}{EA}$; (2) $\Delta l=0$

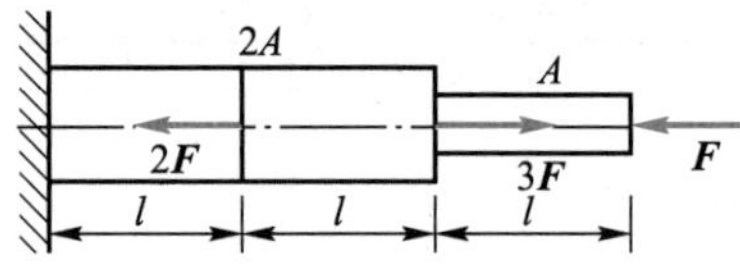

Fig.P7-5

7-6 The bolt shown in Fig.P7-6 has an axial deformation of $\Delta l=0.1$ mm when tightened. It is known that $d_1=8$ mm, $d_2=6.8$ mm, $d_3=7$ mm, $l_1=6$ mm, $l_2=29$ mm, $l_3=8$ mm and the modulus of elasticity of the material $E=210$ GPa. Find the preload force F.

Answer: $F=12.7$ kN

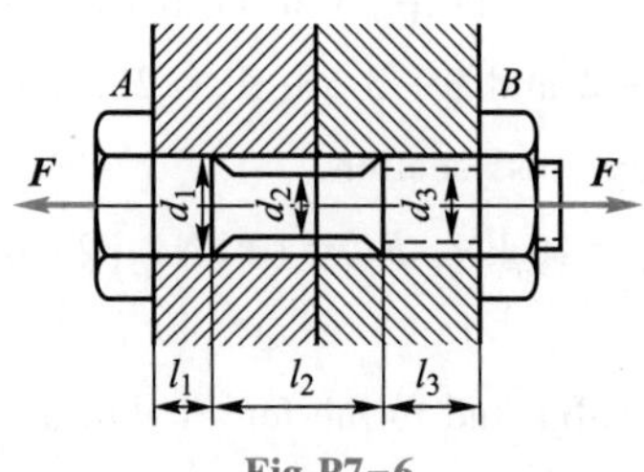

Fig.P7-6

7-7　A type of low-carbon steel is known to have a modulus of elasticity $E = 210$ GPa, a yield stress $\sigma_s = 220$ MPa and a ultimate strength $\sigma_b = 400$ MPa. In a tensile test, when the axial stress in the specimen is 300 MPa, the axial strain $\varepsilon = 3.5\times10^{-3}$ is measured. Find the elastic strain ε_e and the plastic strain ε_p along the axial direction of the specimen at this time.

Answer: $\varepsilon_e = 1.5\times10^{-3}$, $\varepsilon_p = 2.0\times10^{-3}$

7-8　Fig.P7-8 shows a rectangular cross section tensile specimen with a width $b = 40$ mm and a thickness $t = 5$ mm. For each 5 kN increase in tensile force, the axial strain $\varepsilon_1 = 150 \times 10^{-6}$ and the transverse strain $\varepsilon_2 = -38 \times 10^{-6}$ are measured using resistance strain gauges. Find the modulus of elasticity E and Poisson ratio μ of the material.

Answer: $E = 208$ GPa, $\mu = 0.267$

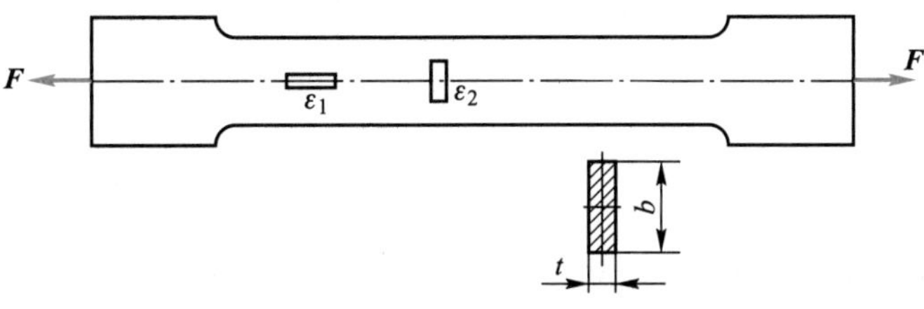

Fig.P7-8

7-9　It shows a truss structure in Fig.P7-9 with each bar consisting of two equilateral angle steels and an allowable stress of the material $[\sigma] = 170$ MPa, try to determine the type of equilateral angle steels required for the bar 1 and bar 2.

Answer: (a) bar 1: 2∟45 mm×45 mm×6 mm, bar 2: 2∟50 mm×50 mm×3 mm; (b) bar 1: 2∟80 mm×80 mm×7 mm, bar 2: 2∟75 mm×75 mm×6 mm

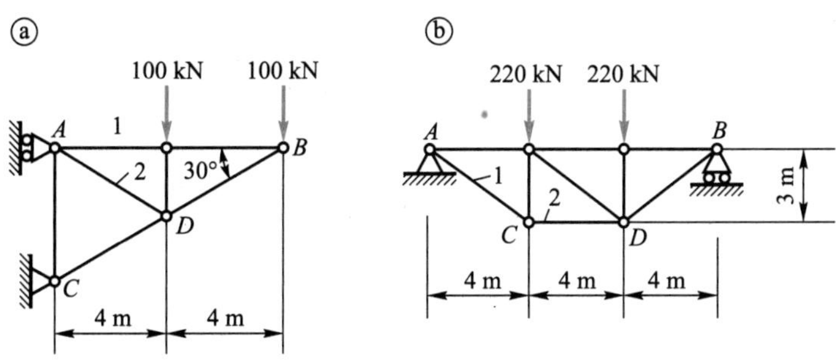

Fig.P7-9

7-10　Fig.P7-10 shows a square structure with a calibrated joint. Each bar is made of cast iron and the ratio of the allowable compressive stress to the allowable tensile stress is $[\sigma^-]/[\sigma^+] = 3$. The cross section area of each bar is A. Find the maximum allowable load F_{max} for the structure.

Answer: $F_{max} = \sqrt{2}A[\sigma^+]$

7-11 The corners of a short wooden column are reinforced with four 40 mm×40 mm×4 mm equilateral angle steels (Fig. P7-11). The allowable stress $[\sigma]_{steel}=160$ MPa and the modulus of elasticity $E_{steel}=200$ GPa for the angle steels are known. the allowable stress $[\sigma]_{wood}=120$ MPa and the modulus of elasticity $E_{wood}=10$ GPa for the timber are also known. Find the allowable load $[F]$.

Answer: $[F]=698$ kN

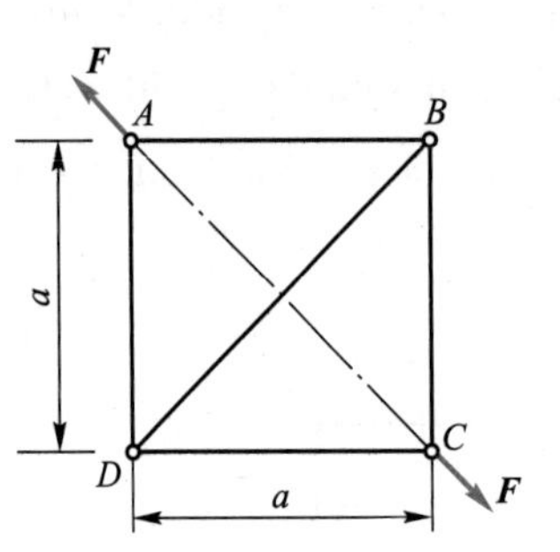

Fig.P7-10

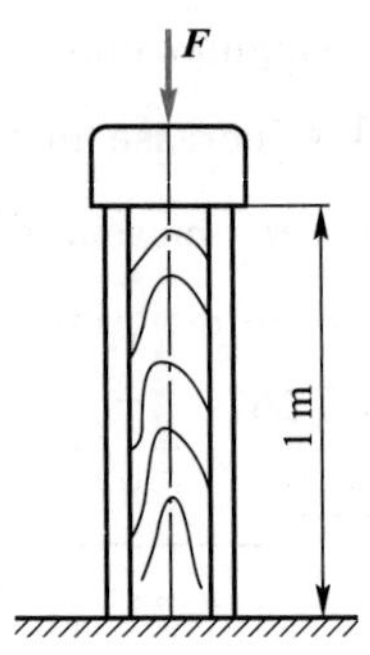

Fig.P7-11

7-12 If the materials of AB and AC are the same (Fig. P7-12). The allowable tensile and compressive stresses are equal. The cross section area of the bar AB is twice that of the bar AC. Find the angle θ at which the structure of the bar system is most reasonable.

Answer: $\theta=60°$

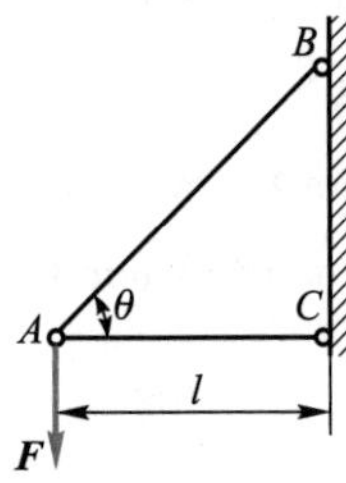

Fig.P7-12

Chapter 8 Shear and Torsion

Teaching Scheme of Chapter 8

Shear and torsion are basic forms of deformation of members. This chapter introduces the characteristics of shear deformation and the practical calculation of shear and bearing of joints, the strength and rigidity caculations of torsion.

8.1 The concept of shear

In engineering, there are often members subjected to a pair of transverse concentrated forces of equal magnitude and opposite direction. The action lines of the porces parallel and close to each other, as shown in Fig.8 - 1a. Under the action of such two external forces, the cross section of the member between the two lines of action will be relatively misshapen, as shown in Fig.8-1b. In engineering, this form of deformation is called shear deformation or direct shear. If the two transverse forces gradually increase, the member will eventually be cut off along the plane of the misalignment. The relative misalignment of the plane is called the shear plane and the member with shear deformation is called shear member. Scissors shear objects is the most typical example of shear damage in everyday life.

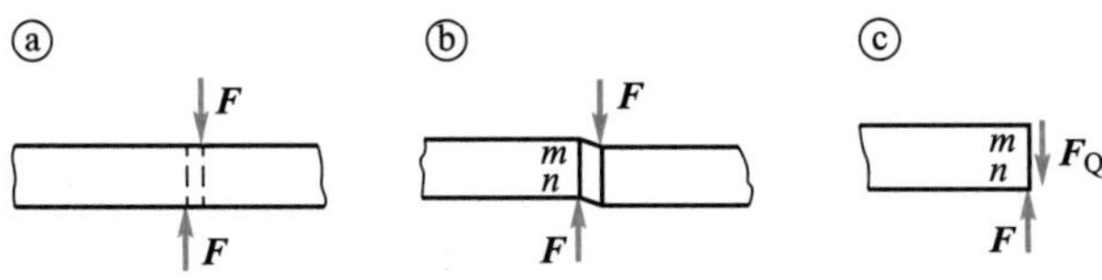

Fig.8-1

Commonly used connections in engineering, such as bolts connecting two steel plates (Fig.8-2), spline connections (Fig.8-3), pins in machinery (Fig.8-4) and the widely used weld joint in steel construction, are all shear members. In addition, people often shear to make the required member shape, such as punching and shear, drilling and cutting. If there is only one shear plane, it is called single shear (Fig.8-2). If there are two planes of shear, it is called double shear (Fig.8-4).

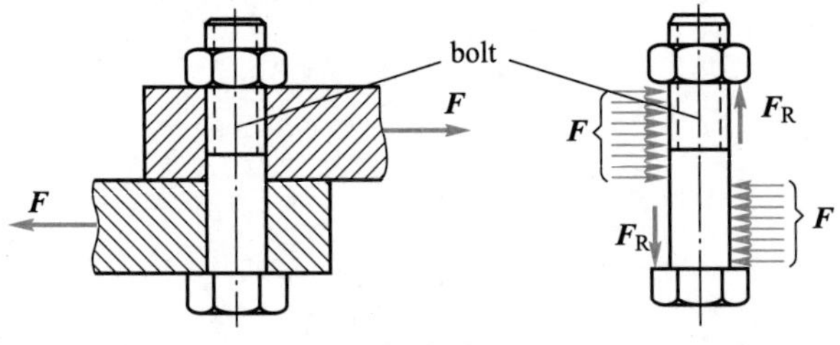

Fig.8-2

The forces and deformations of shear members under load are complex. In shear connectors with various deformations, tensile and bending deformations are usually negligible, but bearing deformations are often not negligible. This chapter discusses the calculation of the shear strength and the bearing strength of members under the action of external forces.

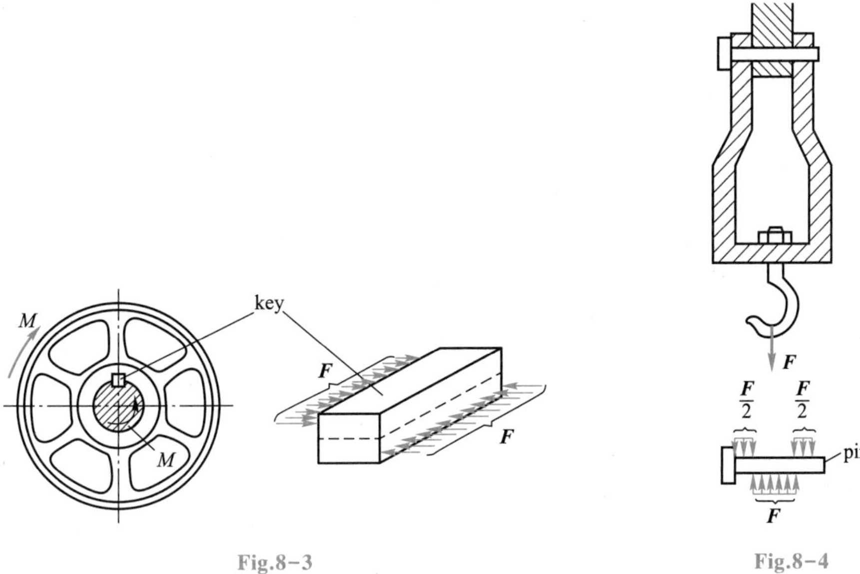

Fig.8-3 Fig.8-4

8.2 Practical calculation of shear

Shear members are very difficult to analyse accurately because of the complexity of forces and deformations.It usually gives some assumptions based on practical experience, and then carries out simplified calculations. This calculation method is called practical calculation. This method is easy to calculate and the results are close to the reality.

Practical calculations for shear mainly include internal force calculations and stress calculations. From the section method, it is known that the internal forces on the shear plane of a shear member should be tangential to the shear surface as shown in Fig.8-1c. This internal force tangent to the cross section is called shear force. It is easily obtained from the equilibrium equation

$$F_Q = F$$

Since the internal force on the shear plane is tangent to the shear plane, the stress on the shear plane should also be tangent to the shear plane. Due to the complexity of the forces and deformation, the distribution of shear stress on the shear surface is also very complex. In order to simplify the calculation, it often assumes that the shear stress τ is uniformly distributed on the shear plane. The practical formula for calculating shear stress is obtained

$$\tau = F_Q / A \tag{8-1}$$

where A is the area of the shear surface.

The shear stress calculated by equation(8-1) is based on assumptions and is not the true shear stress, which is usually referred to the nominal shear stress. When the shear stress τ on the shear plane reaches a certain value, the shear member will be damaged by shear. In order to ensure that the shear member has sufficient strength, it is required that the shear stress should not exceed the allowable shear stress $[\tau]$ of the material.

$$\tau = F_Q/A \leqslant [\tau] \tag{8-2}$$

This is the shear strength condition. The allowable shear stress $[\tau]$ of the material must be determined by shear experiments. In the experiment, we make the stress of the test specimen similar to the actual working condition of the member. Measure the ultimate load required when the test specimen is cut off and then use the practical calculation equation (8-1) to calculate the ultimate shear stress τ_u of the materia. Finally the shear stress is divided by the appropriate factor of safety n to obtain the allowable shear stress $[\tau] = \tau_u/n$. The allowable shear stress of various common materials can be found in relevant material manuals or design specifications. Experimental results show that the shear ultimate strength of the material has an approximate proportional relationship with the tensile (compressive) ultimate strength.

Plastic materials: $[\tau] = (0.6 \text{ to } 0.8)\ [\sigma]$;

Brittle materials: $[\tau] = (0.8 \text{ to } 1.0)\ [\sigma]$.

Based on this relationship, the value of the tensile allowable stress $[\sigma]$ is often used in engineering to estimate the value of the shear allowable stress $[\tau]$.

Example 8-1 The pin connection structure is shown in Fig.8-5a. The load is known to be F = 15 kN. The thickness is t = 8 mm, the diameter of the pin is d = 20 mm and the pin allowable shear stress is $[\tau]$ = 30 MPa. Check the shear strength of the pin.

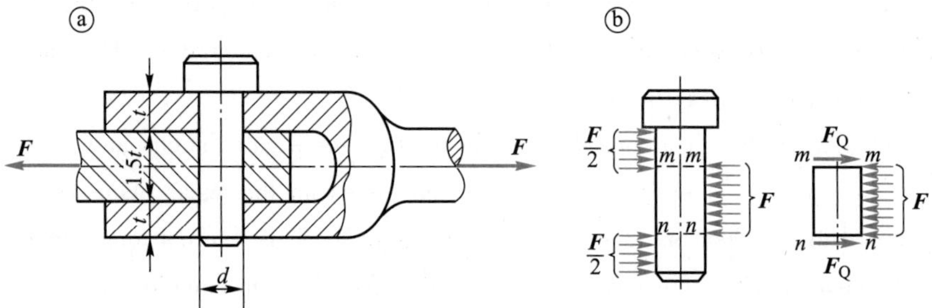

Fig.8-5

Solution The pin is subjected to the forces shown in Fig.8-5b. It can be seen that the pin has two shear planes, which are double shear. From the section method it is easy to find

$$F_Q = \frac{F}{2}$$

The shear stress on the shear plane of the pin is thus

$$\tau = \frac{F_Q}{A} = \frac{15\times10^3}{2\times\frac{\pi}{4}\times(20\times10^{-3})^2} \text{ Pa} = 23.9 \text{ MPa} < [\tau]$$

Therefore the pin meets the strength requirements.

Example 8 – 2 Two steel plates are lap-welded together as shown in Fig. 8 – 6. The plate thickness is $t = 12$ mm and the alllowable shear stress in the weld $[\tau] = 120$ MPa. If the load on the plate is $F_P = 90$ kN, find the length l of the weld.

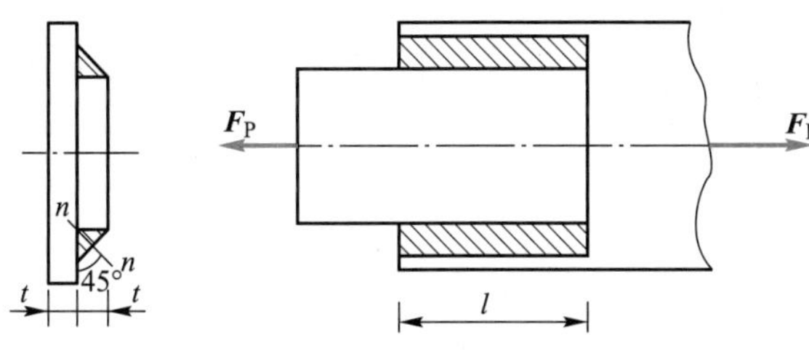

Fig.8–6

Solution Practice has shown that lap welds are often sheared badly at the section n—n with the smallest weld area. When the angle between the weld surface and the steel plate surface is 45°, the shear surface area is

$$A = lt\cos 45°$$

The member is double shear with a shear of

$$F_Q = \frac{F_P}{2}$$

Substituting the above two equations into the shear strength condition, we have

$$\tau = \frac{F_Q}{A} = \frac{F_P}{2lt\cos 45°} \leqslant [\tau]$$

and

$$l \geqslant \frac{F_P}{2t\cos 45°[\tau]} = \frac{90 \times 10^3}{\sqrt{2} \times 12 \times 120}\ \text{mm} = 44.2\ \text{mm}$$

Considering the poor quality of the weld ends, the actual length is usually determined by adding 10 mm to the calculated length, so that $l = 55$ mm. The above algorithm is only suitable for rough calculations of general structural welds.

All of the above are issues that require shear strength to be guaranteed, but somtimes in engineering, such as prevention of mechanical overload and safety pins, shear "breaks" are useful. When overloaded, the safety pin or safety block will be sheared off along the shear plane to protect other important parts. In addition, problems such as the punching of materials and dropouts require that the shear stress τ on the component reaches the ultimate stress τ_u of the material.

$$\tau = \frac{F_Q}{A} \geqslant \tau_u \tag{8-3}$$

Example 8–3 A steel plate with a thickness $t = 5$ mm and an ultimate strength $\tau_u = 320$ MPa is to be punched with a circular hole of diameter $d = 15$ mm, as shown in Fig. 8 – 7a. Calculate the minimum punching shear force F_P required.

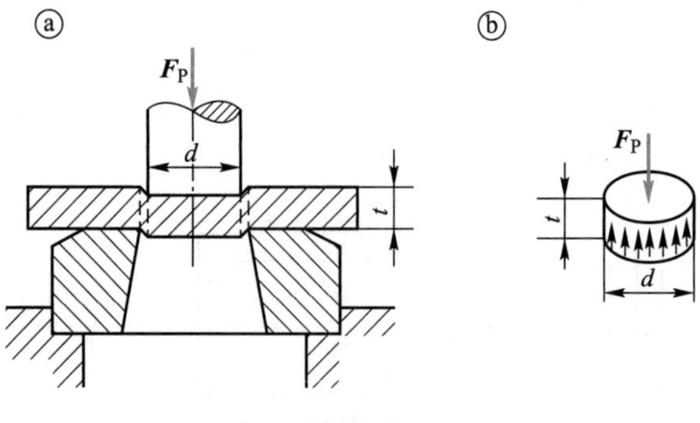

Fig.8-7

Solution The shear plane is a cylindrical surface of diameter d and height t, as shown in Fig.8-7b, and its area is

$$A=\pi dt=\pi\times 15\times 5\ \text{mm}^2=235.5\ \text{mm}^2$$

The shear forces distributed on this cylindrical surface are

$$F_Q=F_P$$

When punching, the working shear stress must reach at least the ultimate shear stress τ_u, so from equation (8-3) we get

$$\begin{aligned}F_P\geqslant\tau_u A&=320\times 10^6\times 235.5\times 10^{-6}\ \text{N}\\&=75.36\times 10^3\ \text{N}\\&=75.36\ \text{kN}\end{aligned}$$

Therefore, the punching shear force F_P needs to be at least 75.36 kN.

8.3 Practical calculations for bearing

In addition to the possibility of shear failure, bearing damage may also occur. Fig.8-8a shows the rivet connection structure, when the rivet and steel plate contact surface pressure is too large. It may make the rivet or steel plate on the rivet hole to produce local plastic deformation, as shown in Fig.8-8b. This pressure acting on the contact surface is called the bearing pressure, denoted by F_{jy}. The contact surface is called the bearing surface and the pressure on the bearing surface is called the bearing stress, denoted by σ_{jy}. Bearing damage can lead to loosening of the connection and affect the normal operation of the member. Therefore, the connection must also be calculated for the bearing strength. In the bearing surface, the distribution of bearing stress is also very complex. It is difficult to get its exact law. In engineering, the bearing problem is still using the practical calculation method, which assume that the distribution of bearing stress is uniform. The bearing stress can be calculated according to the following formula

$$\sigma_{jy}=\frac{F_{jy}}{A_{jy}} \tag{8-4}$$

where A_{jy} is the bearing surface area.

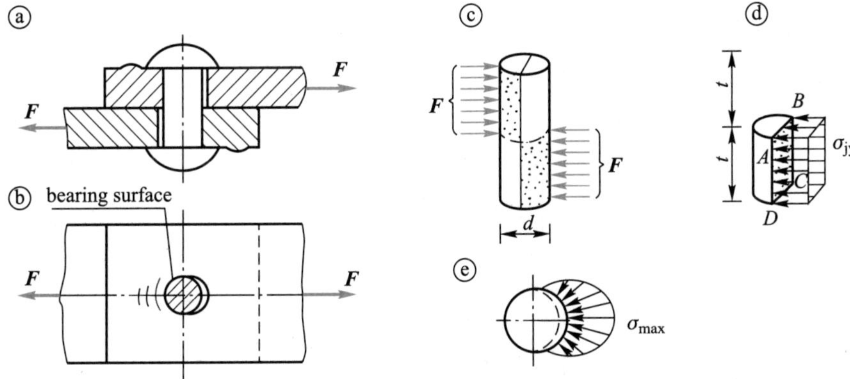

Fig.8-8

The bearing stress obtained by equation (8-4) is not the true stress and is usually referred to as the nominal bearing stress. The calculation of the area of the extruded surface is discussed in two cases as follows.

(1) When the contact surface is flat, as the spline connection shown in the Fig.8-3, the area of the extruded surface for calculation is the actual contact surface area, i.e. the shaded area in the Fig.8-9: $A_{jy}=\frac{1}{2}hl$.

(2) When the contact surface is a semi-cylindrical surface, as the riveted joint shown in Fig.8-8c, the area of the bearing surface for calculation is the the diameter projection area of the actual contact surface. The area of the rectangle *ABCD* in Fig.8-8d is $A_{jy}=dt$. In this way, the nominal bearing stress calculated in accordance with equation (8-4) and the actual maximum bearing stress produced (σ_{max} in Fig.8-8e) are very similar.

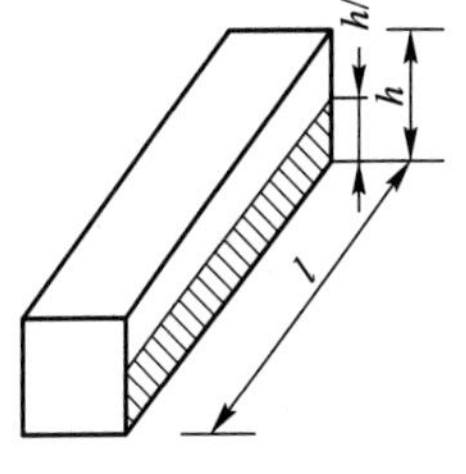

Fig.8-9

To prevent bearing damage, the maximum bearing stress should not exceed the allowable bearing stress $[\sigma_{jy}]$ of the material, i.e.

$$\sigma_{jy}=\frac{F_{jy}}{A_{jy}}\leqslant[\sigma_{jy}] \tag{8-5}$$

This is the bearing strength condition. The allowable beearing stress $[\sigma_{jy}]$ of various commonly used materials can be found in the relevant material handbooks or design codes. From the experimental results, it can be seen that the allowable bearing stress and the allowable tensile stress $[\sigma]$ have the following relationship between.

Plastic materials: $[\sigma_{jy}]=(1.5\sim2.5)\ [\sigma]$;

Brittle materials: $[\sigma_{jy}]=(0.9\sim1.5)\ [\sigma]$.

If the two contacting members are of different materials, the calculation should be made for the member with the weaker bearing strength.

Example 8-4 A riveted joint structure is shown in Fig.8-10a with a known load $F = 100$ kN, a rivet diameter $d = 16$ mm, Width of the steel plate is $b = 90$ mm, an allowable tensile stress $[\sigma] = 160$ MPa for the steel plate. The allowable shear stress is $[\tau] = 130$ MPa for the rivet and the allowable bearing stress is $[\sigma_{jy}] = 320$ MPa for the plate and rivet. Check the strength of the structure.

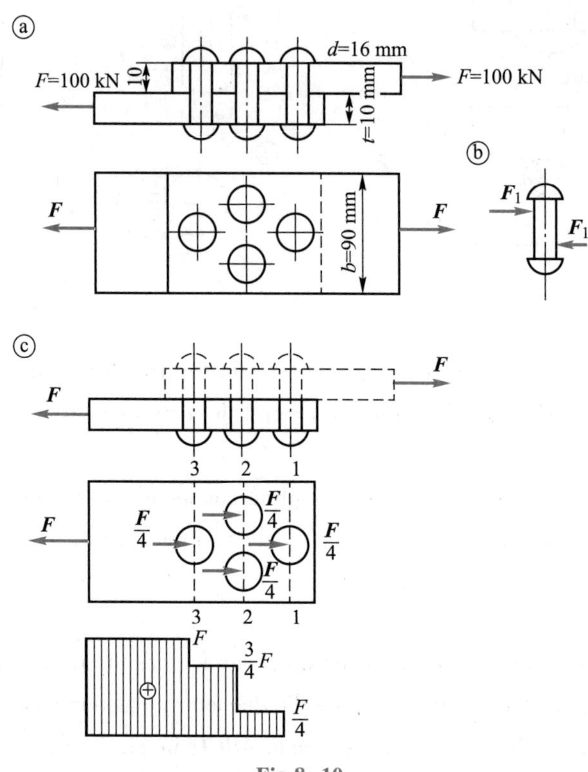

Fig.8-10

Solution There are three possible forms of damage to a riveted joint structure under the action of a tensile force F: damage to the rivet due to shear; damage to the rivet or steel plate due to bearing; and damage to the steel plate due to tension. The following strength checks are carried out according to the possible forms of damage to the structure.

(1) Check the shear strength of the rivet

The rivets are subjected to the forces shown in Fig.8-10b. Each rivet with the same diameter, is of the same material and is uniformly distributed along the axis. So it can be assumed that each rivet is subjected to the same shear force. The tensile force F is equally distributed over each rivet. The force on each rivet is

$$F_1 = \frac{F}{4} = 25 \text{ kN}$$

Therefore, the shear force on the shear plane of the rivet is

$$F_Q = F_1 = 25 \text{ kN}$$

The shear stress in the rivet is thus

$$\tau=\frac{F_Q}{A}=\frac{25\times10^3}{\frac{\pi}{4}\times16^2\times10^{-6}}\ \text{Pa}=124\ \text{MPa}<[\tau]$$

(2) Check the bearing strength of the rivet

The squeezing force of the rivet is

$$F_{jy}=F_1=25\ \text{kN}$$

Since the contact surface is cylindrical, the bearing stress is

$$\sigma_{jy}=\frac{F_{jy}}{A_{jy}}=\frac{25\times10^3}{10\times16\times10^{-6}}\ \text{Pa}=156.25\ \text{MPa}<[\sigma_{jy}]$$

(3) Check the tensile strength of the steel plate

Take the lower piece of steel plate as the object of study, draw its force diagram. Use the cross section method to find the internal force of each section of the steel plate, draw the axial force diagram as shown in the Fig.8-10c. It is clear that the dangerous sections of this steel plate are section 2—2 and section 3—3.

Cross sections 2—2:

$$\sigma_2=\frac{F_{N2}}{(b-2d)t}=\frac{\frac{3}{4}\times100\times10^3}{(90-2\times16)\times10\times10^{-6}}\ \text{Pa}=129\ \text{MPa}<[\sigma]$$

Cross sections 3—3:

$$\sigma_3=\frac{F_{N3}}{(b-d)t}=\frac{100\times10^3}{(90-16)\times10\times10^{-6}}\ \text{Pa}=135\ \text{MPa}<[\sigma]$$

Therefore, the entire structure meets the strength requirements.

Example 8-5 A pulley made of cast iron is connected to a shaft by means of a flat spline as shown in the Fig.8-11. If the allowable shear stress is $[\tau]=60$ MPa for the spline and the allowable bearing stress is $[\sigma_{jy}]=80$ MPa for the cast iron, check the strength of the spline connection.

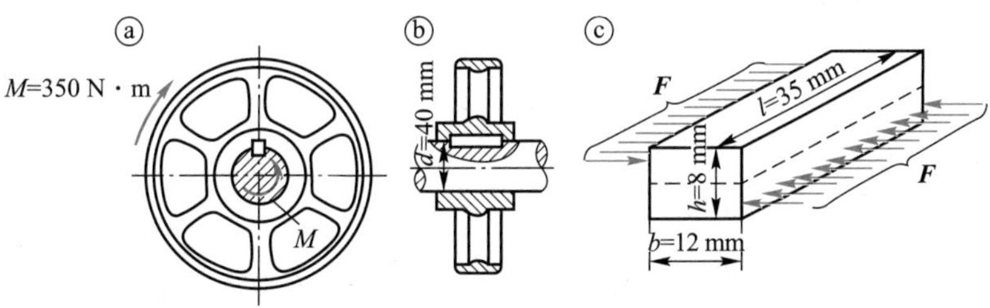

Fig.8-11

Solution For the axis, according to the equilibrium relationship

$$F\frac{d}{2}-M=0$$

We have

$$F=\frac{2M}{d}$$

(1) Check the shear strength of the bond

The shear force on the spline is

$$F_Q=F=\frac{2M}{d}$$

The shear plane area of the spline is

$$A=bl$$

Therefore there is

$$\tau=\frac{F_Q}{A}=\frac{2M}{bld}=\frac{2\times350}{12\times35\times40\times10^{-9}}\text{ MPa}=41.7\text{ MPa}<[\tau]=60\text{ MPa}$$

It can be seen that the splines meet the shear strength requirements.

(2) Check the bearing strength of the wheel

As the spline and shaft are made of steel, the pulley is made of cast iron which is less resistant to bearing than steel. The bearing strength of the wheel should be calibrated.

The squeezing force on the wheel is

$$F_{jy}=F=\frac{2M}{d}$$

Calculate the area of the extruded surface as

$$A_{jy}=\frac{1}{2}hl$$

Therefore there is

$$\sigma_{jy}=\frac{F_{jy}}{A_{jy}}=\frac{4M}{hld}=\frac{4\times350}{8\times35\times40\times10^{-9}}\text{ MPa}=125\text{ MPa}>[\sigma_{jy}]=80\text{ MPa}$$

It can be seen that the bearing strength of the pulley is not sufficient. Thus, the length of the spline should be redesigned.

From the beearing strength condition

$$\sigma_{jy}=\frac{4M}{hld}\leqslant[\sigma_{jy}]$$

The spline length should then be

$$l\geqslant\frac{4M}{hd[\sigma_{jy}]}=\frac{4\times350}{8\times40\times10^{-6}\times80\times10^{6}}\text{ m}=0.054\ 7\text{ m}$$

Final selection of $l=55$ mm.

8.4 The concept of torsion

In engineering practice, there are many bars subjected to moments, such as the steering shaft of a car

(Fig.8-12a), the tapping tap (Fig.8-12b), the main shaft of a hydroelectric generator (Fig.8-12c), etc. Their common feature is that a pair of couples with equal magnitude and opposite direction is applied at the ends of the bar. The couple plane is perpendicular to the axis of the bar. Any two cross sections of the bar rotate relative to each other around the axis of the bar. This form of deformation of the bar is called torsional deformation. According to the section method, when torsional deformation occurs to the rod, the internal force on the cross section is only the moment of the couple located on the face. It is called torque. The bars subjected to torsion is called shafts. This chapter focuses on the torsion of straight shafts with uniform circle section.

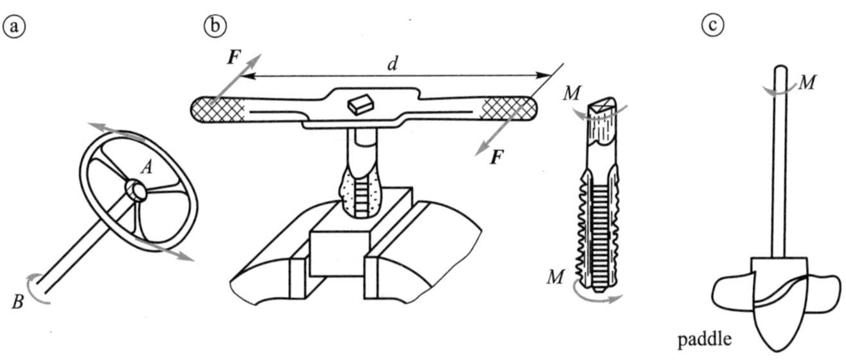

Fig.8-12

8.5 Torque and torque diagram

8.5.1 Calculation of the external moment of couple

In engineering practice, what can be known about shaft are only the power and speed it transmits. Thus, before analysing the internal forces, the external couple moments of the shaft should first be derived from the power and speed transmitted by the shaft.

It is known from theoretical mechanics that the work (i.e. power) P done by a couple per unit of time is equal to the product of its moment M and angular velocity ω.

$$P = M\omega \tag{8-6}$$

The common unit of power in engineering is the kilowatt (kW) and the common unit of speed is the revolutions per minute (r/min). Therefore, if the power is expressed in P and the rotational speed in n, then $\omega = \frac{2n\pi}{60}$. Substituting into the above equation, we have

$$P \times 1000 = M \cdot n \times \frac{2\pi}{60}$$

The result is

$$\{M\}_{\mathrm{N\cdot m}}=\frac{9\ 549\{P\}_{\mathrm{kW}}}{\{n\}_{\mathrm{r/min}}} \tag{8-7}$$

8.5.2 Torque and torque diagrams

When all the external moments applied to the shaft have been found, the internal forces in any cross section can be calculated by the section method. The circular shaft shown in Fig.8-13 is used as an example. Assume that the circular axis is divided into two sections along the n—n section, and study any one. For example, the equilibrium of the left section as following (Fig.8-13b).

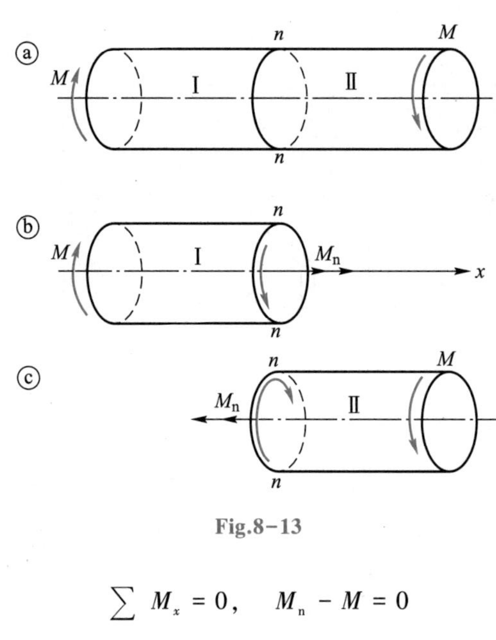

Fig.8-13

$$\sum M_x = 0, \quad M_n - M = 0$$

and

$$M_n = M$$

where M_n is called the torque on section n—n. It is the combined moment of the distributed internal force system of the two parts Ⅰ and Ⅱ interacting on the n—n section.

Similarly, if the right section is the subject of study (Fig.8-13c), the torque M_n on section n—n can also be found. Its value is still M, but its steering is the opposite of that shown in Fig.8-13b. In order to obtain the same positive and negative signs of the torque on the same section from the two sections of the bar, the sign of the torque can be specified as follows: the torque M_n is expressed as a vector according to the right-handed screw rule. When the direction of the vector is the same as the direction of the outer normal of the section, the torque M_n is positive. In this way, the torques M_n shown in Fig.8-13b and c are all positive. Before establishing the equilibrium equation, let the torque be positive. If the result is negative, it means that the direction is opposite to the assumed one.

Torque is an internal moment, which is related to the external moment of couple. When a shaft is divided into two segments using the section method, the torques and the external couple moment form a balanced force system. From this, the magnitude and direction of the torque can be calculated from the external moment of couple. If only the external moment of couple is applied at the ends of the shaft, the torque is equal to its magnitude. If there are more than two external moments acting on the shaft, the torques in each section are not equal. In this case, it is necessary to calculate the torque in each segment using the section method.

A graphical representation of the variation of torque M_n in the direction of the axis is called a torque diagram. Torque diagrams are drawn in a similar way to axial force diagrams.

Example 8-6 On the shaft shown in Fig.8-14a, the active wheel A is connected to the prime mover and the driven wheels B, C and D are connected to the machine tool. The input power of wheel A is known to be $P_A = 50$ kW, the outputs of wheels B, C and D are $P_B = P_C = 15$ kW and $P_D = 20$ kW, respectively. The speed of the shaft is $n = 300$ r/min. Try to find the torque in each section of the shaft and draw a torque diagram.

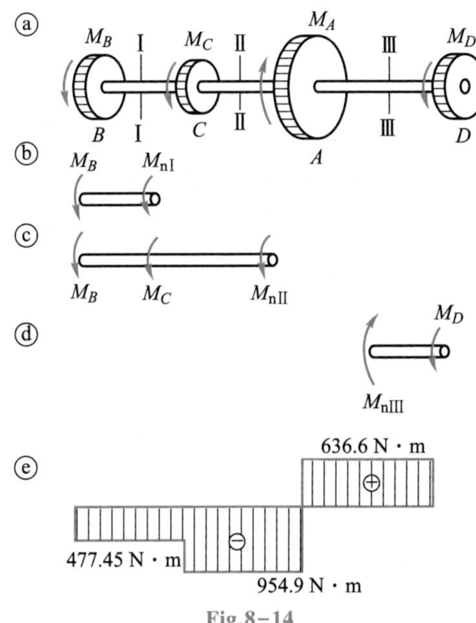

Fig.8-14

Solution (1) Calculate the external moment of couple

$$M_A = 9\ 549 \frac{P_A}{n} = 9\ 549 \times \frac{50}{300}\ \text{N} \cdot \text{m} = 1\ 591.5\ \text{N} \cdot \text{m}$$

$$M_B = M_C = 9\ 549 \frac{P_B}{n} = 9\ 549 \times \frac{15}{300}\ \text{N} \cdot \text{m} = 477.45\ \text{N} \cdot \text{m}$$

$$M_D = 9\ 549 \frac{P_D}{n} = 9\ 549 \times \frac{20}{300}\ \text{N} \cdot \text{m} = 636.6\ \text{N} \cdot \text{m}$$

(2) Calculate torque

BC section: Cut the shaft along cross section I — I within the section BC and set the torque M_{nI} on cross section I — I to be positive (as in Fig.8-14b). From the equilibrium equation

$$M_{nI} + M_B = 0$$

We get

$$M_{nI} = -M_B = -477.45 \text{ N} \cdot \text{m}$$

A negative result indicates that the actual direction of the torque I — I is opposite to the direction set. The torque on each section within the *BC* section is constant, so the torque diagram within this section is a horizontal line (Fig.8−14e).

CA section: From Fig.8−14c it can be seen that

$$M_{n\text{II}} + M_C + M_B = 0$$

There is

$$M_{n\text{II}} = -M_C - M_B = -954.9 \text{ N} \cdot \text{m}$$

Section *AD*: From Fig.8−14d it can be seen that

$$M_{n\text{III}} - M_D = 0$$

There is

$$M_{n\text{III}} = M_D = 636.6 \text{ N} \cdot \text{m}$$

(3) Making torque diagram

Based on the results of the calculations, the torque variation along the axis in each section is represented in Fig.8−14e as a torque diagram. As can be seen from the graph, the maximum torque occurs in the *CA* section with an absolute value of $|M_{nmax}| = 954.9 \text{ N} \cdot \text{m}$.

8.6 Torsion of thin-walled cylinders

In the previous discussion of shear deformation, practical calculations of shear were mainly discussed. But theory of shear has not yet been studied. Now, some important properties concerning shear are discussed based on a simple twist of a thin-walled cylinder.

8.6.1 Stress and deformation in thin-walled cylinders during torsion

Fig.8−15a represents a thin-walled cylinder of equal thickness. A square grid is drawn on the surface with circumferential and longitudinal lines before being subjected to couple. After applying an external moment M_n at each end, the cylinder produces a torsional deformation (Fig.8−15b). At this point the following phenomena can be observed:

(1) The shape, size and spacing of the circumferential lines on the surface of the cylinder remain unchanged, and just rotate relatively around the axis.

(2) Each longitudinal line is inclined at the same angle γ, and the tiny rectangle formed by the longitudinal and circumferential lines becomes a parallelogram.

If a micro-section is intercepted using two sections *m*—*m* and *n*—*n* with a distance of d*x*, and a stress element *abcd* is intercepted from the barrel using two radial longitudinal sections clamped by a tiny center angle (Fig.8−15c and d). It is deduced from the above observed phenomena that the element has neither axial nor circumferential linear strain. Just relative parallel misalignment occurring between adjacent cross sections (ab and cd), i.e. only shear strain. The shear strain of each element on the circumference is the same, which is γ.

From the correspondence between stress and strain, it is clear that only the corresponding shear stress τ exists at each point on the cross section. The direction of the shear stress is along the circumference and is of the same magnitude. Since the cylinder wall is very thin, it can be assumed that the shear stress is uniformly distributed along the wall thickness direction. In summary, when a thin-walled cylinder is twisted, the shear stress in its cross section should be uniformly distributed along the circumference and tangential to the circumference.

Let the average radius of a thin-walled cylinder be R. The wall thickness is t. The micro-area $\mathrm{d}A = tR\mathrm{d}\theta$ is taken in the cross section, and the micro-internal force on it is $\tau\mathrm{d}A$ (Fig.8-15e). The micro-internal moment of the micro-internal force to the center of the circle is $R\tau\mathrm{d}A$. It follows from statics that the sum of all these micro-internal moments over the entire cross section is the torque M_{n} on the cross section.

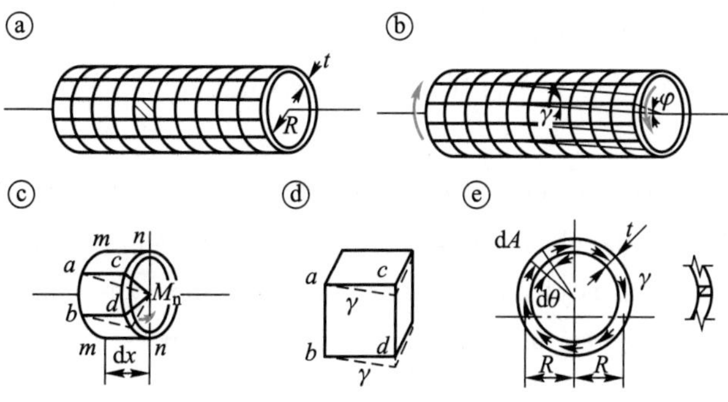

Fig.8-15

$$M_{\mathrm{n}} = \int_A R\tau\mathrm{d}A = \tau RA = \tau \cdot 2\pi R^2 t \tag{8-8}$$

So

$$\tau = \frac{M_{\mathrm{n}}}{2\pi R^2 t} \tag{8-9}$$

or

$$\tau = \frac{M_{\mathrm{n}}}{RA} = \frac{M_{\mathrm{n}}}{2A_0 t} \tag{8-10}$$

where $A = 2\pi Rt$ is the area of the thin-walled cylinder cross section; $A_0 = \pi R^2$ is the area enclosed by the mid line of the cylinder wall on the cross section.

Let l and φ be the length and the relative angle of twist at both ends of the thin-walled cylinder respectively. As seen in Fig.8-15b, the relationship between φ and the shear strain γ is

$$\gamma = \frac{R\varphi}{l} \tag{8-11}$$

8.6.2 Theorem of complementary shearing stresses

Further investigate of the element (tiny positive hexahedron) is shown in Fig.8−16. Let the three side lengths of the element be dx、dy and t respectively. The left and right sides of the element are part of the cross section of the thin-walled cylinder. There are only shear stresses τ on these two sides. According to the equilibrium conditions $\sum F_y = 0$ of the element, the values of shear stress on the left and right sides are equal, but the directions are opposite. The combined force $\tau t\mathrm{d}y$ of the shear stress on these two sides will be synthesized into a couple and its moment is $\tau t\mathrm{d}y \cdot \mathrm{d}x$. As the element is in equilibrium, in the top and bottom surface of the element, there must also be shear stress τ', which synthesis into another couple moment $\tau' t\mathrm{d}y \cdot \mathrm{d}x$ to balance the aforementioned couple. From the equilibrium condition $\sum M_z = 0$, we get

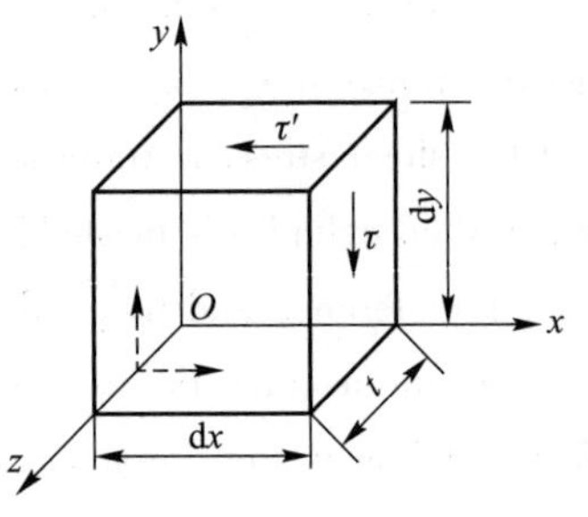

Fig.8−16

$$\tau t\mathrm{d}y \cdot \mathrm{d}x = \tau' t\mathrm{d}x \cdot \mathrm{d}y$$

$$\tau = \tau' \tag{8-12}$$

The above equation shows that shear stress must exist in pairs with equal values on the two planes perpendicular to each other in the element. The shear stresses are both perpendicular to the intersection of the two planes. The direction is of pointing to or deviating from this intersection consistently. This relationship is known as the theorem of complementary shearing stresses.

The stress element shown in Fig.8−16 is called pure shear element. In the pure shear state, the element body only changes its angle, but not its side length. Although the theorem of complementary shearing stresses is derived in pure shear state, it is a general theorem when there is normal stress.

8.6.3 Hook law in shear

Using the torsion of a thin-walled cylinder, a pure shear test can be achieved. The gradually increasing external moment M and the corresponding angle of twist φ were measured in this test. The shear stress τ and shear strain γ can be calculated using equations (8 − 10) and (8 − 11), respectively. The relationship curve between shear stress τ and shear strain γ can be made (i.e. the $\tau-\gamma$ curve). The $\tau-\gamma$ curve for low carbon steel is shown in Fig.8−17. Pure shear tests show that when the shear stress does not exceed the shear proportional limit τ_{p} of the material, the shear stress is proportional to the shear strain, which is the Hooke law in shear. It can be written as

$$\tau = G\gamma \tag{8-13}$$

where G is a constant of proportionality, known as the shear modulus of elasticity. Because γ is dimensionless, G has the same unit as the τ. Commonly used unit is GN/m^2(1 GN/m^2 = 10^9 N/m^2).

"Hooke law in tension and compression", "Hooke law in shear", and "theorem of complementary shearing stresses" are the fundamental theorems of material mechanics, which are often applied in

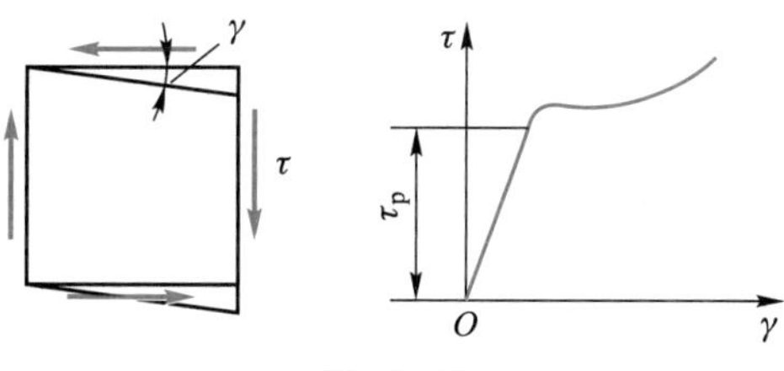

Fig.8-17

theoretical analysis and experimental research. The tensile modulus of elasticity E, the shear modulus of elasticity G, and the Poisson ratio μ are the three elastic constants of a material. For isotropic elastic materials, the following relationships exist between them.

$$G=\frac{E}{2(1+\mu)} \tag{8-14}$$

That is, only two of the three elastic constants are independent.

8.6.4 Energy of shear deformation

From the thin-walled cylinder torsion test, it can be seen that when the shear stress does not exceed the shear proportional limit of the material, the angle of twist φ is proportional to the external torque M. The work done by the external moment is

$$W=\frac{1}{2}M\varphi$$

Under static load, there is a functional principle that all the work done by the external moment during torsion is transferred to strain energy stored inside the thin-walled cylinder. The energy of shear deformation is denoted by U, the

$$U=W=\frac{1}{2}M\varphi$$

Strain energy per unit volume is the strain energy density u. The value of u should be equal to the shear strain energy U divided by the volume V of the thin-walled cylinder. So

$$u=\frac{U}{V}=\frac{1}{2}\cdot\frac{M\varphi}{2\pi Rlt}=\frac{1}{2}\cdot\frac{M}{2\pi R^2 t}\cdot\frac{R\varphi}{l}=\frac{1}{2}\tau\gamma$$

Then using Hook Law for shear, we get

$$u=\frac{1}{2}\tau\gamma=\frac{\tau^2}{2G}=\frac{1}{2}G\gamma^2 \tag{8-15}$$

8.7 Stress and deformation during torsion of circular shafts

8.7.1 Stress during torsion of a circular shaft

When performing torsional strength calculations for circular shafts, the distribution of stresses in the cross section must be further investigated and the maximum stress in the cross section must be found. Similar to the derivation of the formula for normal stress in tension (compression), a comprehensive analysis must be carried out in terms of the geometrical, physical and static relationships of the deformation.

1. Geometric relation

The deformation of circular shaft is like that of a thin-walled cylinder (Fig.8–18). The shape, size and spacing of the circumferential lines don't change and only rotate relatively around the axis. The longitudinal lines are tilted by the same small angle γ. It can be seen that the cross section of a circular shaft remains flat after torsion. Its shape and size remain the same, the radius remains straight and the distance between two adjacent cross sections remains the same. This leads to the assumption that when a circular shaft is twisted, the cross sections are rotated by a small angle around the axis, just like a rigid plane. This is the plane section hypothesis. It is valid because the results derived from it can be confirmed experimentally and have also been proven by the theory of elasticity.

In Fig.8–19a, a micro-section of length $\mathrm{d}x$ is intercepted from the circular shaft by two adjacent cross sections l—l and n—n in Fig.8–19b. Let the torque on the two end faces of the micro-section be m. Take l—l as the reference plane, the section n—n twists by an angle $\mathrm{d}\varphi$ with respect to the section l—l. According to the plane hypothesis, the cross section only makes a rigid rotation during deformation. The radius Oa also turns through an angle $\mathrm{d}\varphi$ to reach Oa'. A small relative misalignment of the ab side with respect to the cd side occurs, the misalignment distance is

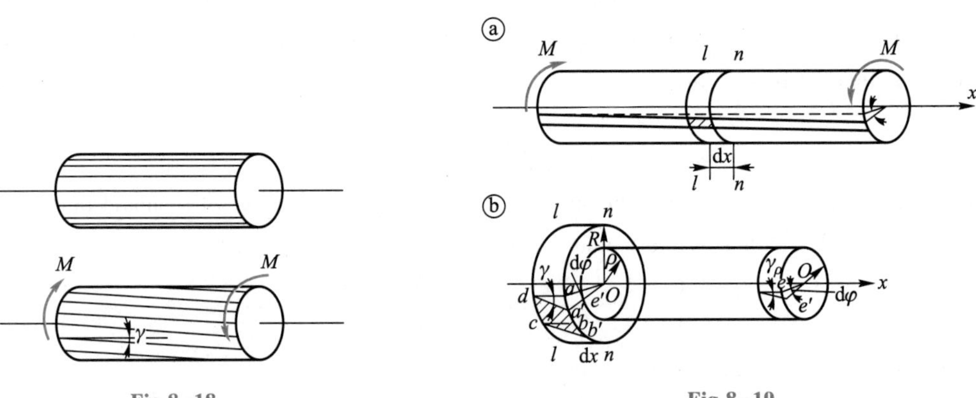

Fig.8–18   Fig.8–19

$$aa' = R \cdot \mathrm{d}\varphi$$

The angle change γ of the original rectangle on the surface of the circular shaft is

$$\gamma \approx \frac{aa'}{ad} = R\frac{\mathrm{d}\varphi}{\mathrm{d}x}$$

The above equation is the formula for shear strain γ at a point on the edge of a circular section. γ occurs in the plane perpendicular to the radius Oa. Similarly, and referring to Fig.8–19c, we can find the shear strain γ_ρ in the cross section at the centre ρ of the circle as

$$\gamma_\rho \approx \rho \frac{d\varphi}{dx} \tag{a}$$

Due to the rigid plane hypothesis, $\frac{d\varphi}{dx}$ is a constant at each point of the same cross section. Therefore, equation (a) shows that: the shear deformation $dx\gamma_\rho$ at any point on the cross section is proportional to ρ. The shear strain γ_ρ and the distance ρ from that point to the centre of the circle are also proportional. γ_ρ occurs in the plane perpendicular to the radius Oe.

2. Physical relation

According to Hooke law in shear, the shear stress at any point with a distance ρ from the center of the circle on the cross section is τ_ρ. It is proportional to the shear strain γ_ρ at that point,

$$\tau = G\gamma \tag{b}$$

Substitute equation (a) into equation (b) to find the shear stress at the distance ρ from the axis as

$$\tau_\rho = G\gamma_\rho = G\rho \frac{d\varphi}{dx} \tag{c}$$

The above formula shows that the shear stress τ_ρ at any point if the cross section is proportional to the distance ρ. The shear stress varies along the radius in a linear fashion, with zero shear stress at the centre of the circle and the maximum shear stress at points on the circumferential edge. Since the shear strain γ_ρ occurs in a plane perpendicular to the radius, the shear stress τ_ρ is also perpendicular to the radius. According to the theorem of complementary shearing stresses, the distribution of shear stresses along the radius in the longitudinal and transverse sections of the solid circular shaft is shown in Fig.8-20.

3. Static relation

Equation (c) gives the law of shear stress distribution, but it still cannot be used to calculate shear stress. As in Fig.8-21, at a point with a distance ρ to the centre, the micro-area dA is taken and there are micro-shear forces $\tau_\rho dA$ on the micro-area dA. The integral of the moment of each micro-shear force is the torque of the cross section M_n.

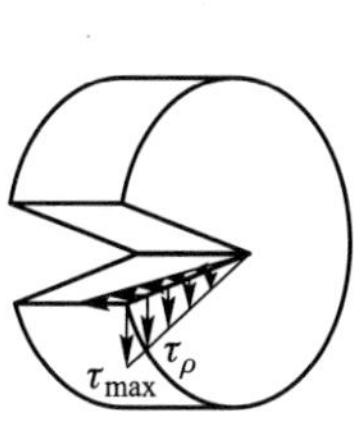

Fig.8-20

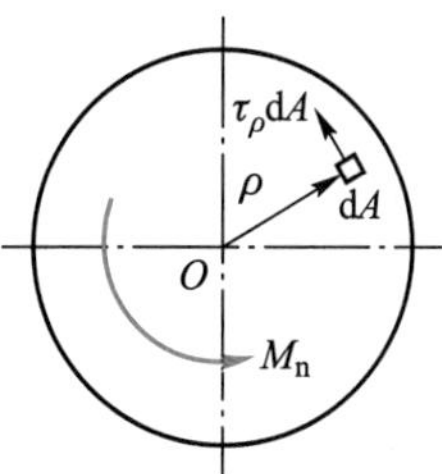

Fig.8-21

$$M_n = \int_A \rho \cdot \tau_\rho dA \tag{d}$$

The integral is carried out over the entire cross section A. Substituting equation (c) into equation

(d) and noting that $\frac{d\varphi}{dx}$ is a constant at the given cross section, we have

$$M_n = \int_A \rho \cdot \tau_\rho dA = G\frac{d\varphi}{dx}\int_A \rho^2 dA \tag{e}$$

let

$$I_p = \int_A \rho^2 dA \tag{8-16}$$

I_p is a quantity related to the geometry and dimensions of the cross section. It is called the polar moment of inertia of the cross section. Thus equation (d) can again be written as

$$M_n = GI_p\frac{d\varphi}{dx} \tag{8-17}$$

Eliminating $\frac{d\varphi}{dx}$ from equations (8-17) and (c) gives

$$\tau_\rho = \frac{M_n\rho}{I_p} \tag{8-18}$$

This is the formula for calculating the shear stress at any point on the cross section when the circular shaft is twisted. When $\rho = R$ (i.e. at each point on the edge of the cross section), the shear stress takes its maximum value.

$$\tau_{max} = \frac{M_n R}{I_p}$$

Citation marks

$$W_n = \frac{I_p}{R} \tag{8-19}$$

W_n is called the section modulus of torsion. This gives

$$\tau_{max} = \frac{M_n}{W_n} \tag{8-20}$$

8.7.2 Angle of twist

The torsional deformation of a circular shaft can be expressed in terms of the relative angle of twist φ of two cross sections. From equation (8-17), we can obtain the relative angle of twist between two cross sections that are separated by dx

$$d\varphi = \frac{M_n}{GI_p}dx$$

Integrating both sides of the above equation gives the relative angle of twist of the two sections separated by l

$$\varphi = \int_l d\varphi = \int_l \frac{M_n}{GI_p}dx \tag{8-21}$$

If the torque M_n between two cross sections G and I_p are constants, the angle of twist between these two sections is

$$\varphi=\frac{M_n l}{GI_p} \tag{8-22}$$

The equation (8−22) is the formula for calculating the torsional deformation of a circular shaft of equal cross section. The larger GI_p, the smaller the angle φ. GI_p reflecting the ability of the circular shaft to resist torsional deformation, known as the torsional rigidity. The sign of the angle φ of twist is specified in the same way as that of torque M_n and its unit is radian (rad).

If the torque or torsional rigidity between two cross sections is variable, the relative torsional angles of the two sections should be calculated by integrating the torsional angles of each section in accordance with equation (8−21) and then summing them algebraically.

8.7.3 Polar moments of inertia and section modulus of torsion

To calculate the polar moments of inertia I_p, it is convenient to use polar coordinates ρ、θ. Take the annular differential area with width $d\rho$ at a distance ρ from that point to the center of the circle: $dA=2\pi\rho d\rho$ (Fig.8−22). Substituting this into equation (8−16), the polar moment of inertia of the circular section is obtained as

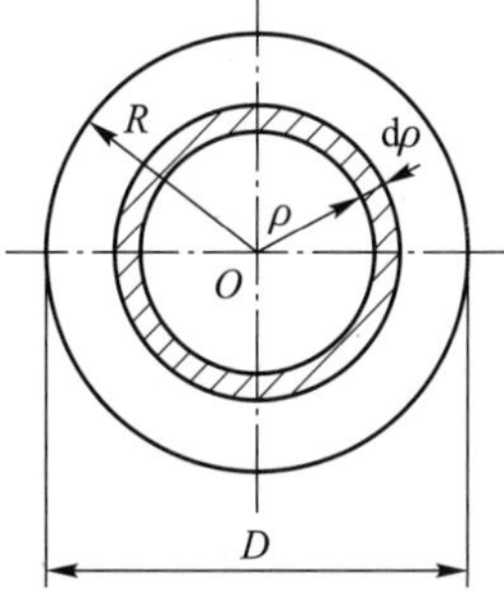

Fig.8−22

$$I_p=\int_A \rho^2 dA=\int_0^{\frac{D}{2}} \rho^2\cdot 2\pi\rho d\rho=\frac{\pi D^4}{32} \tag{8-23}$$

where D is the diameter of the circular section. From equation (8−19) we have

$$W_n=\frac{I_p}{\frac{D}{2}}=\frac{\pi D^3}{16} \tag{8-24}$$

The unit of I_p is the fourth power of the length and the dimension of W_n is the third power of the length.

In the case of a hollow circular shaft, shear stresses on the cross section of the hollow circular shaft should distribute as in Fig.8−23. Therefore, the definite integral in equation (e) should not include the hollow part either. Thus, for the hollow circular shaft we have

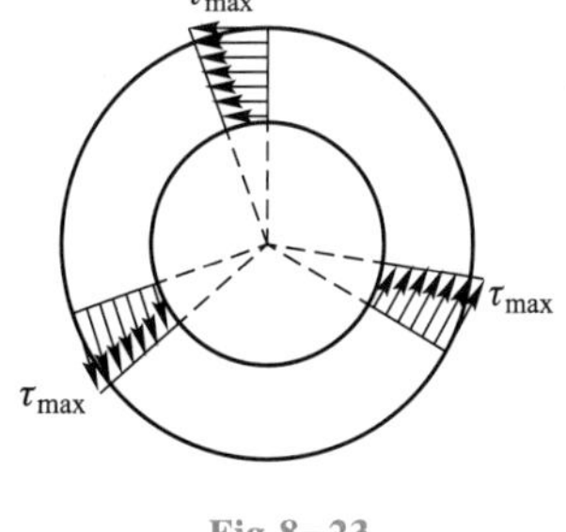

Fig.8−23

$$\left.\begin{aligned} I_p&=\int_A \rho^2 dA=2\pi\int_{\frac{d}{2}}^{\frac{D}{2}} \rho^3 d\rho=\frac{\pi(D^4-d^4)}{32}=\frac{\pi D^4}{32}(1-\alpha^4)\\ W_n&=\frac{I_p}{R}=\frac{I_p}{16D}(D^4-d^4)=\frac{\pi D^3}{16}(1-\alpha^4)\end{aligned}\right\} \tag{8-25}$$

where $\alpha=\frac{d}{D}$, D and d are the outer and inner diameters of the hollow circular section respectively, and R is the outer radius.

8.7.4 Limitations

The above stress and deformation equations are derived based on the rigid plane hypothesis. Tests and elastic theory have shown that rigid plane hypothesis is correct only for circular shaft of equal cross section. Therefore, these formulas are only applicable to isotropic circular bars. When the circular cross section changes slowly along the axis, it can also be approximated by the above formulas. I_p and W_n are also changing along the axis at the same time. In addition, formulas are only applicable when τ_{max} does not exceed the shear proportional limit τ_ρ of the material.

8.8 Torsional strength and rigidity

8.8.1 Strength condition

The strength requirement for torsion of a circular shaft is that the maximum working stress τ_{max} does not exceed the allowable shear stress $[\tau]$ of the material. The strength condition is

$$\tau_{max} \leqslant [\tau]$$

For straight circular shafts of equal cross section, the maximum working stress occurs at points around the perimeter of the cross section in which M_{nmax} is located (absolute terms). From equation (8-20) the above equation can then be written as

$$\tau_{max} = \frac{M_{nmax}}{W_n} \leqslant [\tau] \qquad (8-26)$$

In the case of stepped axes, τ_{max} does not necessarily occur in the section where M_{nmax} is located, as W_n are different in every segment. τ_{max} can be determined when torque and section modulus of torsion are considered comprehensively. In this case, the strength condition is

$$\tau_{max} = \left(\frac{M_n}{W_n}\right)_{max} \leqslant [\tau] \qquad (8-27)$$

The tests indicate that allowable shear stress $[\tau]$ and allowable tensile stress $[\sigma]$ has a certain relationship. For the commonly used plastic materials, $[\tau] = (0.5 \sim 0.7)[\sigma]$. Considering factors such as robbery load, the allowable shear stress taken is generally lower than the allowable shear stress under static load.

8.8.2 Rigidity calculation

Shaft parts should meet not only the strength requirements but also certain conditions of restriction on their deformation, which is called rigidity condition. Too large angle of twist of lathe spindle will cause greater vibration, affecting the accuracy and finish of the workpiece. For precision machinery, rigidity requirements often play a major role. As can be seen from equation (8-22), the angle of twist φ is related to the length l of the shaft. To eliminate the effect of length, the angle θ of torsion per unit length is commonly used in engineering to indicate the degree of torsional deformation. From equation (8-17) θ is

$$\theta=\frac{d\varphi}{dx}=\frac{M_n}{GI_p} \tag{8-28a}$$

To ensure the rigidity of the shaft, it is usually specified that θ_{max} should not exceed the specified allowable value $[\theta]$. The rigidity condition for torsion are thus obtained as

$$\theta_{max}=\frac{M_{nmax}}{GI_p}\leqslant[\theta] \tag{8-28b}$$

In engineering, the unit of $[\theta]$ is customarily used in degrees/meter [noted as (°)/m]. Considering that 1 rad$=\frac{180°}{\pi}$, the rigidity condition is therefore

$$\theta_{max}=\frac{M_{nmax}}{GI_p}\times\frac{180°}{\pi}\leqslant[\theta] \tag{8-29}$$

The value of $[\theta]$ can be found in the relevant brochures.

Example 8-7 Conditions as in Example 8-6 (as shown in Fig.8-14). The drive shaft has a shear modulus of elasticity $G=80$ GPa, an allowable shear stress $[\tau]=30$ MN/m^2 and an allowable angle $[\theta]=0.3°$/m of torsion per unit length. Try to design the diameter d of the shaft according to the strength condition and the rigidity condition.

Solution The largest absolute torque is in the section CA of the shaft

$$M_{nmax}=954\ \text{N}\cdot\text{m}$$

From the strength condition we have

$$\tau_{max}=\frac{M_{nmax}}{W_n}=\frac{M_{nmax}}{\frac{\pi}{16}d^3}\leqslant[\tau]$$

then

$$d\geqslant\sqrt[3]{\frac{16M_{nmax}}{\pi[\tau]}}=\sqrt[3]{\frac{16\times954}{\pi\times30\times10^6}}\ \text{m}=0.054\ 5\ \text{m}$$

From the rigidity condition we have

$$\theta_{max}=\frac{M_{nmax}}{GI_p}\times\frac{180°}{\pi}=\frac{M_{nmax}}{G\times\frac{\pi}{32}d^4}\times\frac{180°}{\pi}\leqslant[\theta]$$

so

$$d\geqslant\sqrt[4]{\frac{32M_{nmax}}{G\pi[\theta]}\cdot\frac{180°}{\pi}}=\sqrt[4]{\frac{32\times954\times180}{80\times10^9\times\pi^2\times0.3}}\ \text{m}=0.069\ 4\ \text{m}$$

Example 8-8 A solid shaft is coupled to a hollow shaft of the same material by means of a tooth-integrated clutch as shown in Fig.8-24. The external moment of couple transmitted by the shaft is

$M=700\ \mathrm{N\cdot m}$. Set the ratio of the inner and outer diameters of the hollow shaft $\alpha=0.5$ and the allowable shear stress $[\tau]=20\ \mathrm{MN/m^2}$. Try to determine the diameter d_1 of the solid shaft and the outer diameter D_2 of the hollow shaft, and compare the cross section areas of the two shafts.

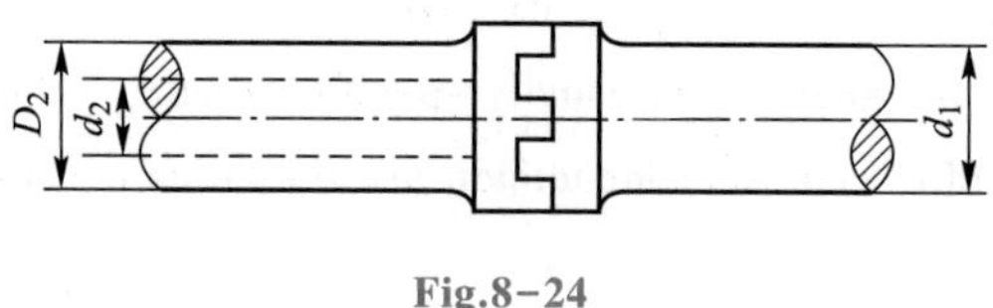

Fig.8-24

Solution From the strength condition we have

$$W_n \geqslant \frac{M_{n\max}}{[\tau]}=\frac{700}{20\times10^6}\ \mathrm{m}=35\times10^{-6}\ \mathrm{m^3}=35\ \mathrm{cm^3}$$

For solid shafts:

$$W_n=\frac{\pi d_1^3}{16}$$

so

$$d_1 \geqslant \sqrt[3]{\frac{16W_n}{\pi}}=\sqrt[3]{\frac{16\times35}{\pi}}\ \mathrm{cm}=5.6\ \mathrm{cm}$$

For hollow shafts:

$$W_n=\frac{\pi D_2^3}{16}(1-\alpha^4)$$

then

$$D_2 \geqslant \sqrt[3]{\frac{16W_n}{\pi}\cdot\frac{1}{1-\alpha^4}}=\sqrt[3]{\frac{16\times35}{\pi}\cdot\frac{1}{1-0.5^4}}\ \mathrm{mm}=5.75\ \mathrm{mm}$$

Inner diameter $d_2=0.5D_2=2.88$ cm. Ratio of the cross section area of a solid shaft to that of a hollow shaft

$$\frac{A_1}{A_2}=\frac{\dfrac{\pi d_1^2}{4}}{\dfrac{\pi D_2^2}{4}(1-\alpha^2)}=\frac{5.6^2}{5.75^2(1-0.5^2)}=1.248$$

Problems

8-1 Fig. P8-1 shows a screw subjected to a tensile force F. The relationship between the allowable shear stress and the allowable tensile stress $[\tau]$ of the material is known to be $[\tau]=0.6[\sigma]$, try to find a reasonable ratio of the screw diameter d to the height h of the screw head.

Answer: $d/h=2.4$

8-2 Clamp shear is shown in Fig.P8-2, $a=30$ mm, $b=150$ mm and the diameter of the pin C $d=5$ mm. A copper wire A of the same diameter as the pin is cut with a force $F=200$ N. Find the average shear stress in the cross section of the copper wire and the pin.

Answer: $\tau_A=51$ MPa, $\tau_C=61.2$ MPa

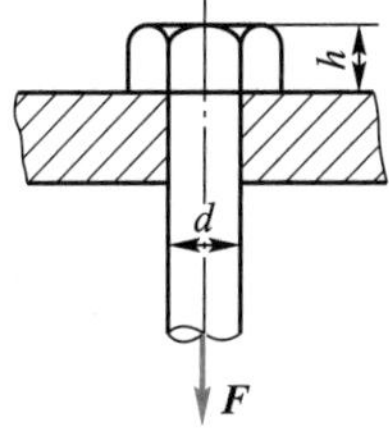

Fig.P8-1

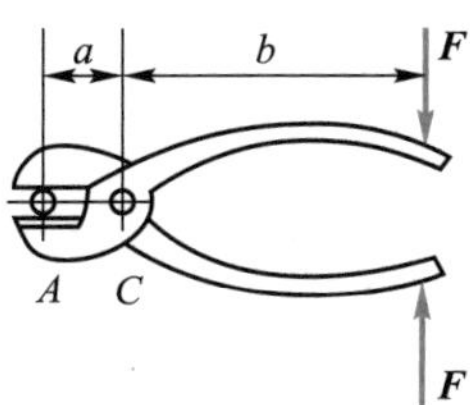

Fig.P8-2

8-3 The spline shaft of the machine tool shown has eight teeth (Fig.P8-3). The length of the fit of the wheel to the shaft is $l=60$ mm, the external torque $M=4$ kN · m and the allowable bearing stress $[\sigma_{jy}]=140$ MPa for the spline shaft.Try to check the bearing strength of the spline shaft.

Answer: $\sigma_{jy}=135$ MPa$<[\sigma_{jy}]$, not safe

8-4 Fig.P8-4 shows a flange coupling which transmits a couple moment of $M=200$ N · m. The flange is connected by four bolts with an inner diameter $d=10$ mm, symmetrically distributed on a circumference of $D=80$ mm. If the allowable shear stress of the bolts $[\tau]=60$ MPa, check the shear strength of the bolts.

Answer: $\tau=15.9$ MPa

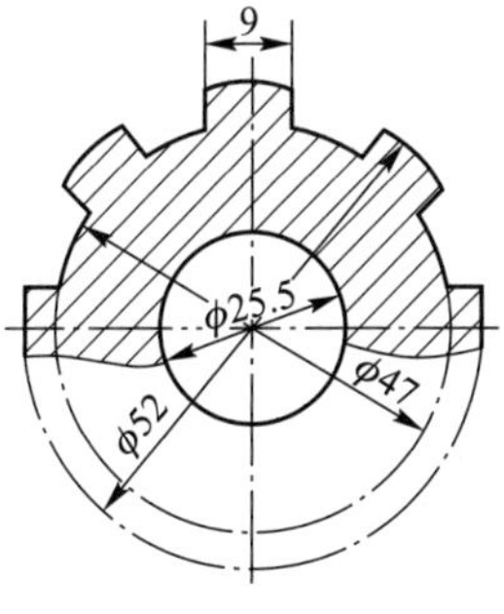

Fig.P8-3

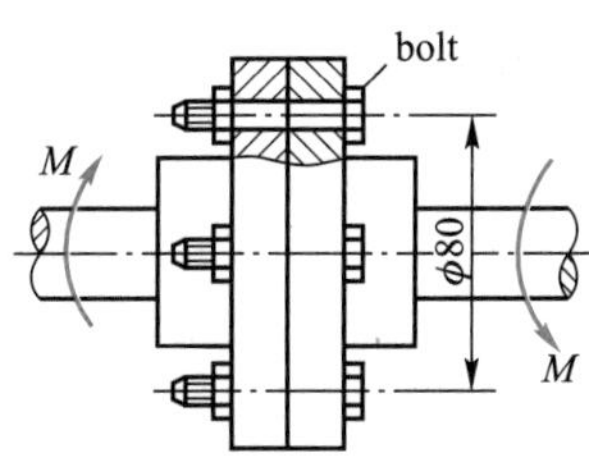

Fig.P8-4

8-5 In the riveted part shown (Fig.P8-5), the diameter of the rivet $d=20$ mm, the allowable shear stress is $[\tau]=140$ MPa and the allowable bearing stress is $[\sigma_{jy}]=320$ MPa. The thickness of the steel plate $t=10$ mm, the width $b=120$ mm, the allowable tensile stress $[\sigma]=100$ MPa and the allowable bearing stress $[\sigma_{jy}]=250$ MPa are all known. Check the strength of the steel plate and rivets under the load $F=50$ kN.

Answer: rivet τ; steel plate $\sigma = 41.6$ MPa; $\sigma_{jymax} = 125$ MPa

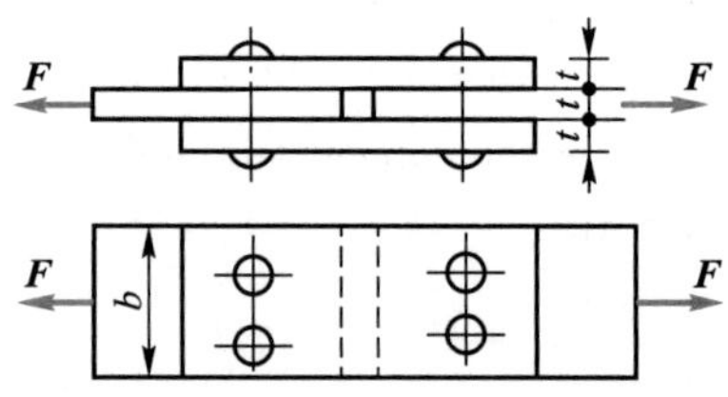

Fig.P8-5

8-6 Fig.P8-6 shows a tie bar fixed to a grating with four rivets of the same diameter. The tie bar and the rivets are made of the same material. The allowable shear stress $[\tau] = 100$ MPa, the allowable bearing stress $[\sigma_{jy}] = 300$ MPa and the allowable tensile stress $[\sigma] = 160$ MPa are known. Check the strength of the rivet and the bar.

Answer: $\tau = 99.5$ MPa, $\sigma_{jy} = 125$ MPa, $\sigma_{max} = 125$ MPa

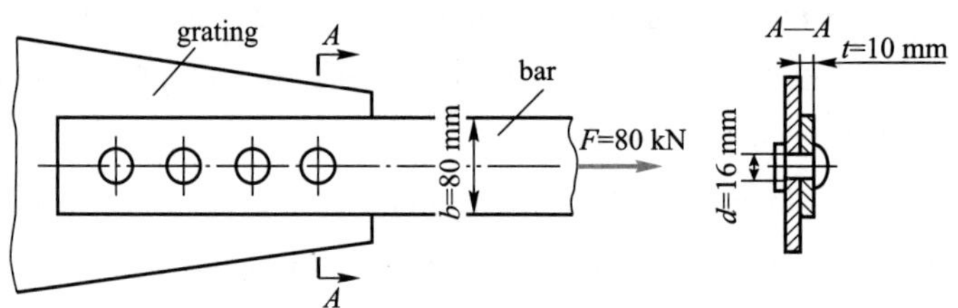

Fig.P8-6

8-7 Fig.P8-7 shows a pin connection. We have known $F = 13$ kN, the thickness of plates $t_1 =$ 8 mm and $t_2 = 5$ mm, a pin of the same material as the plate, with an allowable shear stress $[\tau] =$ 60 MPa and an allowable bearing stress $[\sigma_{jy}] = 200$ MPa. Try to design the diameter d of the pin.

Answer: $d = 1\ 808$ mm

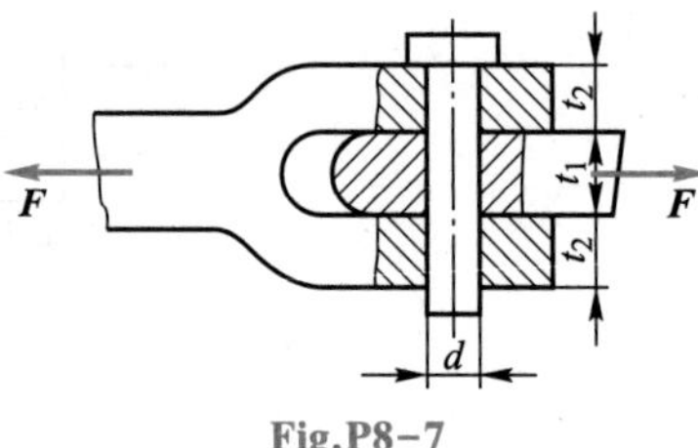

Fig.P8-7

8-8 Fig.P8-8 shows a dowel joint. It has been given that $a = b = 12$ cm, $h = 35$ cm, $c = 4.5$ cm and $F = 40$ kN. Find the shear and compression stresses in the joint.

Answer: $\tau = 0.952$ MPa, $\sigma_{jy} = 7.41$ MPa

8-9 Make the torque diagram of each bar (Fig.P8-9).

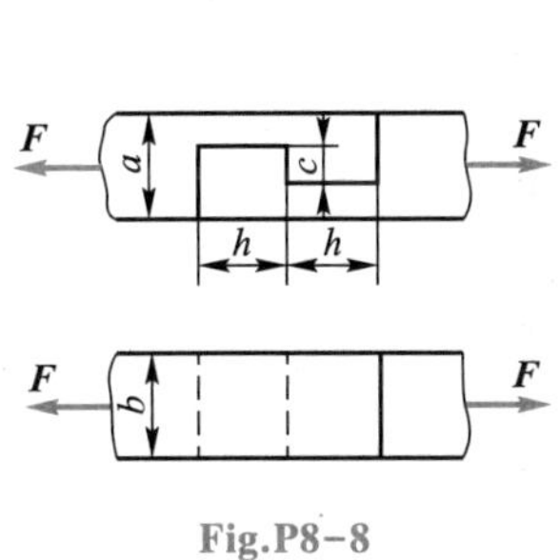

Fig.P8-8

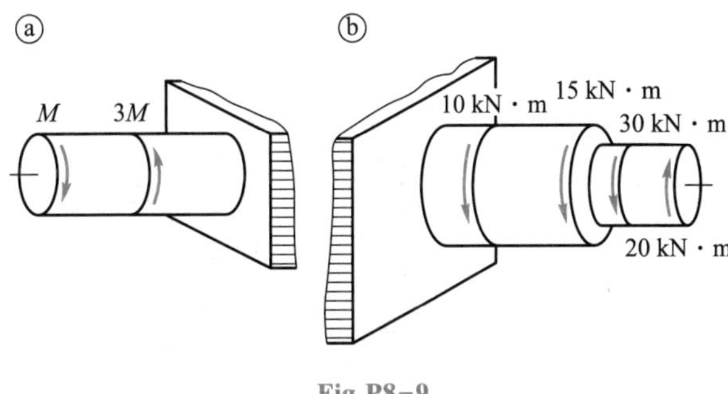

Fig.P8-9

8-10 A circular shaft of diameter $D = 5$ cm is subjected to a torque $M_n = 2.15$ kN · m. Try to find the shear stress at a distance from the shaft centre 1 cm and find the maximum shear stress in the shaft section.

Answer: $\tau_p = 35$ MN/m^2, $\tau_{max} = 87.6$ MN/m^2

8-11 Stepped circular shaft of diameter $d_1 = 4$ cm, $d_2 = 7$ cm is equipped with three pulleys as shown in Fig.P8-11. It is known that the power input by the third pulley is $P_3 = 30$ kW, the power output by the first pulley is $P_1 = 13$ kW. The speed $n = 200$ r/min, allowable shear stress $[\tau] = 60$ MN/m^2, $G =$ 80 GN/m^2 and $[\theta] = 2°$/m are known. Try to check the strength and rigidity of the shaft.

Answer: $\tau_{ACmax} = 49.4$ MN/m^2, $\tau_{DB} = 421.3$ MN/m^2, $\theta_{max} = 1.77°$, safe

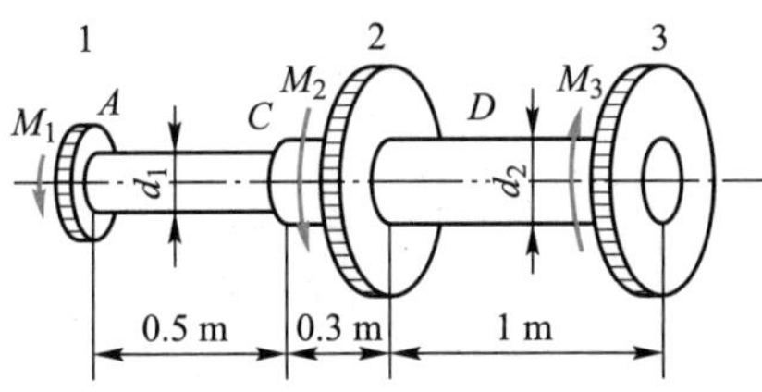

Fig.P8-11

8-12 The speed of the drive shaft (Fig.P8-12) is $n = 500$ r/min. We have the input power $P_1 =$ 500 kW of the active wheel 1, the output $P_2 = 200$ kW and $P_3 = 300$ kW of the driven wheel 2 and 3, respectively. We also known$[\tau] = 70$ MN/m^2, $[\theta] = 1°$/m, $G = 80$ GN/m^2. (1) Try to determine the diameter d_1 of the segment AB and the diameter d_2 of the segment BC. (2) If the same diameter is chosen for the two segments BC and AB, try to determine the diameter d. (3) What is the most appropriate arrangement for the driving and driven wheels?

Answer: (1) $d_1 \geqslant 84.6$ mm, $d_2 \geqslant 74.5$ mm; (2) $d \geqslant 84.6$ mm; (3) The Driving Wheel 1 is reasonably placed between the driven wheels 2 and 3

8-13 A drilling rig has a power 10 kW, a speed $n = 180$ r/min, and a depth $l = 40$ cm. If the resistance of the soil to the drill pipe can be regarded as a uniformly distributed couple, try to find the concentrate t of this distributed couple and make a torque diagram of the drill pipe (Fig.P8-13).

Answer: $t = 0.013\ 3$ (kN · m)/m

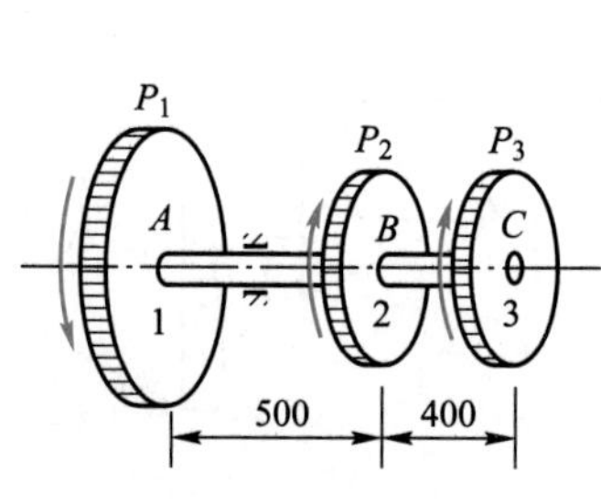

Fig.P8-12

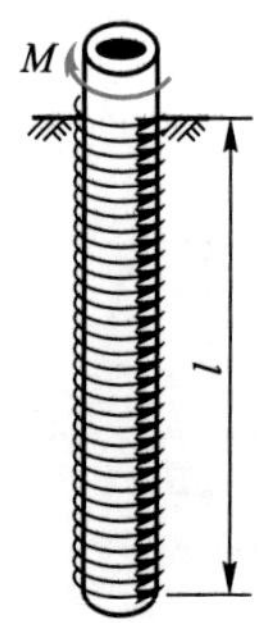

Fig.P8-13

8-14 A steel round bar of diameter $d=25$ mm is subject to axial tension 60 kN. The extend length is 0. 113 mm of the standard distance of 200 mm. When it is twisted by a pair of external moments of couple 0.2 kN · m, it is twisted by an angle 0.73 2°. Try to find the elastic constants E、G and μ of the steel.

Answer: $E=216$ GN/m^2, $G=81.8$ GN/m^2, $\mu=0.32$

8-15 The structure is loaded as shown in Fig.P8-15. We have known $d_1=100$ mm, $d_2=50$ mm, $G=82$ GPa, $[\tau]=40$ MPa, and $[\theta]=0.5°/\text{m}$. Check the strength and rigidity of the shaft and find the angle of twist of the section C.

Answer: Section AB: $\tau_{max}=81.5$ MN/m^2, $\theta_{max}=0.36°/\text{m}$

Section BC: $\tau_{max}=81.5$ MN/m$^2>[\tau]$, $\theta_{max}=2.28°/\text{m}$, $\varphi_C=0.37\times10^{-2}$ rad

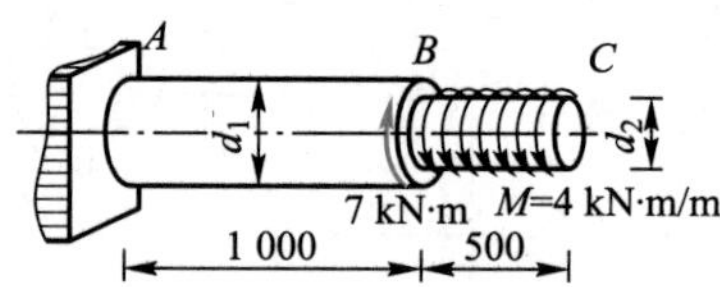

Fig.P8-15

Chapter 9 Geometric Properties of Cross Sections

Teaching Scheme of Chapter 9

Some geometry of the cross section is used when calculating the stress and deformation of the member. Such as tensile (press) deformation used in the cross-sectional area A, torsional deformation used in the extreme moment of inertia I_p, etc. They are only related to the size and shape of the section and are called geometric properties of the section. In bending deformation, some other geometric properties of the section are used, such as static moment, moment of inertia, product of inertia, etc. This chapter will concentrate on their definition and calculation methods.

9.1 Static moment of area and centroid

9.1.1 Static moment of area

As shown in Fig.9−1, the section area is A. The coordinate (y, z) is the core coordinate of the microelement area dA. The ydA and zdA are called the static moment for the z and y axes respectively, and sometimes called area moments. The integral across the entire section

$$\left.\begin{aligned} S_z &= \int_A y\mathrm{d}A \\ S_y &= \int_A z\mathrm{d}A \end{aligned}\right\} \tag{9-1}$$

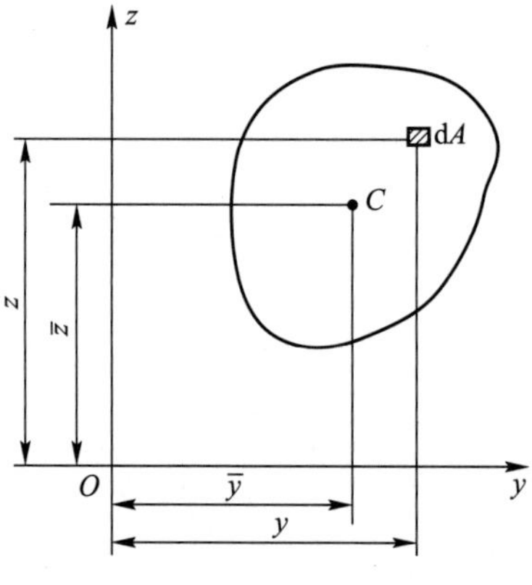

Fig.9−1

Static moments are for certain axes and will vary for different axes for the same section. The value of the static moment can be positive, negative, or zero, and its dimension is L^3.

9.1.2 Relationship between static moment of area and centroid

If the cross section is considered as a homogeneous and equal-thickness sheet, its centre of gravity is the centroid. According to the position of the centre of gravity, the coordinates of the centroid C in the zOy coordinate system $\bar{z}$, $\bar{y}$ can be obtained, i.e.

$$\bar{z} = \frac{\int_A z\mathrm{d}A}{A} = \frac{S_y}{A}, \quad \bar{y} = \frac{\int_A y\mathrm{d}A}{A} = \frac{S_z}{A} \tag{9-2}$$

When the position of the centroid is known, the static moment can beobtained

$$S_z = A\bar{y}, \quad S_y = A\bar{z} \tag{9-3}$$

The coordinate axis passing through the centroid within a plane is called the centroidal axis. From equation (9-3), it can be seen that the static moment of the cross-sectional pattern to its mandrel is equal to zero. Conversely, if the static moment of a section to an axis is zero, this axis must be the centroidal axis of the center of the section.

Example 9-1 Try to find the static moments S_y and S_z of the semicircular section shown in Fig.9-2 and the coordinates of centroid.

Solution (1) Since the z axis is the axis of symmetry and must pass through the centroid of the section, we have

$$\bar{y}=0, \quad S_z=0$$

(2) For the purpose of calculating S_y take a narrow strip parallel to the y axis as the micro area

$$\mathrm{d}A = 2R\cos\theta\mathrm{d}z$$

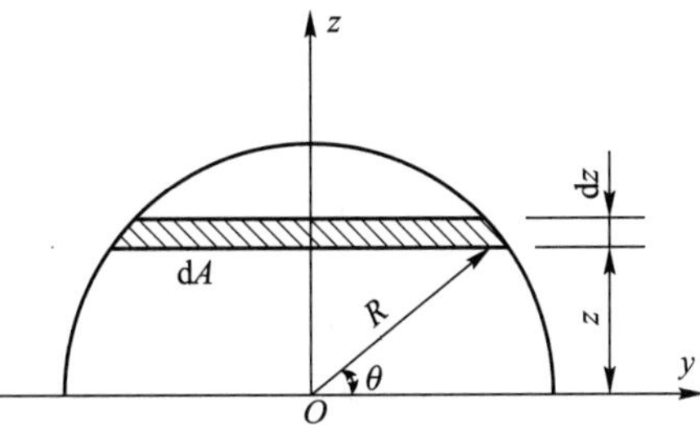

Fig.9-2

and

$$z=R\sin\theta, \quad \mathrm{d}z=R\cos\theta\mathrm{d}\theta$$

then

$$\mathrm{d}A = 2R^2\cos^2\theta\mathrm{d}\theta$$

Substituting above epuation into equation (9-1) gives

$$S_y = \int_A z\mathrm{d}A = \int_0^{\frac{\pi}{2}} R\sin\theta \cdot 2R^2\cos^2\theta\mathrm{d}\theta = \frac{2}{3}R^3$$

Substituting S_y into equation (9-2), we get

$$\bar{z}=\frac{S_y}{A}=\frac{\frac{2}{3}R^3}{\frac{1}{2}\pi R^2}=\frac{4R}{3\pi}$$

9.1.3 Centroid of composite area

A section consisting of simple picutres (e.g. circles, rectangles, triangles, etc.) is called a composite section. The static moment of a composite section for a given axis is equal to the algebraic sum of the static moments of the simple picutres, i.e.

$$S_z = \sum_{i=1}^{n} A_i \bar{y}_i, \quad S_y = \sum_{i=1}^{n} A_i \bar{z}_i \tag{9-4}$$

where A_i and $(\bar{y}_i, \bar{z}_i)$ are the area and centroidal coordinates of each simple picutre in the composite section, and n is the number of simple picutres. Substituting equation (9-4) into equation (9-3), we can obtain the equation for the coordinates of the centroid of the composite section

$$\bar{y} = \frac{\sum_{i=1}^{n} A_i \bar{y}_i}{\sum_{i=1}^{n} A_i}, \quad \bar{z} = \frac{\sum_{i=1}^{n} A_i \bar{z}_i}{\sum_{i=1}^{n} A_i} \tag{9-5}$$

The centroid and center of gravity of a homogeneous object are the same. The centroid is an intrinsic property of geometric structures and is independent of coordinates.

9.2 Moment of inertia, product of inertia and radius of gyration

9.2.1 Moment of inertia

In the Fig.9-3, $y^2 \mathrm{d}A$ and $z^2 \mathrm{d}A$ are called the moments of inertia of the micro-area $\mathrm{d}A$ with respect to the z and y axes, respectively. The integral over the entire cross-sectional area A

$$\left.\begin{aligned} I_z &= \int_A y^2 \mathrm{d}A \\ I_y &= \int_A z^2 \mathrm{d}A \end{aligned}\right\} \tag{9-6}$$

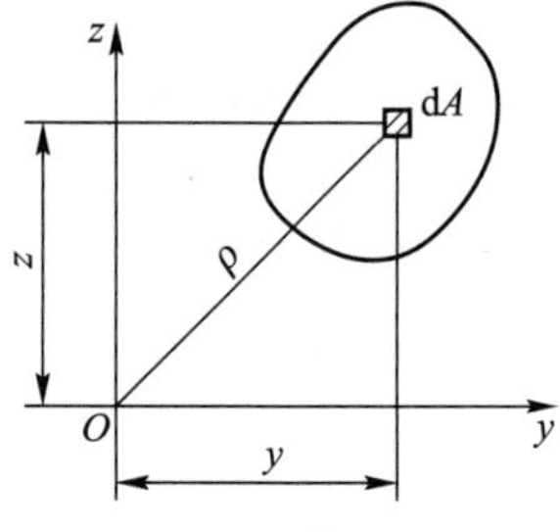

Fig.9-3

It is defined as the moment of inertia or second moment for the section with respect to the z and y axes, respectively. The moment of inertia of the same section varies for different axes, but it is always positive. The dimension of the moment of inertia is L^4.

9.2.2 Product of inertia

In the Fig.9-3, the product $yz\mathrm{d}A$ is called the inertial product of the micro-area $\mathrm{d}A$ over the z, y axis. The integral over the entire cross-sectional area A

$$I_{yz} = \int_A yz \mathrm{d}A \tag{9-7}$$

It is defined as the product of inertia for the section with respect to the z, y axes. The product of inertia for a section varies for different axes. The product of inertia may be positive, negative or zero. The dimension of the product of inertia is L^4.

9.2.3 Radius of gyration

The moment of inertia is sometimes expressed as the product of the cross-sectional area A and a certain length squared, i.e.

$$I_y = A i_y^2,\quad I_z = A i_z^2 \tag{9-8a}$$

$$i_y = \sqrt{\frac{I_y}{A}},\quad i_z = \sqrt{\frac{I_z}{A}} \tag{9-8b}$$

where i_y, i_z is defined as the radius of gyration of the section with respect to the z and y axis, respectively, with the dimension of length.

9.2.4 Polar moments of inertia

The distance from the area of the micro-element dA to the origin O of the coordinates is denoted by ρ.

Define the integral

$$I_p = \int_A \rho^2 \mathrm{d}A$$

It is the polar moment of inertia of the section with respect to the origin of the coordinates.

Because

$$\rho^2 = y^2 + z^2$$

We have

$$I_p = \int_A \rho^2 \mathrm{d}A = \int_A (y^2 + z^2)\mathrm{d}A = \int_A y^2 \mathrm{d}A + \int_A z^2 \mathrm{d}A$$

and

$$I_p = I_y + I_z \tag{9-9}$$

From each of these definitions, we can get the following conclusions.

(1) The same section has different moments of inertia and products of inertia for different coordinate axes.

(2) The sum of the moments of inertia I_y and I_z is constantly equal to the moment of inertia I_p of the section at the intersection of these two axes.

(3) I_y, I_z and I_p are constantly positive, while the product of inertia I_{yz} may have a negative or zero value. They all have the dimension of L^4.

(4) The product of inertia with respect to two orthogonal axes is equal to zero if the axes is the axis of symmetry of the section.

Example 9-2 Calculate the moment of inertia of a rectangular section shown in Fig.9-4 with respect to the y and z axes of symmetry.

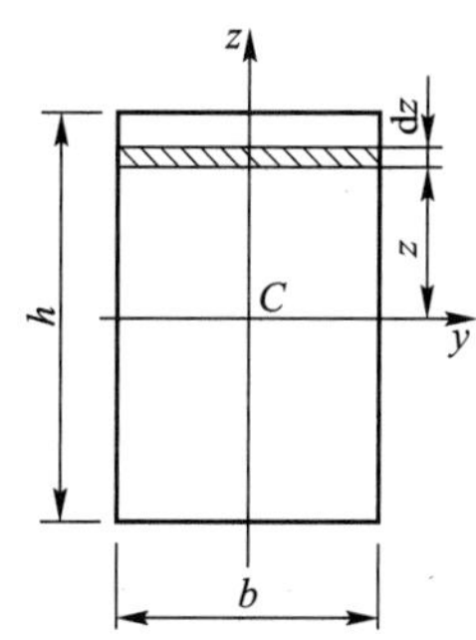

Fig.9-4

Solution Take a narrow rectangle parallel to the y axis to be the micro-area dA. Then

$$\mathrm{d}A = b\mathrm{d}z$$

The moment of inertia of the section with respect to the$y$ axis is

$$I_y = \int_A z^2 \mathrm{d}A = \int_{-\frac{h}{2}}^{\frac{h}{2}} bz^2 \mathrm{d}z = \frac{bh^3}{12}$$

A similar moment of inertia to the z axis is

$$I_z = \frac{b^3 h}{12}$$

Example 9-3 Calculate the moment of inertia of a circular section (shown in Fig.9-5) with respect to its centroidal axis.

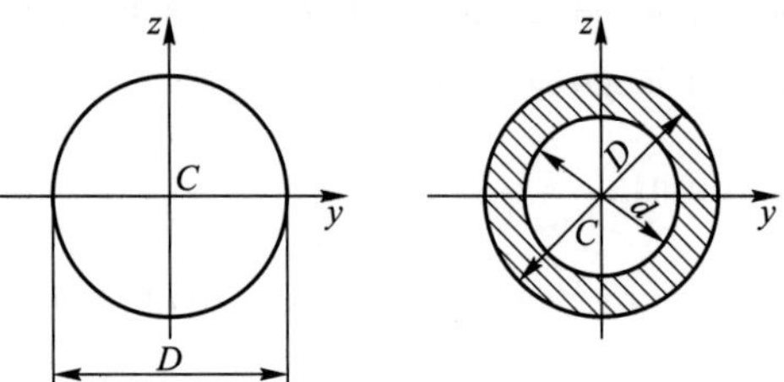

Fig.9-5

Solution Since the circular section is symmetric with respect to any centroidal axis, there is $I_y = I_z$.

The polar moment of inertia ofthe circular section to its centroid is

$$I_{\mathrm{p}} = \frac{\pi D^4}{32}$$

From equation (9-9) we have

$$I_y = I_z = \frac{1}{2} I_{\mathrm{p}} = \frac{\pi D^4}{64}$$

For a composite section, the moment of inertia of the composite section for a certain axis should be equal to the sum of the moments of inertia of each constituent section for the same axis, i.e.

$$I_y = \sum_{i=1}^{n} I_{yi}, \quad I_z = \sum_{i=1}^{n} I_{zi} \tag{9-10}$$

For a hollow circular section,

$$I_y = I_z = \frac{\pi D^4}{64} - \frac{\pi d^4}{64} = \frac{\pi D^4}{64}(1 - \alpha^4)$$

where $\alpha = d/D$.

9.3 Parallel axis theorem

The same section has different moments of inertia and products of inertia for different axes. The relationship between the moment of inertia and the product of inertia for two pairs of parallel axes (one of which is a centroidal axis) is discussed below.

In the Fig.9-6, y_C and z_C are orthogonal axes passing through the centroid of the section, and are parallel to the y and z axes respectively. The distances between the two parallel axes are a and b, respectively. We have

$$y=y_C+b,\quad z=z_C+a$$

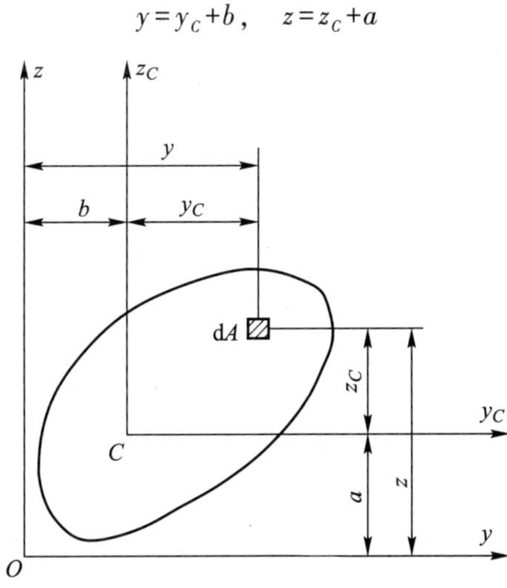

Fig.9-6

From the definition of moment of inertia

$$I_y=\int_A z^2\mathrm{d}A=\int_A (z_C+a)^2\mathrm{d}A=\int_A (z_C^2+2az_C+a^2)\mathrm{d}A\pi=I_{y_C}+a^2A \tag{9-11}$$

Since point C is the centroid, $S_{y_C}=0$. Similarly, we have

$$I_z=I_{z_C}+b^2A$$

$$I_{yz}=I_{y_Cz_C}+abA$$

The above equation is known as the parallel axis theorem.

Example 9-4 Calculate the moment of inertia of the T-shaped section shown in the Fig.9-7 for the y_C and z_C centroidal axes.

Solution Position of the section's centroid

$$\bar{y}=0,\quad \bar{z}=0.046\ 7\ \mathrm{m}$$

To calculate the moments of inertia I_{y_C} and I_{z_C}, firstly calculate the moments of inertia of the two rectangles Ⅰ and Ⅱ for the y_C and z_C axes respectively

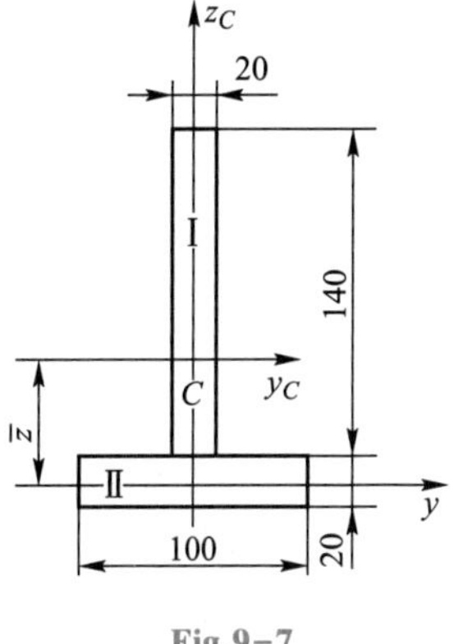

Fig.9-7

$$I_{y_C}^{\mathrm{I}}=\frac{1}{12}\times0.02\times0.14^3\ \mathrm{m}^4+(0.07+0.01-0.046\ 7)^2\times0.02\times0.14\ \mathrm{m}^4$$

$$=7.69\times10^{-6}\ \mathrm{m}^4$$

$$I_{y_C}^{\mathrm{II}}=\frac{1}{12}\times0.1\times0.02^3\ \mathrm{m}^4+0.046\ 7^2\times0.1\times0.02\ \mathrm{m}^4$$

$$=4.43\times10^{-6}\ \mathrm{m}^4$$

$$I_{z_C}^{\mathrm{I}}=\frac{1}{12}\times 0.14\times 0.02^3\ \mathrm{m}^4=0.09\times 10^{-6}\ \mathrm{m}^4$$

$$I_{y_C}^{\mathrm{II}}=\frac{1}{12}\times 0.02\times 0.1^3\ \mathrm{m}^4=1.67\times 10^{-6}\ \mathrm{m}^4$$

So the moment of inertia of the whole section with respect to the y_C and z_C axes is

$$I_{y_C}=I_{y_C}^{\mathrm{I}}+I_{y_C}^{\mathrm{II}}=7.69\times 10^{-6}\ \mathrm{m}^4+4.43\times 10^{-6}\ \mathrm{m}^4=12.1\times 10^{-6}\ \mathrm{m}^4$$

$$I_{z_C}=I_{z_C}^{\mathrm{I}}+I_{z_C}^{\mathrm{II}}=0.09\times 10^{-6}\ \mathrm{m}^4+1.67\times 10^{-6}\ \mathrm{m}^4=1.76\times 10^{-6}\ \mathrm{m}^4$$

9.4 Transformation equation and principal moment of inertia

Now, we discuss the relationship between the moment of inertia and the product of inertia when the coordinate system is rotated about the origin.

9.4.1 Transformation equation

The moment of inertia and product of inertia shown in Fig.9-8 are I_y, I_z and I_{yz}. Rotating the zOy coordinate system by an angle of α around point O to obtain the new coordinate system z_1Oy_1. α is specified as positive for counterclockwise rotation.

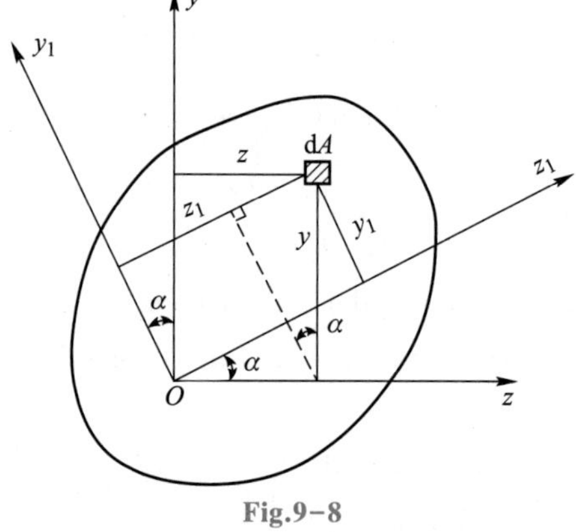

Fig.9-8

For transformation of coordinates by the axis of rotation, we have

$$z_1=z\cos\alpha+y\sin\alpha$$

$$y_1=y\cos\alpha-z\sin\alpha$$

so

$$I_{z_1}=\int_A y_1^2\mathrm{d}A=\int_A(y\cos\alpha-z\sin\alpha)^2\mathrm{d}A$$

$$I_{y_1}=\int_A z_1^2\mathrm{d}A=\int_A(z\cos\alpha+y\sin\alpha)^2\mathrm{d}A$$

$$I_{y_1z_1}=\int_A y_1z_1\mathrm{d}A=\int_A(z\cos\alpha+y\sin\alpha)(y\cos\alpha-z\sin\alpha)\mathrm{d}A$$

Expand and get

$$\left.\begin{aligned}I_{z_1}&=\frac{I_z+I_y}{2}+\frac{I_z-I_y}{2}\cos 2\alpha-I_{yz}\sin 2\alpha\\I_{y_1}&=\frac{I_z+I_y}{2}-\frac{I_z-I_y}{2}\cos 2\alpha+I_{yz}\sin 2\alpha\\I_{y_1z_1}&=\frac{I_z-I_y}{2}\sin 2\alpha+I_{yz}\cos 2\alpha\end{aligned}\right\}\tag{9-12}$$

These are transformation equations. From the above expressions, we can obtain a property independent of coordinates:

$$I_{y_1}+I_{z_1}=I_y+I_z=I_{\mathrm{p}}$$

9.4.2 Principal axis of inertia and principal moment of inertia

If the product of inertia to a pair of orthogonal axes is zero, this pair of axes is called the principal axis. The moment of inertia about a principal axis is called the principal moment of inertia. When the principal axis is the centroidal axis, it is called the centroidal principal axis. The moment of inertia with respect to the centroidal principal axis is referred to as the centroidal principal moment of inertia.

If the section has an axis of symmetry, this axis is one of the principal axes of the centroid. The other principal axis of the centroid is the axis passing through the section centroid and perpendicular to this axis of symmetry.

For the cross section without axis of symmetry, the position of the principal axis is determined by calculation.

Substituting $\alpha=\alpha_0$ into the third equation of equation (9-12) and making $I_{y_1z_1}=0$, we have

$$\frac{I_z-I_y}{2}\sin 2\alpha_0+I_{yz}\cos 2\alpha_0=0$$

and

$$\tan 2\alpha_0=-\frac{2I_{yz}}{I_z-I_y}\tag{9-13}$$

By substituting the equation into the first two equations of equation (9-12), we obtain I_{y_0} and I_{z_0}. From equation (9-13), we obtain $\sin 2\alpha_0$ and $\cos 2\alpha_0$. Substituting into equation (9-12) and finally obtaining the general formula for the principal moment of inertia is

$$\left.\begin{aligned}I_{z_0}&=\frac{I_z+I_y}{2}+\sqrt{\left(\frac{I_z-I_y}{2}\right)^2+I_{yz}^2}\\I_{y_0}&=\frac{I_z+I_y}{2}-\sqrt{\left(\frac{I_z-I_y}{2}\right)^2+I_{yz}^2}\end{aligned}\right\}\tag{9-14}$$

The maximum and the minimum value of the moment of inertia with respect to all axes are the principal moments of inertia. Tthe principal axes and the principal moments of inertia are intrinsic properties of geometric structures and independent of coordinates.

Example 9-5 Find the position of the centroidal principal axis of the Z-shaped section shown in the Fig.9-9 and find the centroidal principal moment of inertia of thesection.

Solution (1) Determine the position of the centroid. Since the Fig.9-9 is antisymmetric, its centre of symmetry C is the centroid.

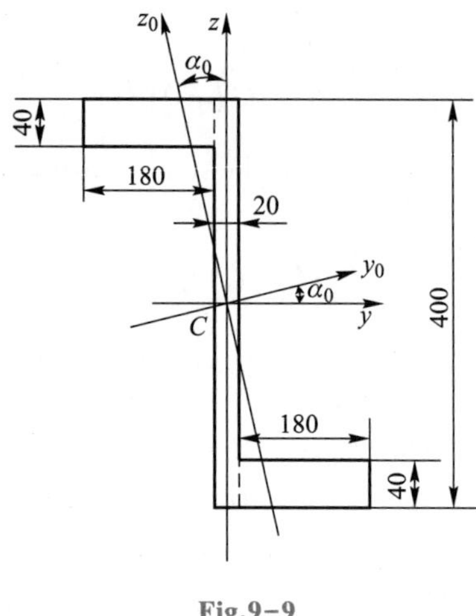

Fig.9-9

(2) The dashed line is now divided into three rectangles to calculate the moment of inertia.

$$I_y=\frac{2}{12}\times 40^3\text{ cm}^4+2\left(\frac{18}{12}\times 4^3+18^2\times 18\times 4\right)\text{ cm}^4=57\ 515\text{ cm}^4$$

$$I_z=\frac{40}{12}\times 2^3\text{ cm}^4+2\left(\frac{4}{12}\times 18^3+10^2\times 18\times 4\right)\text{ cm}^4=18\ 315\text{ cm}^4$$

$$I_{yz}=-2(18\times 10\times 18\times 4)\text{ cm}^4=-25\ 920\text{ cm}^4$$

(3) Determining the centroidal principal axis.

Determine the centroidal principal axis y_0, z_0 and the principal moments I_{y_0}, I_{z_0} from I_y, I_z, I_{yz}.

$$\tan 2\alpha_0=-\frac{2I_{yz}}{I_y-I_z}=-\frac{2(-25\ 920)\text{ cm}^4}{(57\ 515-18\ 315)\text{ cm}^4}=1.322\ 4$$

$$2\alpha_0=52.90°,\quad \alpha_0=26.45°$$

(4) The centroidal principal moment of inertia is now determined.

$$\left.\begin{cases}I_{y_0}\\ I_{z_0}\end{cases}\right\}=\frac{I_y+I_z}{2}\pm\sqrt{\left(\frac{I_y-I_z}{2}\right)^2+I_{yz}^2}=\begin{cases}70\ 411\text{ cm}^4\\ 5\ 419\text{ cm}^4\end{cases}$$

It can be seen that $I_{y_0}=I_{max}=70\ 411\text{ cm}^4$ and $I_{z_0}=I_{min}=5\ 419\text{ cm}^4$.

If the plane is the cross section of the bar, the plane defined by the principal centroidal axis of inertia and the axis of the bar is called the principal centroidal plane of inertia. The principal centroidal axis of inertia, the principal centroidal moment of inertia and the principal centroidal plane of inertia are of importance in the theory of bending of bars. The product of inertia for the symmetry cross section is equal to zero and the cross-sectional centroid must be on the axis of symmetry. Therefore, the symmetry axis of the cross section is the principal axis of inertia.

Problems

9-1 Calculate the static moment S_z about the z axis for the shaded part of the section (Fig.P9-1).

Answer: $S_z = \frac{b}{2}\left(\frac{h^2}{4} - y^2\right)$

9-2 Calculate the moment of inertia I_z for the box section shown about the horizontal centroidal z axis (Fig.P9-2).

Answer: $I_z = 1.55 \times 10^{10}\ \text{mm}^4$

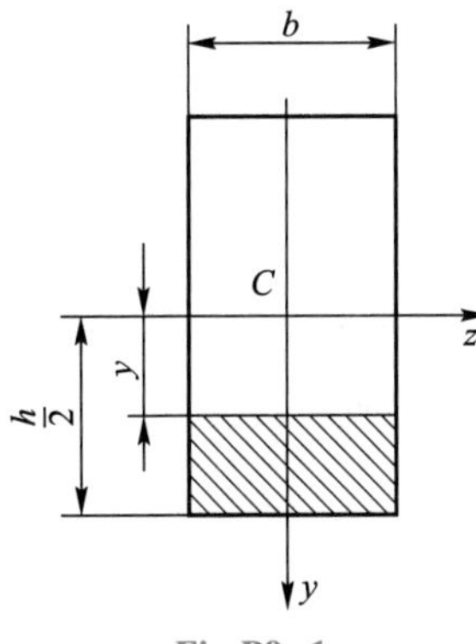

Fig.P9-1

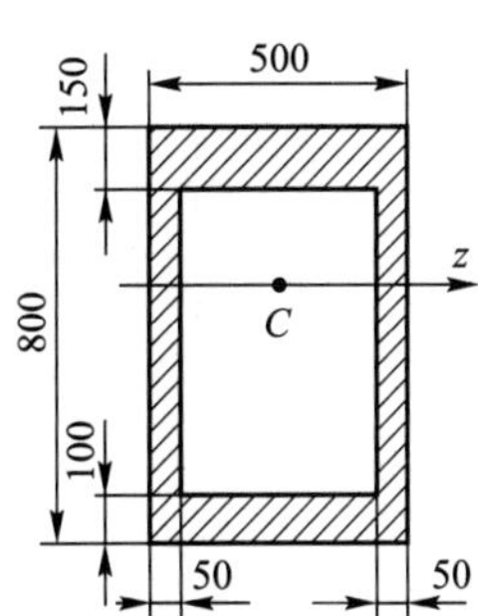

Fig.P9-2

9-3 The static moments for square section are $S_z = S_y = 0$ (Fig.P9-3), Is the moment of inertia I_z and I_y of the square section zero? Try to calculate I_z and I_y.

Answer: $I_y = I_z = \frac{a^4}{12}$

9-4 The ratio of the rectangle's height to its width is $\frac{h}{b} = \frac{3}{2}$ (Fig.P9-4). If a semicircle of diameter $d = 0.5h$ is cut from each of the left and right sides, try to find:

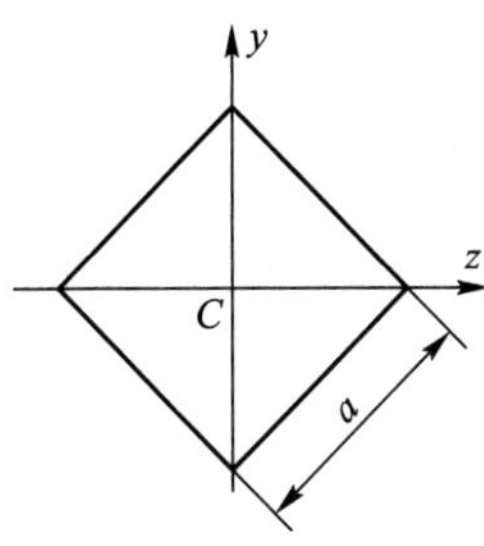

Fig.P9-3

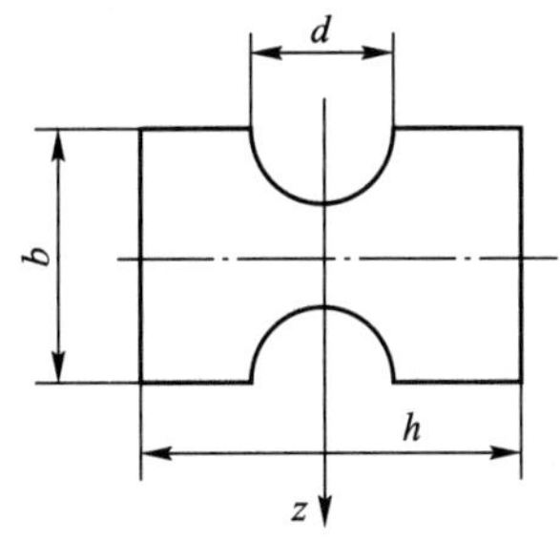

Fig.P9-4

(1) The percentage of the area cut off to the original area.

(2) The ratio of the moment of inertia I'_z after cutting away to the moment of inertia I_z of the original rectangle.

Answer: (1) 29.4% ; (2) 94%

9-5 Find the principal centroidal moments of inertia for each section shown (Fig.P9-5).

Answer: (a) $I_y = 23\ 808\ \text{cm}^4$, $I_z = 18\ 800\ \text{cm}^4$;

(b) $\alpha = 22.2°$, $I_{z_0} = 353.9\ \text{cm}^4$, $I_{y_0} = 59.2\ \text{cm}^4$

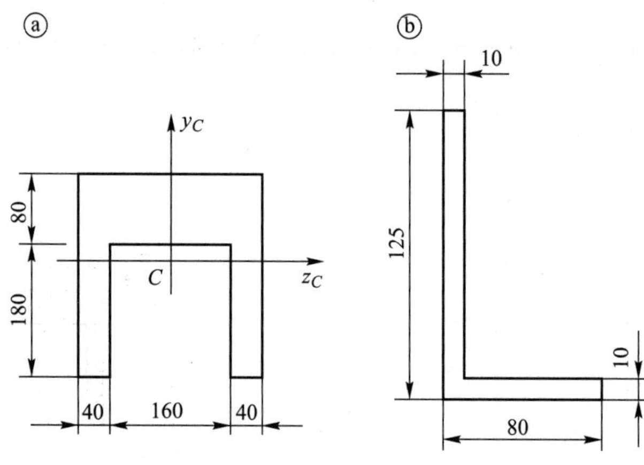

Fig.P9-5

9-6 Find the principal centroidal moment of inertia I_{y_C} of the semicircle shown in Fig.P9-6 and the moment of inertia I_y with respect to the centre of the circle O. It is known that $d = 2$ m and O is the centre of the semicircle.

Answer: $I_y = \dfrac{\pi d^4}{128} = 0.392\ 7\ \text{m}^4$, $I_{y_C} = \dfrac{\pi d^4}{128} - \dfrac{d^4}{18} = 0.109\ 7\ \text{m}^4$

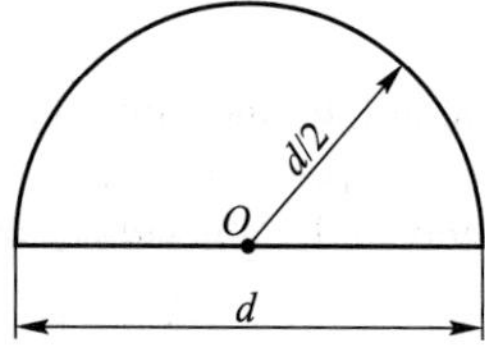

Fig.P9-6

Chapter 10 Plane Bending

Teaching Scheme of Chapter 10

10.1 Introduction

10.1.1 Plane bending of beams

Bending is a basic form of deformation in engineering practice. The crane beam shown in Fig. 10 – 1a and the axle shown in Fig. 10 – 1b are both members of bending deformation. The common feature of these members is that they can all be simplified to a bar, which in the plane passing through the axis is subjected to an external force (transverse force) perpendicular to the axis of the bar or an external couple. Under the action of such an external force, the axis of the bar will be bent into a curve as shown in Fig. 10 – 1a. This form of deformation is called bending. In engineering, the members of bending deformation are usually called beams.

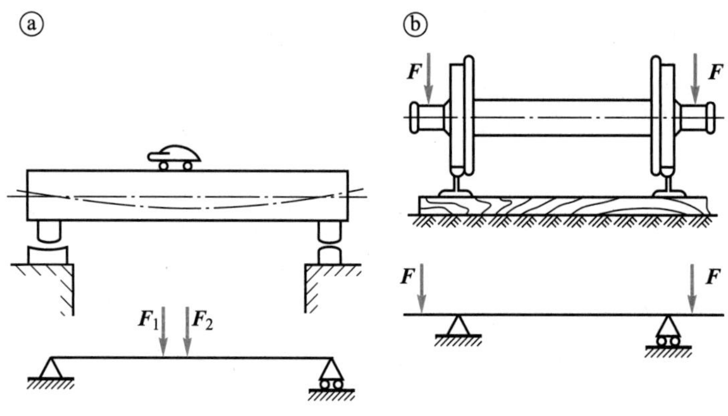

Fig.10–1

(a) Diagram of the crane crossbeam; (b) Schematic diagram of the axle

In engineering, most beams have an axis of symmetry in their cross section, as shown in Fig.10–2. The plane passing through the axis of the beam and the axis of symmetry of the section is called the longitudinal plane of symmetry. When the external load on the beam is located in the longitudinal plane of symmetry, the axis of the beam will bend into a plane curve located in the same plane. This is called plane bending. It is the most common situation in engineering practice and is discussed in this chapter.

10.1.2 Supports and loads

For the straight beam, when the external forces act in the longitudinal plane of symmetry, the calculation sketch of the beam can be represented in its axis. It is necessary to simplify the support and external forces to create a calculation sketch.

The external forces acting on a beam, including loads and support reactions, can be simplified to the following three types:

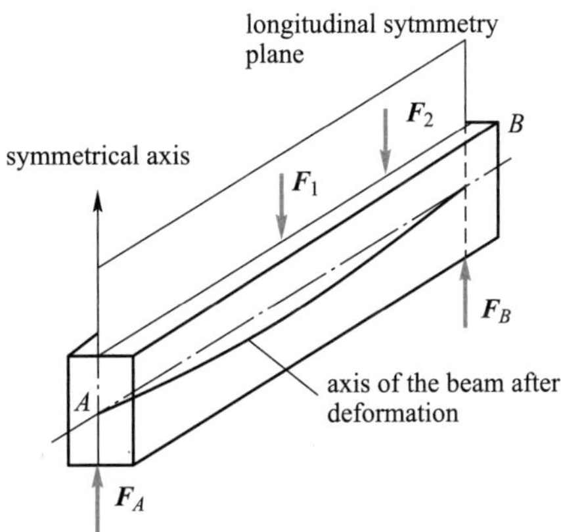

Fig.10-2

(1) Concentrated forces. Such as pressure of wheel on a highway bridge and the pressure of a train wheel on an axle. The range of these action is much smaller than the length of a highway bridge or axle. It can be considered to be concentrated at one point. This type of force is called concentrated force or concentrated load, and its unit is often expressed in Newton (N).

(2) Distributed load. A distributed load is a transverse force distributed continuously along the entire length or part of the length of a beam. The load of uniform distribution is called the uniform load. The intensity of uniform load is defined by the load per unit length expressed in q, and its unit is often expressed in Newton/m (N/m).

(3) Concentrated couple. When F is applied parallel to the axis, the load will become a concentrated force F acting in the direction of the axis and a concentrated couple $M_e = Fr$ acting in the plane of the beam axis after moving F to a point on the axis. The units of the couple are often expressed in Newton · meters (N · m).

The way in which the beams are supported can be simplified to the following three forms:

(1) Roller support: The calculation sketch for this type is shown in Fig.10-3a or b. It allows the beam end not only to rotate freely around the hinge center, but also to move along the plane of the support. Roller support only restricts the movement of the beam at the support in the direction perpendicular to the plane of the support. Therefore, there is only one unknown support reaction force, i.e., the support reaction force F_R perpendicular to the bearing plane and through the hinge center, as shown in Fig.10-3c. The sliding bearing and radial rolling bearing can be simplified as a roller support.

(2) Fixed hinge support (pin support): The calculation sketch for this type of bearing is shown in Fig.10-3d or e. It allows the beam end to rotate only about the hinge center without any free movement. Therefore, the support reaction force passes through the hinge center, but its magnitude and direction are unknown. The support is usually represented by two mutually perpendicular components, the support reaction force F_H and F_R, as shown in Fig.10-3f. Thus, this type of bearing has two unknown support reactions. For example, thrust bearings and radial thrust ball bearings can be

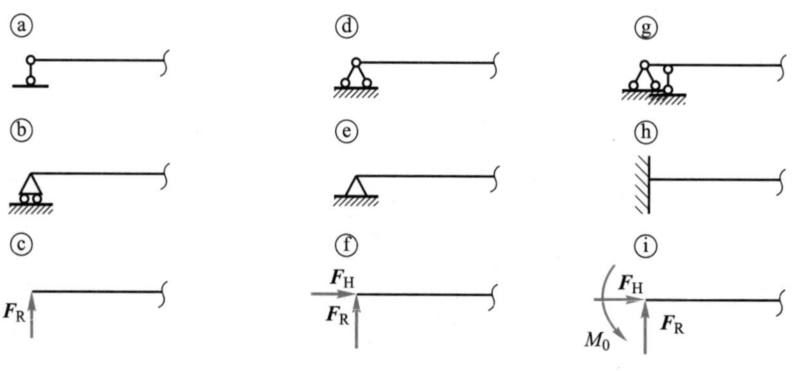

Fig.10-3

simplified as pin support.

(3) Fixed end support (fixed support) : The calculation sketch is shown in Fig.10-3g or h. It prevents any movement or rotation of the beam within the fixed end. Therefore, the magnitude, direction and point of action of the support reaction force are unknown. And it is usually simplified to two vertical forces F_H and F_R acting at the center of the cross section of the fixed end and a support reaction couple M_0 as shown in Fig.10-3i. Thus, such a bearing has three unknown support reactions. A long bearing can be reduced to a fixed end support.

In the case of plane bending, the external forces acting on the beam (including load and support reaction) are a plane force system. When the bearing reaction forces of a beam can be fully derived from the static equilibrium equations, the beam is called statically determinate beams.

According to the support condition of the beam, the common static beams in engineering practice have the following three forms:

(1) Simply supported beam. Pin support at one end of the beam, roller support at the other end, as shown in Fig.10-4a.

(2) Cantilever beam. One end of the beam is fixed and the other end is free, as shown in Fig.10-4b.

(3) Overhanging beam. The beam is supported by a pin support and a roller support, with one or both ends of the beam extending freely, as shown in Fig.10-4c.

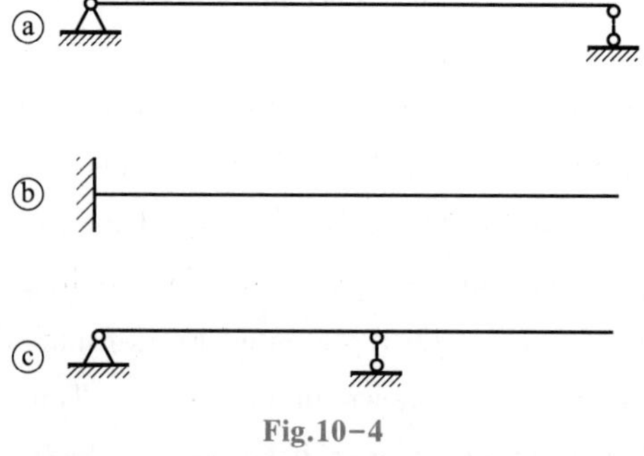

Fig.10-4

10.2 Internal force in beams

According to the equations of static equilibrium, all the external forces acting on the beam are known. The internal forces of the beam can be further investigated. The internal forces of a beam are the forces that interact between one segment of the beam and another under the action of an external force. In order to calculate the stresses and deformations in a beam, the internal forces in the cross section must first be investigated.

10.2.1 Section method for calculating internal forces in beams

When all external forces on the beam are known, the section method can be used to determine the internal forces in any cross section of the beam. The simply supported beam AB shown in Fig.10-5a is analyzed for internal forces at cross section $n—n$ at a distance x from end A. Taking the whole beam as the object of study, the static equilibrium equations can be used to find the support reactions F_{RA} and F_{RB} of the beam. Cut the beam into left and right segments hypothetically at the cross section $n—n$ and take the left section as the object of study (Fig.10-5b). In order to maintain the equilibrium of the left section, i.e. to satisfy the equilibrium conditions $\sum F_y = 0$ and $\sum M_O = 0$ of the left section, there must be two internal force components at cross section $n—n$: the internal force F_Q parallel to the cross section and the internal couple M in the load plane. Internal force F_Q is called shear force and internal couple M is called bending moment. According to the equilibrium condition of the left section beam $\sum F_y = 0$, we get

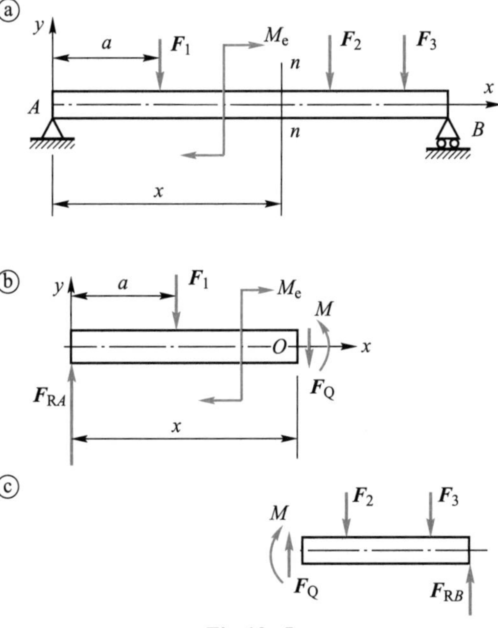

Fig.10-5

$$F_Q = F_{RA} - F_1 \tag{10-1}$$

From $\sum M_O = 0$, we get

$$M = F_{RA}x - F_1(x-a) + M_e \tag{10-2}$$

Similarly, F_Q and M of section $n—n$ can be obtained from the right beam according to its equilibrium conditions, which are equal in magnitude but opposite in direction to those obtained from the left beam.

10.2.2 Sign conventions

In order that the shear forces and bending moments obtained separately from the left and right section of section $n—n$ have the same values and uniform signs, the sign of them must be specified in accordance with the deformation of the beam as follows:

(1) The sign of shear force: If the shear force F_Q tends to rotate clockwise around the micro-section, it is positive. In the opposite direction, it is negative. As shown in Fig.10-6.

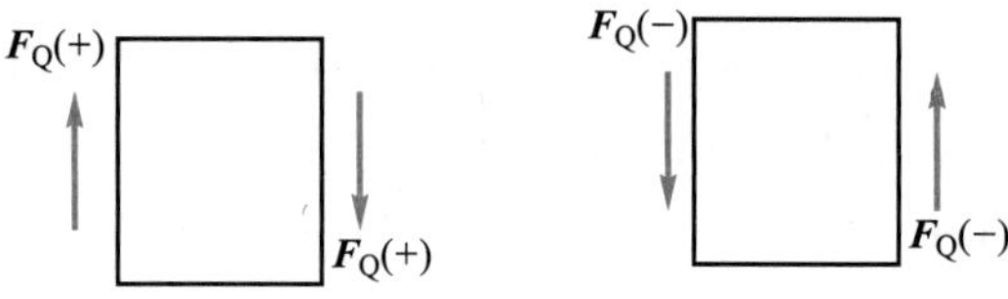

Fig.10-6

(2) The signs of the bending moment: If the bending moment M causes a downward convex deformation of the micro-section, with the upper part under compression and the lower part under tension, the bending moment M is positive. In the opposite direction, it is negative. As shown in Fig.10-7.

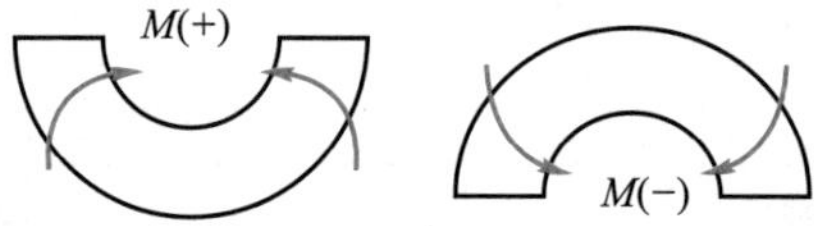

Fig.10-7

The shear force F_Q and the bending moment M in this cross section must be of equal magnitude and the same sign, regardless of whether the left or the right segment is selected.

10.2.3 Calculation of internal force in beams

From equation (10-1) and (10-2), we get:

(1) The shear force in a cross section is numerically equal to the algebraic sum of all lateral forces in the beam to the left (or right) of the section. An upward force on the left-hand beam (or a downward force on the right-hand beam) produces a positive shear force; the opposite produces a negative shear force.

(2) The bending moment in a section is numerically equal to the algebraic sum of the moments of all external forces to the left (or right) of the section with respect to the centroid of the section. External forces acting upwards (either to the left or to the right of the section) produce a positive moment, while external forces acting downwards produce a negative moment. The external couple on the left side of the section produces a positive bending moment when it is clockwise and negative when it is counterclockwise. While the external couple on the right side of the section produces a positive bending moment when it is counterclockwise and negative when it is clockwise.

The above rule can be summarized in the recipe "Left up, right down, shear force is positive; left down, right up, bending moment is positive".

Example 10-1 Find the internal forces in cross section 1—1 and 2—2 of shown in Fig.10-8.

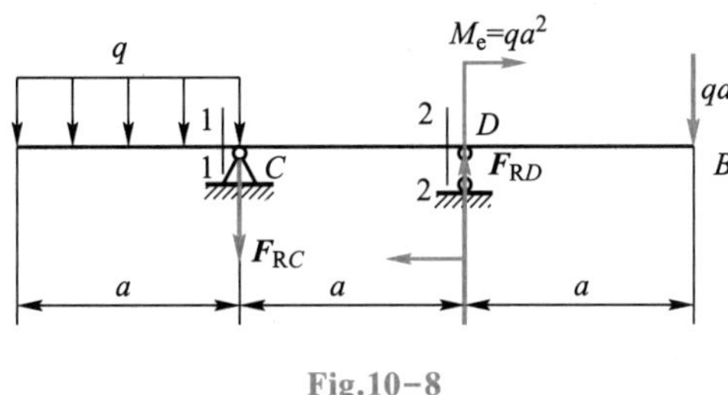

Fig.10-8

Solution Firstly find the support reaction force from

$$\sum M_C = 0, \quad F_{RD} = 2.5qa$$

$$\sum M_D = 0, \quad F_{RC} = 0.5qa$$

From section 1—1, the external forces on the left-hand beam are directly derived

$$F_{Q1} = -qa, \quad M_1 = -qa \cdot 0.5a = -0.5qa^2$$

As can be seen, the two calculations yield identical results.

From section 2—2, the external forces on the right-hand beam are calculated directly

$$F_{Q2} = qa - F_{RD} = qa - 2.5qa = -1.5qa$$

$$M_2 = -qa \cdot a - M_e = -qa^2 - qa^2 = -2qa^2$$

10.3 Shear force and bending moment diagrams

10.3.1 Shear force and bending moment equations

Analysis of the previous examples shows that the shear forces and bending moments are generally different in different cross sections of the beam, i.e. the shear forces and bending moments are variable along the beam axis. The previous section focused on the method of calculating the internal forces of a beam at a given cross section. In practice, what is of interest and what is required is the maximum value of the internal forces and its location. Therefore, in addition to the method of calculating the internal forces in a given cross section, it is also necessary to know the variation of the internal forces in each cross section along the beam axis.

If the position of the cross section on the beam axis is expressed in transverse coordinates, the shear force and bending moment in each cross section of the beam can be expressed as a function of x,

$$F_Q = F_Q(x), \quad M = M(x) \tag{10-3}$$

These two equations are known as the shear force equation and the bending moment equation, respectively. These two equations are usually written with the left end of the beam as the origin of the coordinates. Sometimes, for convenience, the origin can also be taken at the right end of the beam.

10.3.2 Shear force and bending moment diagrams

The shear force equation and the bending moment equation are not intuitive enough to understand the variation of shear force and bending moment with the position of the section. In order to clearly represent the variation of internal forces in the cross section along the beam axis, the variation of shear forces and bending moments along the beam axis can be represented by graphical lines. When graphing, select x as the abscissa, F_Q or M as the vertical coordinate, to draw $F_Q(x)$ and $M(x)$ of the graph line. The lines are called shear and moment diagrams, respectively, referred to as F_Q and M diagrams. The maximum values of shear force and bending moments in the beam can be determined, as well as the location of the corresponding cross section.

Example 10-2 The cantilever beam shown in Fig.10-9a is subjected to a concentrated force F at its free end. Try to list the shear equation and bending moment equation of the beam and make a shear force diagram and bending moment diagram.

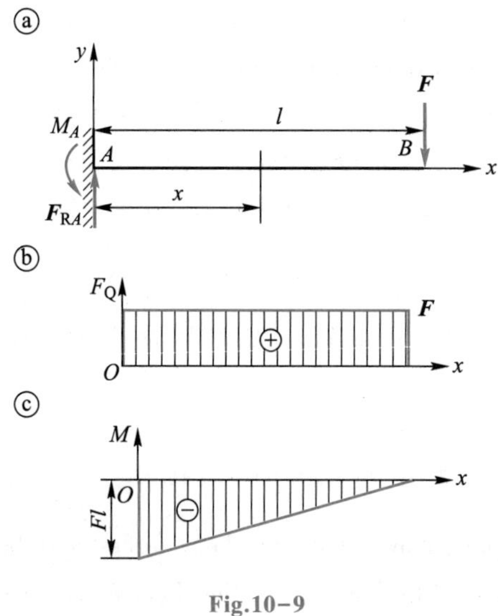

Fig.10-9

Solution (1) Find the equation of shear and the equation of bending moment

Using the fixed support A as the origin, the coordinate system is chosen as shown in Fig.10-9a. When writing the shear force and moment equations, the right-hand side of any cross section is selected to avoid calculating the support reactions F_{RA} and M_A at the fixed end.

Using the method of superposition of external forces, the equations of shear and bending moment are derived from the (10-3) equation as

$$F_Q(x)=F \quad (0<x<l) \tag{10-4}$$

$$M(x)=-F(l-x) \quad (0<x\leqslant l) \tag{10-5}$$

(2) Make shear force and bending moment diagrams

Equation (10-4) shows that the shear force in each cross section of the beam is the same F. The resulting shear force diagram is a straight line located on the upper side of the x axis and parallel to the x axis, as shown in Fig.10-9b. From equation (10-5) we know that the bending moment is a linear function of x, and therefore the bending moment diagram is an inclined straight line. The bending moment diagram as shown in Fig.10-9c can be drawn by determining two points, such as $M=-Fl$ at $x=0$ and $M=0$ at $x=l$.

As seen in Fig.10-9b and c, the shear force is the same in all cross sections of the beam, and the bending moment is maximum at the fixed end section.

$$|M|_{max}=Fl$$

Example 10-3 A simply supported beam shown in Fig.10-10a, is subjected to a uniform load q. Find the shear force and bending moment equations, and then make a shear force diagram and bending moment diagram.

Solution (1) Find the support reaction force

Since the structure and loads are symmetrical at midpoint of the span length, it is easy to find the two support reaction forces as

$$F_{RB}=\frac{ql}{2}, \quad F_{RA}=\frac{ql}{2}$$

(2) Find the equations of shear force and bending moment

With point A as the origin of coordinates, the coordinate system is chosen as shown in Fig.10-10a. Taking an arbitrary cross section at a distance x from the left end (at the origin of coordinates), the left segment of the beam is studied. The shear force equation and bending moment equation are derived from equation (10-3) using the method of superposition as

$$F_Q(x)=\frac{ql}{2}-qx \quad (0<x<l) \tag{10-6}$$

$$M(x)=\frac{ql}{2}x-\frac{qx^2}{2} \quad (0\leqslant x\leqslant l) \tag{10-7}$$

(3) Make shear force and bending moment diagrams

From equation (10-6), the shear force diagram is an inclined straight line and the shear force diagram of the beam can be drawn by determining two points on it, as shown in Fig.10-10b. From equation (10-7), the moment diagram is a quadratic parabola. In order to draw this parabola, at least three or four points must be determined, for example, $x=0$ at $M=0$, $M=3ql^2/32$ at $x=l/4$, $M=\frac{ql^2}{8}$ at $x=l/2$, $x=l$ at $M=0$. The moment diagram of the beam is made through these points, as shown in Fig.10-10c.

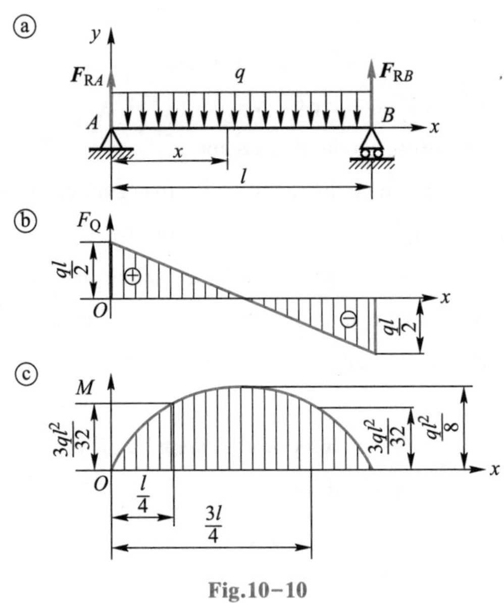

Fig.10-10

From the shear force and bending moment diagrams, it can be seen that the maximum shear force is on the inner cross section of the two supports with a value of $|F_Q|_{max}=ql/2$ and the maximum bending moment is on the cross section at the mid-point of the span with a value of $|M|_{max}=ql^2/8$, while the shear force $F_Q=0$ in this section.

Example 10-4 A simply supported beam shown in Fig.10-11a, is subjected to a concentrated force F. Find the shear force and bending moment equations, and then make a shear force diagram and bending moment diagram.

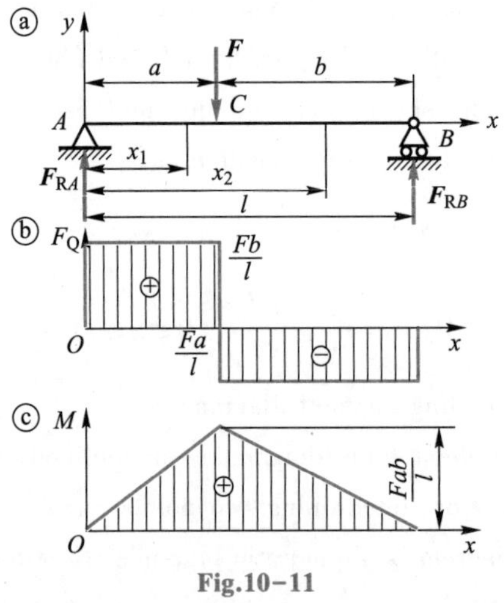

Fig.10-11

Solution (1) Find the support reaction force

Derived from the equilibrium conditions $\sum M_B = 0$ and $\sum M_A = 0$

$$F_{RA} = \frac{b}{l}F$$

$$F_{RB} = \frac{a}{l}F$$

(2) Find the shear force equation and bending moment equation

The beam is divided into AC and CB, because of the concentrated force F at point C. The shear force and bending moment equations cannot be expressed in the same equation, so they need to be written in different segments.

Take an arbitrary section at x_1 from the origin in the AC section and select the left side beam. The shear force equation and the bending moment equation can be derived from equation (10-3).

$$F_Q(x_1) = \frac{Fb}{l} \quad (0<x_1<a) \tag{10-8}$$

$$M(x_1) = \frac{Fb}{l}x_1 \quad (0<x_1 \leqslant a) \tag{10-9}$$

Take an arbitrary section at x_2 from the origin in the section CB and select the right-hand side of the beam. The shear force equation and the bending moment equation are

$$F_Q(x_2) = -\frac{Fa}{l} \quad (a<x_2<l) \tag{10-10}$$

$$M(x_2) = \frac{Fa}{l}x_2 \quad (a<x_2 \leqslant l) \tag{10-11}$$

(3) Make shear force and bending moment diagrams

From equations (10-8) and (10-10), the shear force diagrams of both sections of the beam are straight lines parallel to the x axis. From equations (10-9) and (10-11), the bending moment diagrams of both sections of the beam are inclined straight lines. The shear force and bending moment diagrams are shown in Fig.10-11b and c.

In Fig.10-11b and c, if $a < b$, the shear force at any cross section of the AC section is maximum, i.e. $F_{Q\max} = pb/l$. The bending moment at the section C where the concentrated force F acts is maximum, i.e. $M_{\max} = Fab/l$ when $a = b = l/2$. So, when the concentrated force F acts at the mid-point of the span, the maximum bending moment will occur at the mid-span section and $M_{\max}$ reaches a maximum value of $M_{\max} = Fl/4$.

In addition, at the point where the concentrated force is applied, the shear force undergoes a sudden change, the value of which is equal to the value of the concentrated force,

$$\left| -\frac{Fa}{l} - \frac{Fb}{l} \right| = \frac{a+b}{l}F = F$$

The shear force and bending moment equations for both sides of the beam at the point of concentrated force are different and should be written in sections; the shear force and bending moment diagrams should also be drawn in sections.

Example 10–5 A simply supported beam shown in Fig.10–12a, is subjected to a couple. Find the shear force and bending moment equations, and make a shear force and bending moment diagram.

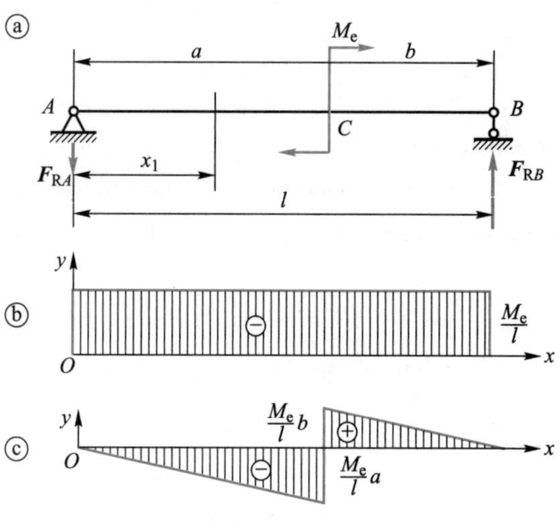

Fig.10–12

Solution (1) Find the support reaction force

From the equilibrium equation it can be obtained that

$$F_{RA}=F_{RB}=\frac{M_e}{l}$$

(2) Find the shear force and bending moment equations

The load acting on this beam is a couple M_e. There is only one shear force equation for the whole beam, but the moment equations for the sections AC and CB are different. The shear force and moment equations are derived from (10–3) as

$$\left.\begin{aligned} F_Q(x)&=-\frac{M_e}{l} \quad (0<x<a)\\ M(x)&=-\frac{M_e}{l}x \quad (0\leqslant x<a)\\ M(x)&=\frac{M_e}{l}(l-x) \quad (a<x\leqslant l)\end{aligned}\right\}$$

(3) Shear force and bending moment diagrams

The shear force and bending moment diagrams are plotted from the above equations and are shown in Fig.10–12b and c. The shear force diagram for the whole beam is a straight line parallel to the x axis, and the shear force values in each cross section of the whole beam are equal to M_e/l.

In the case of $a<b$, the bending moment is maximum at the section slightly to the right of point C, $|M|_{max}=M_eb/l$.

From the bending moment diagram, it can be seen that at the action of the concentrated couple M_e, the bending moment changes abruptly and the value of the abrupt change is the value of this concentrated couple moment, i.e.

$$\left|\frac{M_e}{l}b-\left(-\frac{M_e}{l}a\right)\right|=\frac{a+b}{l}M_e=M_e$$

The moment equations are different on each side of the couple and should be written in sections, and the moment diagram should also be drawn in sections.

From the above examples, the steps for making shear force and bending moment diagrams can be summarized as follows.

(1) Find the support reaction force.

(2) Segmenting the beam where the concentrated forces (including concentrated loads and support reactions), concentrated couples, and the set of distributed loads change.

(3) Determine the origin of the coordinates, then write the equations of shear force and bending moment for each section of the beam.

(4) Draw the shear force and bending moment diagrams for each section of the beam based on the shear force and bending moment equations.

10.4 Relations among shear force, bending moment and loads

The internal forces are caused by the loads and there are some relationships between the internal forces and the loads. Recognizing and understanding these relationships are important not only in drawing or examining the shear force and bending moment diagrams of a beam, but also in solving other problems associated with beams.

10.4.1 Differential relationship

Analyzing the shear force and bending moment equations of the examples in Section 10.3, it can be seen that by taking the derivative of the bending moment equation (10-7) with respect to the variable x, we obtain

$$\frac{\mathrm{d}M}{\mathrm{d}x}=\frac{ql}{2}-qx$$

This is exactly the shear force equation (10-6). If the shear force equation (10-6) is derived with respect to the variable$x$, we get

$$\frac{\mathrm{d}F_{\mathrm{Q}}}{\mathrm{d}x}=-q$$

This is exactly the load set q. A negative sign indicates a downward direction of load intensity.

The above differential relationships between load intensity, shear force and bending moment is common in straight beams. This relationship is derived from the general case below.

As shown in Fig.10-13a, consider an arbitrary beam subjected to a load. The distributed load set $q(x)$ is a continuous function and is specified to be positive upwards. The origin of the coordinates is now taken to be at the left end of the beam, and a section of length dx is taken from the beam using two adjacent cross sections of coordinate x and $x+dx$, as shown in Fig. 10-13b. In the section with coordinates x, the bending moment and shear force are $M(x)$ and $F_Q(x)$, respectively. In the section with coordinates $x+dx$, the bending moment and shear force are $M(x)+dM(x)$ and $F_Q(x)+dF_Q(x)$, respectively. The bending moments and shear forces are assumed to be positive and the variation of the load set along the length of dx is omitted. According to the conditions of static equilibrium, the load set is given by

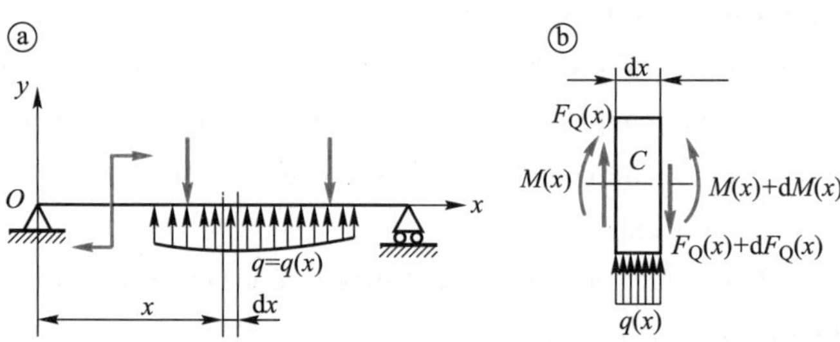

Fig.10-13

$$\sum F_y = 0, \quad F_Q(x) + q(x)dx - [F_Q(x) + dF_Q(x)] = 0$$

Got

$$\frac{dF_Q(x)}{dx} = q(x) \tag{10-12}$$

then by

$$\sum M = 0, \quad -M(x) - F_Q(x)dx - q(x)\cdot dx\cdot\frac{dx}{2} + M(x) + dM(x) = 0$$

Omitting the higher order minima, the collation gives

$$\frac{dM(x)}{dx} = F_Q(x) \tag{10-13}$$

Substituting equation (10-13) into equation (10-12), we get

$$\frac{d^2M(x)}{dx^2} = q(x) \tag{10-14}$$

The above three equations are the differential relationships between load intensity, shear force and bending moment. It essentially represents the equilibrium conditions of a micro-segmented beam. The relations (10-12), (10-13) and (10-14) show that the bending moment $M(x)$ is of order one higher than the shear force $F_Q(x)$, which in turn is of order one higher than the load set $q(x)$.

It is noting that the load intensity is positive in the upward direction and the x axis is positive in the rightward direction. The shear force $F_Q(x)$ and the bending moment $M(x)$ are both continuous functions of x in the interval. There is neither a concentrated force nor a concentrated couple in the interval.

10.4.2 Law of Internal force diagram

Based on the differential relationships, the above questions can be composite to summarize some of the laws of internal force diagrams.

(1) The shear force equation is different at the concentrated force and at the end of the distributed load. So it should be written in sections. The shear force diagram should also be drawn in sections. The bending moment equations are different at the point of concentrated force, at the point of couple and at the end of the distributed load. It should be written in sections. The bending moment diagram should also be drawn in sections.

(2) If there is no load on a section of the beam, i.e. $q(x)=0$, then the shear force on the section is constant, i.e. $F_Q(x)=c$. And the bending moment is a linear function of x, i.e. $M(x)=cx+b$. The shear force diagram on the section of the beam is a horizontal line and the bending moment diagram is generally a tilted straight line. The slope of which is equal to the value of the shear force F_Q. If $F_Q>0$, the bending moment diagram is a tilted straight line on the right; if $F_Q<0$, the bending moment diagram is a tilted straight line on the right.

(3) A section of the beam is subjected to a uniform load $q(x)=c$. The shear force on this section is a linear function of x, i.e. $F_Q(x)=cx+b$. The bending moment is a quadratic function of x, i.e. $M(x)=\frac{1}{2}cx^2+bx+d$. Therefore, the shear force diagram is a sloping straight line and the bending moment diagram is a quadratic parabola. If the direction of the uniform load is upwards, i.e. $q(x)>0$, the shear force diagram is a straight line inclined to the right and the bending moment diagram is convex downwards. This means that the bending moment diagram is convex in the opposite direction to the direction of the uniform load.

(4) If the shear force $F_Q(x)=0$, i.e. $\frac{\mathrm{d}M}{\mathrm{d}x}=0$, the slope of the moment diagram is zero. The moment in that section is an extreme value. However, note that the maximum value of the bending moment may be at $F_Q=0$, or at the point of concentrated force and concentrated couple.

(5) At the point of concentration, the shear force diagram changes abruptly and the change is equal to the value of the concentration force. In the transition from the left adjacent section to the right adjacent section, the direction of the abrupt change in the shear force diagram corresponds to the direction of the concentrated force. The abrupt changing of shear force diagram means that there is an abrupt change in the slope of the moment diagram, so that the moment diagram forms a turning point (e.g. Fig.10-14).

At the place where the couple acts, the bending moment diagram changes abruptly and the value of the abrupt change is equal to the value of that couple moment. But the shear force diagram does not change. At the place where the couple in the clockwise direction, the bending moment diagram is abruptly changed upward at the transition from the left-neighboring section to the right-neighboring

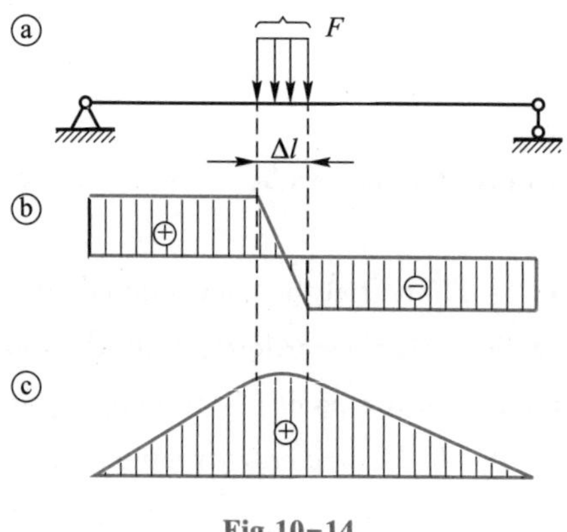

Fig.10-14

section. While the bending moment diagram abruptly changed downward at the transition from the left-neighboring domain to the right-neighboring domain under action of the couple in the counterclockwise direction.

In fact, the concentrated forces are distributed over a micro-segment of the beam, and the shear force and bending moment diagrams should vary continuously over this micro-segment. For the sake of simplicity, the distributed forces on the micro-segment Δl are considered as concentrated forces (i.e. $\Delta l \to 0$). A similar explanation can be given for the abrupt changes in the moment diagram at the couple.

Example 10-6 Using the relationships between bending moment, shear force and distributed load intensity, draw the shear force and bending moment diagrams of the overhanging beam shown in Fig.10-15a and check.

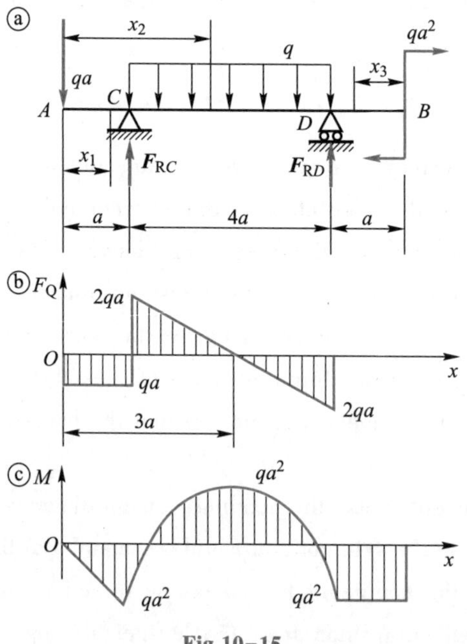

Fig.10-15

Solution (1) Find the support reaction force.

From the equilibrium equation $\sum M_D = 0$ and $\sum M_C = 0$, find $F_{RC} = 3qa$, $F_{RD} = 2qa$.

(2) Draw shear force and bending moment diagrams

Shear force diagram should be drawn in three sections AC, CD and DB.

For the section AC, the shear force diagram is a horizontal line. This horizontal line can be drawn simply by calculating the shear force on any cross section of the section AC. Using the method of superposition of external forces, we get

$$F_{QAright} = -qa$$

Similarly, we can know that the shear force diagram of section DB is also a horizontal line, and we only need to calculate the shear force on any cross section of section DB to draw this horizontal line. Using the method of superposition of external forces, we get

$$F_{QBleft} = 0$$

For section CD, the shear force diagram is diagonally straight because of the uniform load acting on this section of the beam. We just need to calculate the shear force on the cross section at both ends of the beam, and then connect these two points. Using the method of superposition of external forces, we get

$$F_{QCrihgt} = 2qa, \quad F_{QDleft} = -2qa$$

The final shear force diagram is shown in Fig.10-15b.

The bending moment diagram should be drawn in three sections, AC, CD and DB.

For the section AC, because the shear force diagram is a horizontal line, the moment diagram should be a straight line. We need to calculate the bending moment out of the section at both ends of the section AC, and then connect these two points.

Using the method of superposition of external forces, we can find

$$F_{QAright} = 0, \quad F_{QCleft} = -qa^2$$

For the section CD, since the shear force diagram is a diagonal straight line, the bending moment diagram should be a quadratic parabola. For the drawing of the parabola, it needs to find two endpoints and the endpoint. The extreme value point is the point where the shear force is zero, i.e. the section corresponding to $x = 3a$. From the method of superposition of external forces, the bending moment on this section can be calculated as

$$M_{x=3a} = qa^2$$

Again, using the external force superposition method, we get

$$M_{Cright} = -qa^2, \quad M_{Dleft} = -qa^2$$

For the section DB, since the shear force is 0, the bending moment diagram is a horizontal line. According to the method of superposition of external forces, find

$$M_{Dright} = -qa^2$$

The final bending moment diagram is shown in Fig.10-15c.

10.5 Normal stress on cross section in pure bending

Shear force and bending moment on the cross section are the resultant force and resultant moment of the distributed internal force system on the cross section. To calculate the strength of the beam, it is necessary to further study the distribution law of the internal forces on the cross section and establish the formula for calculating the stress of the beam.

10.5.1 Pure bending

When a straight beam is bent in plane, there are generally both bending moment and shear force in the cross section, which make the beam produce bending deformation and shear deformation at the same time. This bending is called transverse-force bending. When there is only bending moment and no shear force in the cross section, the bending of the beam is called pure bending. On the simply supported beam shown in Fig.10–16a, there are two external forces F acting symmetrically in the longitudinal symmetry plane of the beam. The shear force and bending moment diagrams of the beam are shown in Fig.10–16b and c, respectively. As can be seen from the diagram, the shear force is equal to zero and the bending moment is constant in each cross section of the section CD. The pure bending occurs in this section. While the shear force and bending moment are not zero in the cross sections of the AC and DB sections, the transverse bending occurs.

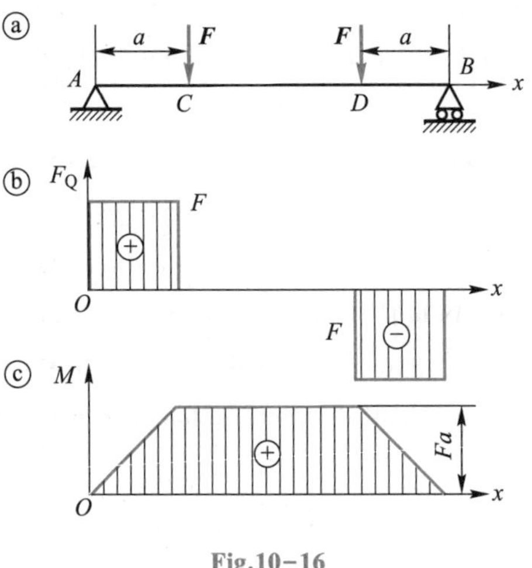

Fig.10–16

10.5.2 Normal stress in pure bending

It is known that the bending moment M is the combined couple moment composed of the normal distributed internal forces in the cross section, while the shear force F_Q is the combined force composed of the tangential distributed internal forces in the cross section. Therefore, the beam generally has both normal stress σ and shear stress τ in the cross section, which are related to the bending moment M and shear force F_Q, respectively. The calculation of the normal stresses in the beam in pure bending is studied.

The determination of the distribution law of the stresses in the cross section and the establishment of the stress calculation formula is a super-static problem, which must take into account the geometric, physical and static relations of the deformation.

1. Pure bending experiments

Pure bending experiments can be easily performed on a material testing machine. The deformation of section CD of a rectangular beam shown in Fig. 10 - 16a is studied. Before the force is applied, longitudinal lines aa and bb parallel to the beam axis and transverse lines m—m and n—n perpendicular to the longitudinal lines are drawn on the sides of the section CD as shown in Fig. 10 - 17a. After deformation, the following phenomena can be observed.

(1) The transverse line remains straight, but is turned by a small angle (Fig.10-17b).

(2) The longitudinal line becomes curved, but remains perpendicular to the transverse line (Fig.10-17b).

(3) The longitudinal line located at the concave edge shortens and the longitudinal line at the convex edge elongates (Fig.10-17b).

(4) In the beam width direction, the upper part elongates and the lower part shortens (Fig.10-17c)

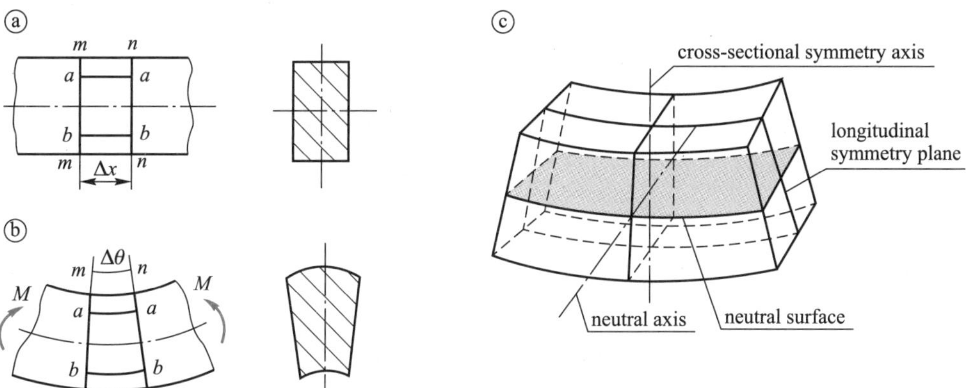

Fig.10-17

Based on the above surface deformation phenomena, the following assumptions are made about the deformation and forces on the beam.

Plane section hypothesis: According to the phenomena (1) and (2), it can be assumed that each cross section of the beam remains as a plane perpendicular to the axis of the deformed beam after deformation. All cross sections only turned by an angle around one of the axes on the cross section.

Uniaxial force assumption: According to the phenomena (3) and (4), it can be assumed that all the fibers parallel to the axis of the beam do not squeeze each other. All the longitudinal fibers are axially stretched or compressed.

Practice shows that the stress and deformation equations derived on the basis of the above assumptions are in line with the actual situation. Also, the same conclusion is obtained from the elasticity theory in the case of pure bending.

From the above assumptions, a deformation model of the beam can be established, as shown in Fig.10−17c. Imagine that the beam is composed of numerous layers of longitudinal fibers. After the beam is deformed, the upper longitudinal fibers are shortened and the lower longitudinal fibers are elongated. While the continuity of the deformation shows that there must be a layer of longitudinal fibers in the beam which is neither elongated nor shortened. This layer of fiber is called the neutral surface. The intersection line between the neutral surface and the cross section is called the neutral axis. When the beam is deformed, the cross section is rotated around its neutral axis.

2. Deformation geometry equation

A micro-segment beam of length dx is now cut from the beam using two adjacent cross sections m—m and n—n (Fig.10−18a). The deformation of the longitudinal fibers a—a at a distance y from the neutral surface O_1O_2 is studied after the beam bending.

According to the plane hypothesis, the cross sections m—m and n—n remain as a plane after beam deformation, but only rotate by an angle dθ relative to each other as shown in Fig.10−18b. Let the radius of curvature of the neutral surface after deformation be ρ, then the original length of the longitudinal fiber b—b is ρdθ, the length of b'—b' after deformation is $(\rho+y)\mathrm{d}\theta$. The normal strain of longitudinal fiber b—b is

$$\varepsilon=\frac{(\rho+y)\mathrm{d}\theta-\rho\mathrm{d}\theta}{\rho\mathrm{d}\theta}=\frac{y}{\rho} \qquad (10-15)$$

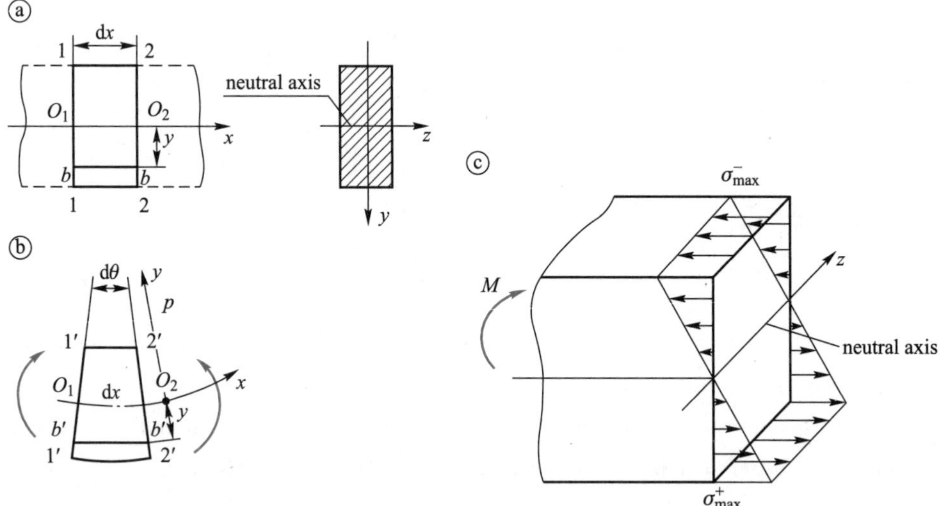

Fig.10−18

This equation shows that the longitudinal normal strain ε at any point on the cross section is proportional to the distance y from that point to the neutral axis.

3. Physical equations

Under the assumption of uniaxial stress, all longitudinal fibers are in axial tension or compression. Therefore, when the stress does not exceed the proportional limit of the material, it is obtained from Hooke law that

$$\sigma = E\varepsilon = E\frac{y}{\rho} \tag{10-16}$$

For a given cross section, E/ρ is a constant. Therefore, it can be seen from equation (10-16) that the normal stress is proportional to the distance from that point to the neutral axis. The normal stress in bending is linearly distributed along the height of the cross section.

4. Static relations

Fig.10-19 shows a section taken from a beam. The y axis is the longitudinal symmetry axis of the section and the z axis is the neutral axis. The micro-internal force acting on the micro-area $\mathrm{d}A$ is $\sigma\mathrm{d}A$, and these micro-forces in the whole cross section form a space-parallel force system perpendicular to the cross section. This force system can be simplified to three internal force components, namely, the axial force F_N, the moments M_y, M_z to the y and z axes respectively. According to the theorem of combined force and combined moment, we get

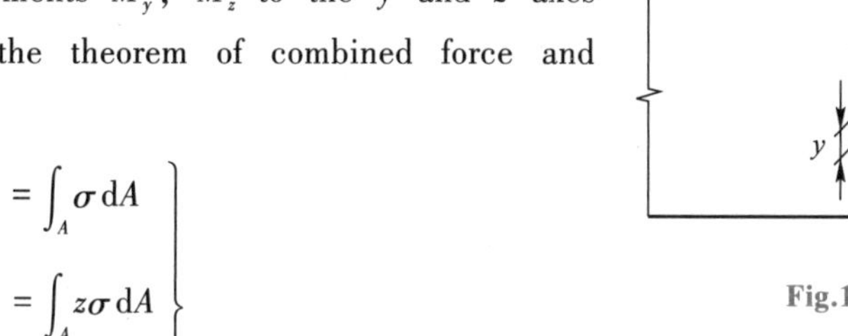

Fig.10-19

$$\left.\begin{aligned} F_N &= \int_A \sigma \mathrm{d}A \\ M_y &= \int_A z\sigma \mathrm{d}A \\ M_z &= \int_A y\sigma \mathrm{d}A \end{aligned}\right\}$$

The axial force F_N in the cross section and the moment M_y on the axis are equal to zero, while the moment M_z to the z axis is equal to the bending moment M. The above several static relations become

$$F_N = \int_A \sigma \mathrm{d}A = 0 \tag{10-17}$$

$$M_y = \int_A z\sigma \mathrm{d}A = 0 \tag{10-18}$$

$$M_z = \int_A y\sigma \mathrm{d}A = M \tag{10-19}$$

Substituting equation (10-16) into equation (10-17) and assuming that the material has the same elastic modulus in tension and compression, we get

$$\int_A E\frac{y}{\rho}\mathrm{d}A = \frac{E}{\rho}\int_A y\mathrm{d}A = 0$$

The integral in equation

$$\int_A y\mathrm{d}A = S_z$$

is the static moment to its neutral axis. For a given cross section, $\frac{E}{\rho}$ is a constant that is not equal to zero. Therefore, there must be

$$\int_A y\mathrm{d}A = S_z = 0$$

This indicates that the neutral axis must pass through the centroid of the cross section.

Substituting equation(10-16) into equation (10-18), we get

$$\int_A z\sigma\mathrm{d}A = \frac{E}{\rho}\int_A yz\mathrm{d}A = 0$$

where the integral

$$\int_A yz\mathrm{d}A = I_{yz}$$

is the product of inertia against the y axis and z axis. Since the y axis is the axis of symmetry, the $I_{yz}=0$ and equation (10-18) is automatically satisfied.

Substituting equation(10-16) into equation (10-19), we get

$$\frac{E}{\rho}\int_A y^2\mathrm{d}A = M$$

where the integral

$$\int_A y^2\mathrm{d}A = I_z$$

is the moment of inertia of the cross section area with respect to the neutral z axis. We get

$$\frac{1}{\rho}=\frac{M}{EI_z} \tag{10-20}$$

This is the basic formula for determining the curvature $1/\rho$ of the neutral surface of the beam. The larger the bending moment M, the larger the neutral surface curvature; the larger the EI_z, the smaller the neutral surface curvature. EI_z is called the flexural rigidity of the beam, which indicates the ability of the beam to resist bending deformation.

Substitute (10-20) back to (10-16) to obtain the formula for calculating the normal stress in cross section of the beam in pure bending

$$\sigma=\frac{My}{I_z} \tag{10-21}$$

The normal stress is proportional to the binding moment and the distance from that point to the neutral axis. Moreover, it is inversely proportional the moment of inertia of the cross section area.

5. Section modulus of bending

From equation (10-21), the maximum normal stress occurs at the farthest distance from the neutral axis, i.e.

$$\sigma_{\max}=\frac{My_{\max}}{I_z}$$

Merge the two geometric quantities I_z and y_{max} of the cross section, we obtain

$$W_z = \frac{I_z}{y_{max}}$$

then there are

$$\sigma_{max} = \frac{M}{W_z} \tag{10-22}$$

W_z is called the section modulus of bending and is a geometric quantity of the beam strength with the dimension of [length]^3.

For a rectangular section (Fig.10-20a)

$$W_z = \frac{I_z}{y_{max}} = \frac{\frac{bh^3}{12}}{\frac{h}{2}} = \frac{bh^2}{6}$$

For a circular section (Fig.10-20b)

$$W_z = \frac{I_z}{y_{max}} = \frac{\frac{\pi d^4}{64}}{\frac{d}{2}} = \frac{\pi d^3}{32}$$

The flexural section modulus of bending of various steel sections can be found in the table of sections.

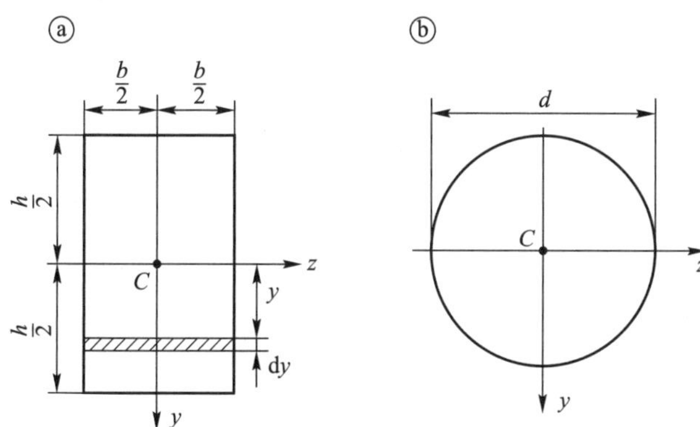

Fig.10-20

Example 10-7 A simply supported beam is shown in Fig.10-21. We have known $l = 3$ m, $q = 40$ kN/m. Try to find the normal stress at points a and b on the dangerous section.

Solution (1) Make a bending moment diagram, as shown in Fig.10-21c. The dangerous section (the section where the maximum bending moment is located) is the span section. The maximum bending moment is

$$M_{max}=\frac{1}{8}ql^2=\frac{1}{8}\times40\times3^2\ \text{kN}\cdot\text{m}=45\ \text{kN}\cdot\text{m}$$

(2) Calculation of moment of inertia I_z.

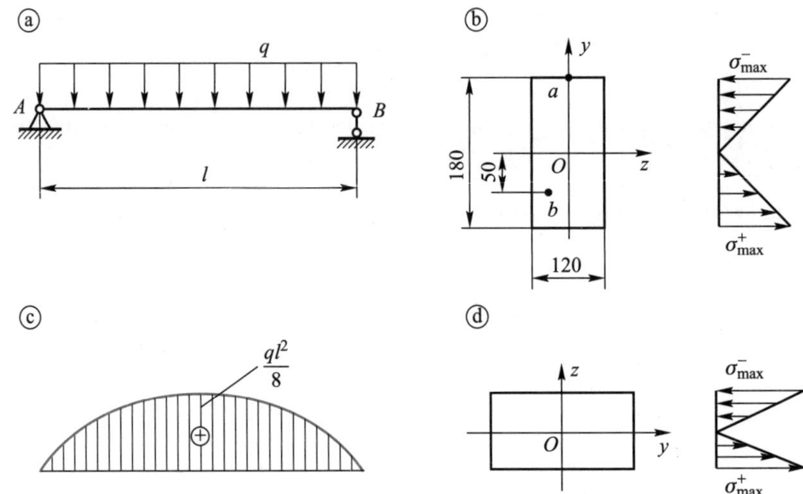

Fig.10-21

$$I_z=\frac{bh^3}{12}=\frac{1}{12}\times120\times180^3\times10^{-12}\ \text{m}^4=58.3\times10^{-6}\ \text{m}^4$$

(3) Find the normal stress at points a, b.

Using equation (10-21), we find

$$\sigma_a=\frac{M_{max}y_a}{I_z}=\frac{45\times10^3\times90\times10^{-3}}{58.3\times10^{-6}}\ \text{Pa}=69.5\times10^6\ \text{Pa}$$

$$\sigma_b=\frac{M_{max}y_b}{I_z}=\frac{45\times10^3\times50\times10^{-3}}{58.3\times10^{-6}}\ \text{Pa}=38.6\times10^6\ \text{Pa}$$

Since the bending moment of this section is positive, the deformation of the beam in this section is convex side down. The fibers below the neutral axis are under tension and above fibers are under compression. So, the compressive stress at point a is determined to be the tensile stress at point b.

10.6 Shear stress in bending

In shear bending, there are both normal and shear stresses in the beam cross section. In general, the normal stress is the main factor causing damage to the beam. However, when the span length of the beam is short, the cross section is high, or the web of the cross section is thin like an I-shaped beam, the value of shear stress may also be quite large, and shear stress strength check is necessary. The shear stressin bending of beams with rectangular cross section is discussed as an example, and the shear stress formulas of several common cross section beams are introduced.

10.6.1 Shear stresses in rectangular beams

The shear force F_Q and bending moment M in a simply supported rectangular beam shown in Fig.10-22 induce shear stresses τ and normal stresses σ, respectively. In deriving the shear stress formula, the following assumptions are made about its distribution law.

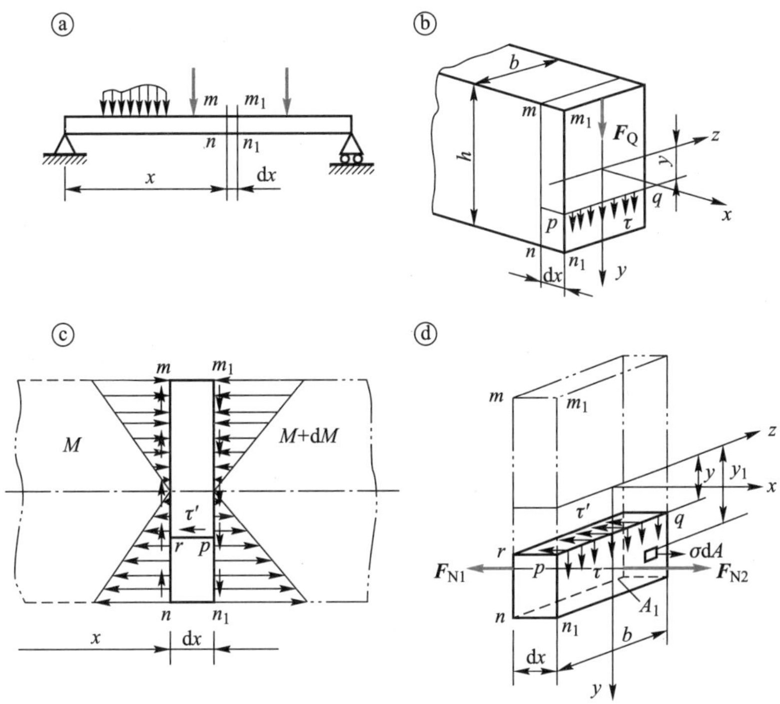

Fig.10-22

(1) The direction of shear stress at any point on the cross section is parallel to the direction of shear force F_Q.

(2) The shear stress is uniformly distributed along the section width. The magnitude of the shear stress is only related to the coordinate y. The shear stress is equal at each point of equal distance to the neutral axis.

When the height h is greater than its width b, the shear stress formula established on the basis of the above assumptions is quite accurate. Now the two cross sections mn and m_1n_1 are used to intercept the micro segment of dx in the beam. In general, the bending moments in these two cross sections are M and M+dM, respectively. To calculate the longitudinal section pr at a distance y from the neutral surface, cut $prnn_1$ from micro segment beam and study the equilibrium of this hexahedron. On the right-hand side pn_1, there is a normal stress σ, caused by the bending moment M+dM. The combined force of the internal force system consisting of the micro-internal force σdA is

$$F_{N2} = \int_{A_1} \sigma \mathrm{d}A = \int_{A_1} \frac{M + \mathrm{d}M}{I_z} y_1 \mathrm{d}A = \frac{M + \mathrm{d}M}{I_z} \int_{A_1} y_1 \mathrm{d}A = \frac{M + \mathrm{d}M}{I_z} S_z^* \tag{10-23}$$

where A_1 is the area covered by the side pn_1 and the area of

$$S_z^* = \int_{A_1} y_1 \mathrm{d}A \tag{10-24}$$

is static moment of the partial cross-sectional area A_1 to the neutral axis. This value varies with the position of the longitudinal section *pr*. Similarly, the combined force of the internal force system on the left side *rn* can be found as

$$F_{N1} = \frac{M}{I_z} S_z^* \tag{10-25}$$

Due to F_{N1} and F_{N2} are not equal, there must be a shear force on the top surface *rp* of the hexahedron, which corresponds to a shear stress τ'. According to the theorem of complementary shearing stresses and the assumptions, it is known that τ' is numerically equal to $\tau(y)$, and is also uniformly distributed along the cross-sectional width. The shear stress composed of τ' is

$$\mathrm{d}F'_Q = \tau' b \mathrm{d}x \tag{10-26}$$

to maintain the equilibrium of the x direction of the lower micro-segment, i.e.

$$F_{N2} - F_{N1} - \mathrm{d}F'_Q = 0 \tag{10-27}$$

Substituting(10−23), (10−25) and (10−26) into equation (10−27)

$$\frac{M+\mathrm{d}M}{I_z} S_z^* - \frac{M}{I_z} S_z^* - \tau' b \mathrm{d}x = 0$$

After simplification, we get

$$\tau' = \frac{\mathrm{d}M}{\mathrm{d}x} \frac{S_z^*}{I_z b} = \frac{F_Q S_z^*}{I_z b}$$

It is numerically equal to $\tau(y)$. The shear stress in the cross section at y from the neutral axis is

$$\tau(y) = \frac{F_Q S_z^*}{I_z b} \tag{10-28}$$

The above equation is the formula for the shear stress in a rectangular beam. F_Q is the shear force in the cross section, b is the section width, and I_z is the moment of inertia of the entire cross section to the neutral axis. S_z^* is the static moment of the area A_1.

For a rectangular section, it is advisable to $\mathrm{d}A = b y_1$, then

$$S_z^* = \int_{A_1} y_1 \mathrm{d}A = \int_y^{\frac{h}{2}} b y_1 \mathrm{d}y_1 = \frac{b}{2}\left(\frac{h^2}{4} - y^2\right)$$

So equation (10−28) can besimplified to

$$\tau = \frac{F_Q}{2I_z}\left(\frac{h^2}{4} - y^2\right) \tag{10-29}$$

From equation (10−29), it can be seen that the shear stress τ varies along the height of the section according to the parabolic law (Fig. 10−23b). At the top and bottom edges $\left(y = \pm\frac{h}{2}\right)$ of the section, the shear stress is $\tau=0$. At points on the neutral axis ($y=0$), the shear stress is maximum, and its value is

$$\tau_{max}=\frac{F_Q h^2}{8I_z}$$

Substituting $I_z=\frac{bh^3}{12}$ into the above equation, we get

$$\tau_{max}=\frac{3}{2}\frac{F_Q}{bh} \qquad (10-30)$$

The maximum shear stress of a rectangular section beam is 1.5 times of the average shear stress. In general, the direction of the shear stress can be judged from the direction of the shear force F_Q, as shown in Fig.10-23b.

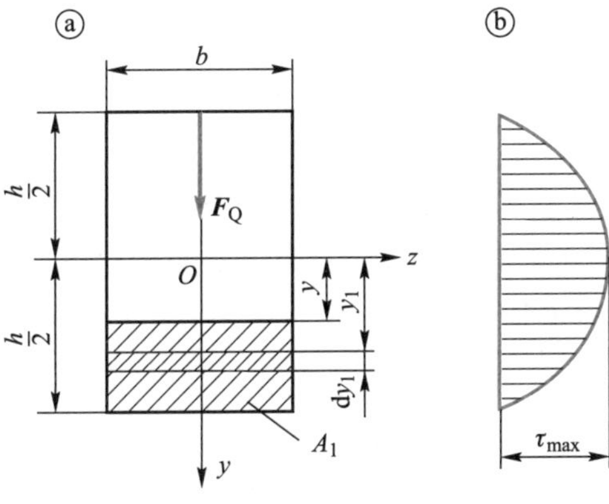

Fig.10-23

10.6.2 Maximum shear stress on other common sections

1. I-shaped section

The shear stress on the web plate can be calculated from equation (10-28), i.e.

$$\tau(y)=\frac{F_Q S_z^*}{I_z b}$$

The maximum shear stress occurs in the neutral axis (Fig.10-24b), and its value is

$$\tau_{max}=\frac{F_Q S_{z\,max}^*}{I_z b} \qquad (10-31)$$

2. Circular, circular cross section

For a circular section, the shear stress is calculated to be the maximum at each point on the neutral axis, where the shear stress is parallel to the$y$ axis, and its value is

$$\tau_{max}=\frac{4}{3}\frac{F_Q}{\pi R^2} \qquad (10-32)$$

The maximum shear stress in the circular section is also on the neutral axis (Fig.10-25). If the circular wall thickness t is much smaller than its average radius R_0, the maximum shear stress can also be calculated by equation (10-28), and finally we obtained

$$\tau_{max} = \frac{F_Q}{\pi R_0 t} = 2\frac{F_Q}{A} \tag{10-33}$$

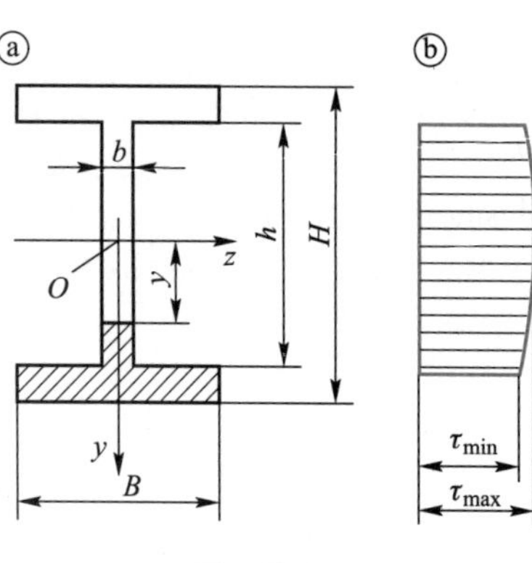

Fig.10-24

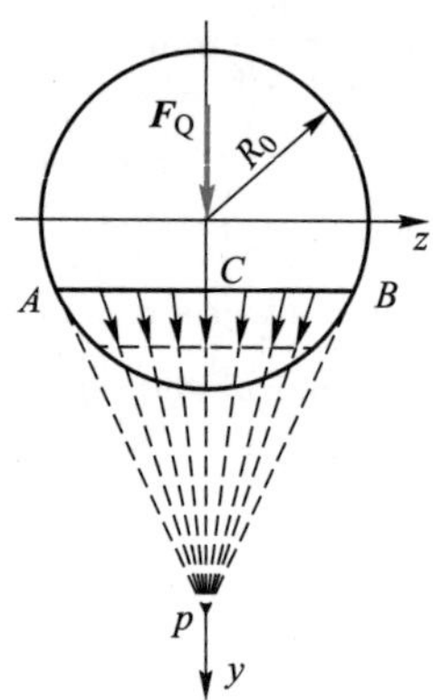

Fig.10-25

Example 10-8 Try to analyze the horizontal shear stress components on the flange of the I-shaped beam section shown in Fig.10-26a.

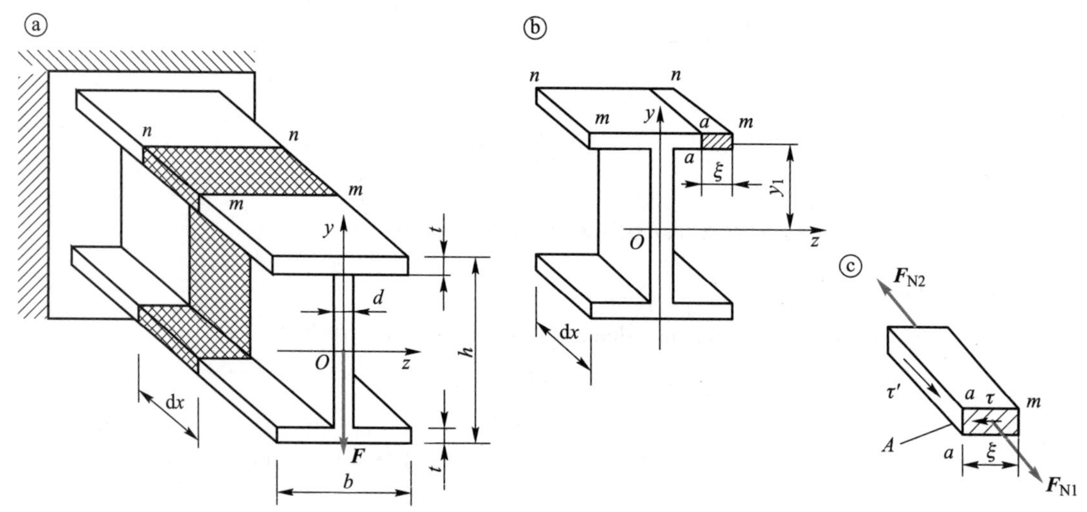

Fig.10-26

Solution The diagram showing a beam with I-shaped beam section subjected to shear bending, the shear force in each section is $F_Q = F$.

To find the horizontal shear stress at the flange a—a, first intercept the micro-segment dx in the beam as shown in Fig.10-26b, then truncate it along the axis direction at a—a. Select the truncated right part as shown in Fig.10-26c, and analyze its force and equilibrium.

The combined forces of the normal stresses on the front and back sides of the selected part are F_{N1} and F_{N2}. Following the analysis of the previous rectangular section beam, we can see that

$$F_{N1} = \int_{A^*} \sigma dA = \int_{A^*} \frac{M}{I_z} y_1 dA = \frac{M}{I_z} S_z^*$$

$$F_{N2} = \int_{A^*} \sigma dA = \int_{A^*} \frac{M + dM}{I_z} y_1 dA = \frac{M + dM}{I_z} S_z^*$$

where A^* is the area of the behind and front plane. S_z^* is the static moment of the area A^* to the neutral z axis.

It can be seen that

$$S_z^* = t\xi\left(\frac{h}{2}-\frac{t}{2}\right)$$

F_{N1} and F_{N2} are not equal. In order to maintain the balance of the right part along the axis, the left part must have a shear force on the right part

$$dF_Q' = \tau' t dx$$

The equilibrium equation is obtained

$$F_{N1} - F_{N2} + dF_Q' = 0$$

or

$$\frac{M}{I_z} S_z^* - \frac{M+dM}{I_z} S_z^* + \tau' t dx = 0$$

have to

$$\tau' = \frac{\frac{dM}{dx} S_z^*}{I_z t} = \frac{F_Q S_z^*}{I_z t} = \frac{F_Q}{I_z t} t\xi\left(\frac{h}{2}-\frac{t}{2}\right) = \frac{F_Q}{I_z}\left(\frac{h}{2}-\frac{t}{2}\right)\xi$$

From the theorem of complementary shearing stresses, the shear stress at the flange a—a of the cross section is

$$\tau = \tau' = \frac{F_Q}{I_z}\left(\frac{h}{2}-\frac{t}{2}\right)\xi$$

From the above equation, the magnitude of horizontal shear stress on the flange is proportional to the distance from the end point ξ. Its distribution is shown in the Fig. 10-27. Its direction can be determined by the magnitude of the forces F_{N1} and F_{N2}.

It can be seen from the Fig.10-27 that the direction of shear stress in the cross section is like a stream of water "flowing" from both sides of the upper flange to the inside, then merging through the web, and finally flwing separately to both sides of the lower flange. This is usually called shear stress flow or shear flow. In practice, the direction of shear stress on the web plate is often determined according to the direction of the section shear force F_Q. While the direction of shear flow is used to determine the direction of shear stress in each part of the flange.

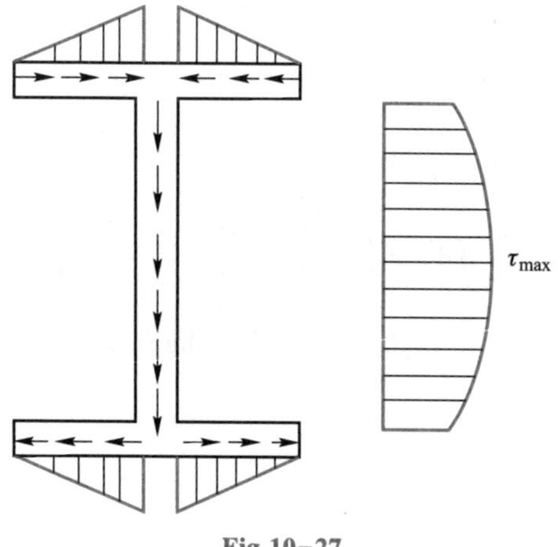

Fig.10-27

10.7 Calculation of strength

In general, both normal stress and shear stress exist in the beam. The maximum normal stress and shear stress occur at different locations in the cross section. Therefore, the normal stress strength condition and shear stress strength condition should be established separately.

10.7.1 Strength condition of bending stress in pure bending

For long and slender beams, the main controlling factor of strength is the normal stress. In order to ensure that the beam can work safely and properly, the maximum normal stress must not exceed the allowable normal stress $[\sigma]$ of the material. Therefore, the normal stress strength condition of the beam is

$$\sigma_{max}=\left(\frac{M}{W_z}\right)_{max}\leqslant[\sigma] \tag{10-34}$$

For a beam of equal section, the normal stress strength condition can be written as

$$\sigma_{max}=\frac{M_{max}}{W_z}\leqslant[\sigma] \tag{10-35}$$

For the materials with equal tensile and compressive strength (e.g. carbon steel), the maximum absolute normal stress in the beam should not exceed the allowable stress. For materials with unequal tensile and compressive strengths (e.g. cast iron), the strength should be calculated separately, i.e.

$$\sigma_{max}^{+}\leqslant[\sigma^{+}]$$

$$\sigma_{max}^{-}\leqslant[\sigma^{-}]$$

where σ_{max}^{+} and σ_{max}^{-} are the maximum tensile stress and maximum compressive stress, respectively. $[\sigma^{+}]$ and $[\sigma^{-}]$ are the corresponding allowable stresses.

The strength conditions of the bending stress can generally be used for strength verification, section design and determination of the allowable load calculation. The general process of strength calculation of normal stress:

(1) Find the support reaction force of the beam and draw the bending moment diagram.

(2) Find out the dangerous section and determine the dangerous point according to the bending moment diagram.

(3) According to the normal stress strength condition, perform the corresponding strength calculation.

10.7.2 Strength condition of shear stress in bending

The maximum shear stress usually occurs at the neutral axis of the cross section, where the normal stress is zero. So, the point of action of the maximum shear stress is in pure shear. Therefore, the maximum shear stress in the beam is required not to exceed the allowable shear stress, and the shear stress strength condition is

$$\tau_{max}=\left(\frac{F_Q S_{z\max}^{*}}{I_z b}\right)_{max}\leqslant[\tau] \tag{10-36}$$

The value of allowable shear stress $[\tau]$ is also specified in the design specifications.

For straight beams of equal section, the maximum shear stress occurs in the section with the highest shear force. At this point the above equation can be rewritten as

$$\tau_{max}=\frac{F_{Q\max}S_{z\max}^{*}}{I_z b}\leqslant[\tau] \tag{10-37}$$

When calculating the bending strength, theoretically, both the normal stress strength condition and the shear stress strength condition should be satisfied. When selecting the cross section of the beam, it is generally chosen according to the normal stress strength condition and then checked according to the shear stress strength condition. For slender solid section beams, the normal stress is the principal stress, so the analysis can be done according to the normal stress strength condition without checking the shear stress.

However, shear stress checks must be performed in the following cases.

(1) The span of the beam is short, or a large load is applied near the support. The bending moment of the beam is small and the shear force is large.

(2) For jointed or welded I-shaped beams, because the web is thin and the height of the section is quite large, the ratio of thickness to height is less than the corresponding ratio for sections.

(3) For beams made by welding or gluing, shear stress calculations are generally performed on the welded or glued surfaces, etc.

(4) Because of the poor shear resistance of wood along the grain, the wood beam is likely to break along the neutral surface when the shear stress is high. So the shear stress should also be calibrated for wood beam.

Example 10-9 The material of the winch reel mandrel is 45 steel, and its allowable stress is $[\sigma]=100$ MPa. The structure and forces of the mandrel is shown in Fig.10-28a, where $F=25.3$ kN. Try to calibrate the normal stress strength of the mandrel.

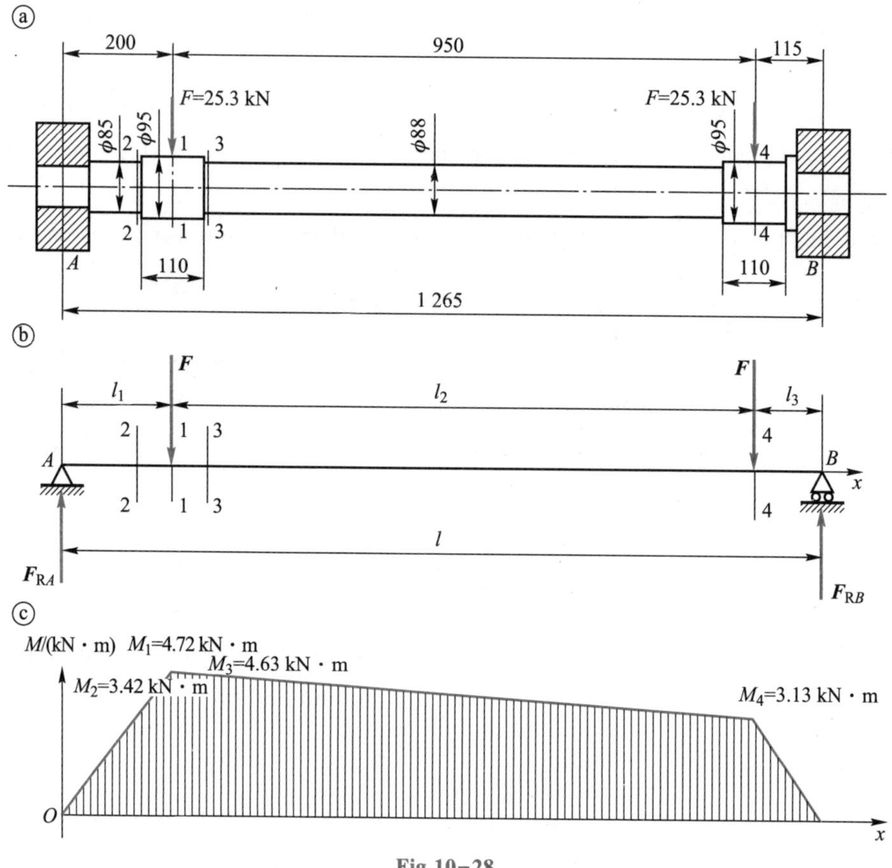

Fig.10-28

Solution According to the structure and force of the mandrel, the force diagram is drawn in Fig.10-28b. From the static equilibrium equation, we find the support reaction force

$$F_{RA} = \frac{F(l_2+l_3)+Fl_3}{l} = 23.6\ \text{kN}$$

$$F_{RB} = \frac{F(l_2+l_1)+Fl_1}{l} = 27.2\ \text{kN}$$

Since there are no distributed loads and concentrated couples, the bending moment diagram of each section consists of straight lines connected by each section. The bending moments of sections A, B, 1—1 and 4—4 are calculated as

$$M_A = M_B = 0$$

$$M_1 = F_{RA} l_1 = 4.72\ \text{kN} \cdot \text{m}$$

$$M_4 = F_{RB} l_3 = 3.13\ \text{kN} \cdot \text{m}$$

Its bending moment diagram is shown in Fig.10-28c. Beam AB is a non-equal section beam. The moment diagram and section dimensions need to be considered comprehensively to determine the dangerous section. Based on the moment diagram and the variation of the mandrel diameter along the axis, the dangerous sections are determined as sections 1—1, 2—2 and 3—3.

The values of bending moments at sections 1—1, 2—2 and 3—3 are

$$M_1 = 4.72 \text{ kN} \cdot \text{m}$$

$$M_2 = F_{RA}(l_1 - 0.055 \text{ m}) = 3.42 \text{ kN} \cdot \text{m}$$

$$M_3 = F_{RA}(l_1 + 0.055 \text{ m}) - F \times 0.055 \text{ m} = 4.63 \text{ kN} \cdot \text{m}$$

Check the normal stress strengths of the sections in which they are located respectively.

$$\sigma_{\max 1} = \frac{M_1}{W_{z1}} = \frac{32M_1}{\pi d_1^3} = \frac{32 \times 4.72 \times 10^6}{\pi \times 95^3} \text{ MPa} = 56.1 \text{ MPa} < [\sigma]$$

$$\sigma_{\max 2} = \frac{M_2}{W_{z2}} = \frac{32M_2}{\pi d_2^3} = \frac{32 \times 3.42 \times 10^6}{\pi \times 85^3} \text{ MPa} = 56.7 \text{ MPa} < [\sigma]$$

$$\sigma_{\max 3} = \frac{M_3}{W_{z3}} = \frac{32M_3}{\pi d_3^3} = \frac{32 \times 4.63 \times 10^6}{\pi \times 88^3} \text{ MPa} = 69.2 \text{ MPa} < [\sigma]$$

It can be seen that the mandrel satisfies the normal stress strength condition.

Example 10-10 The simply supported beam with $F = 100$ kN, $l = 2$ m, $a = 0.2$ m is shown in Fig.10-29a. Allowable stress of the material is $[\sigma] = 160$ MPa and $[\gamma] = 100$ MPa. Try to select a suitable type of I-shaped beam.

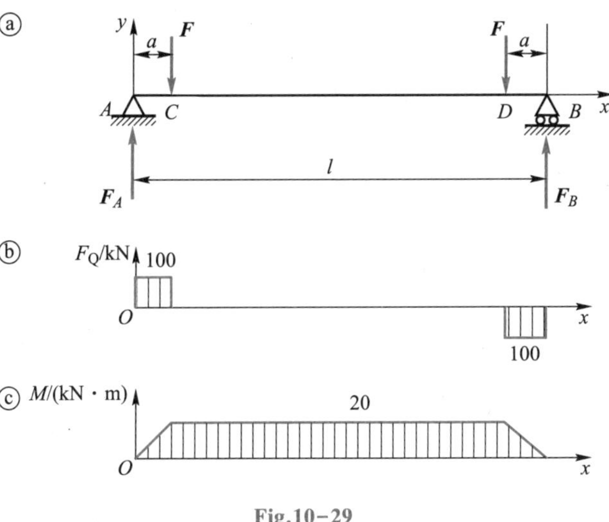

Fig.10-29

Solution The shear force and bending moment diagrams are made according to the forces, as shown in Fig.10-29b and c.

With normal stress strength condition

$$\sigma_{\max} = \left(\frac{M}{W_z}\right)_{\max} = \frac{M_{\max}}{W_z} \leqslant [\sigma]$$

There are

$$W_z \geqslant \frac{M_{\max}}{[\sigma]} = \frac{20 \times 10^6}{160} \text{ mm}^3 = 125 \times 10^3 \text{ mm}^3 = 125 \text{ cm}^3$$

Choose gauge I-shaped beam 16 with $W_z = 141\ \text{cm}^3$, $I_z/S^*_{max} = 13.8$ cm, $b_0 = 6$ mm. Check the shear stress strength on the ends AC and DB:

$$\tau_{max} = \frac{F_Q S^*_{z\,max}}{I_z b_0} = \frac{100\times10^3}{138\times6}\ \text{MPa} = 120.8\ \text{MPa} > [\tau]$$

Therefore, the shear stress strength condition is not satisfied. It is needed to reselect again according to the shear stress strength condition. By

$$\tau_{max} = \frac{F_Q S^*_{z\,max}}{I_z b_0} \leqslant [\tau]$$

We get

$$\frac{I_z b_0}{S^*_{z\,max}} \geqslant \frac{F_Q}{[\tau]} = \frac{100\times10^3}{100}\ \text{mm}^2 = 10^3\ \text{mm}^2 = 10\ \text{cm}^2$$

Try to choose the I-shaped beam 18 with $I_z/S^*_{max} = 15.4$ cm, $b_0 = 6.5$ mm.

$$\frac{I_z b_0}{S^*_{z\,max}} = 15.4\times0.65\ \text{cm} = 10.01\ \text{cm}^2 > 10\ \text{cm}^2$$

Therefore, I-shaped beam 18 can meet the strength conditions of shear stress in bending. If the trial selection of the model still can not meet the requirements, you can choose the next model in turn and then calculate, until the strength conditions are met. In the principle of strength enough and economic, we should not jump to choose the model.

10.8 Displacement and rigidity condition of beams

10.8.1 Displacement of beams

Take the simply supported beam shown in the Fig.10-30 as an example. The axis of the beam before deformation is set as x axis and the left end is set as the origin of the coordinate y axis. After the beam is deformed, its axis will be bent into a curve in the xOy plane, which is called the deflection curve of the beam.

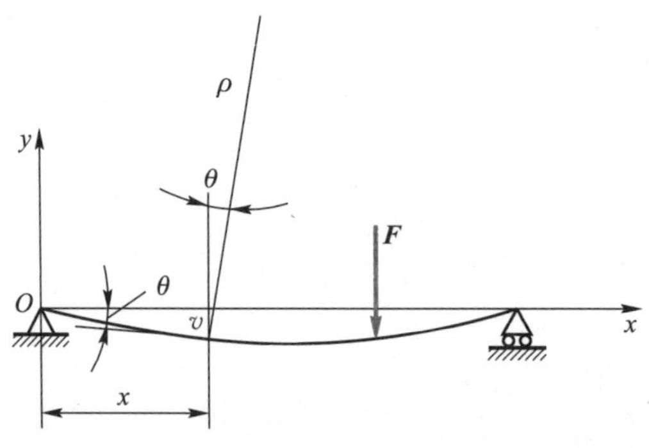

Fig.10-30

The linear displacement v of the point on the beam axis (i.e. the section centroid) in the direction perpendicular to the x axis (the deflection of that point) and the angular displacement θ of the cross section around its neutral axis (the angle of rotation of the section) are used to measure the displacement of a beam. As for the displacement of the section center in the x direction, it is much smaller than the deflection under the small deformation condition, and can be omitted.

The deflection of each point on the beam varies with x. This variation rule is expressed in the deflection curve equation as

$$v=f(x) \tag{10-38}$$

Angle θ is also equal to the angle between the tangent line of the flexure curve at that section and the axis (Fig.10-30). Because the deflection curve is a very flat, θ is a very small angle. We get

$$\theta \approx \tan\theta = \frac{\mathrm{d}v}{\mathrm{d}x} = f'(x) \tag{10-39}$$

10.8.2 Rigidity condition

In order to ensure the normal operation of the beam, we not only require the beam to have sufficient strength, but also need to control the deformation of the beam. So, its maximum deflection and maximum angle of rotation are within the specified permitted range. The rigidity condition of the beam is

$$f_{max} \leqslant [f] \tag{10-40}$$

$$\theta_{max} \leqslant [\theta] \tag{10-41}$$

where f_{max} and θ_{max} are the maximum deflection and maximum angle of rotation of the beam. The $[f]$ and $[\theta]$ are the allowable deflection and allowable angle of rotation. Their specific values are determined by the working conditions and can be found in the code.

For specific problems, there are different requirements. Sometimes the angle of rotation is also used to establish the rigidity condition . That is

$$|\theta|_{max} \leqslant [\theta] \tag{10-42}$$

It is worth pointing out that $|f_{max}|$ and $|\theta_{max}|$ determined by equation (10-40) are not necessarily really the maximum deflection and maximum angle of rotation in the beam. They may be the deflection and angle of rotation of a particular section where the displacement needs to be limited.

Be similar with strength conditions, there are three aspects of rigidity calculation, which are rigidity check, section design and determination of allowable loads

10.9 Deflection of beams

10.9.1 Approximate differential equations of deflection curve

In order to obtain the equation of the deflection curve, it is necessary to establish the relationship between the deformation and the external force, i.e., equation (10-20), where I_z is abbreviated as I, then

$$\frac{1}{\rho}=\frac{M}{EI} \tag{10-43}$$

This formula was derived in the study of purely bending beams. In transverse bending, the shear force F_Q deformation is small and can be omitted in most cases. The above equation is approximated as the basic equation of deformation in shear bending. Its accuracy meets the engineering needs. Since the bending moment M is function of x, the curvature at different locations along the bar length is different. As long as substitute the bending moment equation $M(x)$ of the beam into deflection curve equation (10-43), the curvature equation can be obtained

$$\frac{1}{\rho(x)}=\frac{M(r)}{EI} \tag{10-44}$$

And from the curvature formula of plane curve in advanced mathematics

$$\frac{1}{\rho(x)}=\pm\frac{v''}{[1+(v')^2]^{\frac{3}{2}}} \tag{10-45}$$

Substituting equation(10-45) into equation (10-44), we get

$$\pm\frac{v''}{[1+(v')^2]^{\frac{3}{2}}}=\frac{M(x)}{EI} \tag{10-46}$$

This is the differential equation for the deflection curve of the beam. Under small deformation $v'\ll 1$, the above equation can be approximated as

$$\pm v''=\frac{M(x)}{EI} \tag{10-47}$$

This equation has a positive or negative sign. But after specifying the sign of the bending moment and selecting the xOy coordinate system, the sign is determined. In the previous section, it is stated that when the bending moment is positive, the beam axis is bent into a downward convex curve; when the bending moment is negative, the beam axis is bent into an upward convex curve. It is also known that when the positive direction of v is specified upward, the second-order derivative v'' is positive for evert point on a downward convex curve; while the second-order derivative v'' is negative for every point on an upward convex curve. Since v'' and $M(x)$ have tne same sign, the equation (10-47) should take a positive sign.

$$v''=\frac{M(x)}{EI} \tag{10-48}$$

As for the positive direction of the x axis, it does not affect selection of sign of v''. Therefore, it also does not affect the sign of equation (10-48). For the variable section beam, the I should be understood as $I(x)$ for variable section beams. For equal-section beams, I is a constant, equation (10-48) is also often written as

$$EIv''=M(x) \tag{10-49}$$

The equation (10-48) is usually referred to as the approximate differential equation of the deflection curve of beams, from which the equation of θ and the equation of f can be derived.

10.9.2 Integration method

According to equation (10-48), we get the angle equation

$$\theta = \frac{\mathrm{d}v}{\mathrm{d}x} = \int \frac{M}{EI}\mathrm{d}x + C \tag{10-50}$$

Then multiply both ends of equation (10-48) by dx and integrate to obtain the equation of the deflection curve

$$v = \int\left(\int \frac{M}{EI}\mathrm{d}x\right)\mathrm{d}x + Cx + D \tag{10-51}$$

where C and D are integration constants.

For a specific beam, the deflection or angle of rotation of some sections is sometimes known. For example, at the fixed end, both the deflection and the angle of rotation are equal to zero (Fig.10-31a). The deflection is equal to zero on the pin support (Fig.10-31b). These conditions are collectively referred to as the boundary conditions. In addition, the deflection curve should be a continuous smooth curve and should not have the case shown in Fig.10-31c and d. Thus, there is a uniquely determined deflection and angle of rotation at any point of the deflection curve. This is the continuity condition. Based on the continuity condition and the boundary condition, the integration constants in equations (10-50) and (10-51) can be determined.

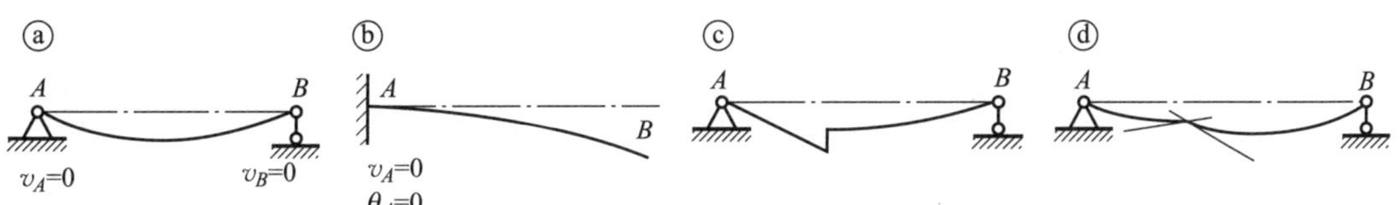

Fig.10-31

Example 10-11 The cantilever beam AB shown in Fig.10-32 is subjected to a concentrated force F at its free end. Determine the maximum deflection and maximum angle of rotation of the beam by establishing the deflection curve and angle of rotation equations. Let the flexural rigidity EI of the beam be a constant.

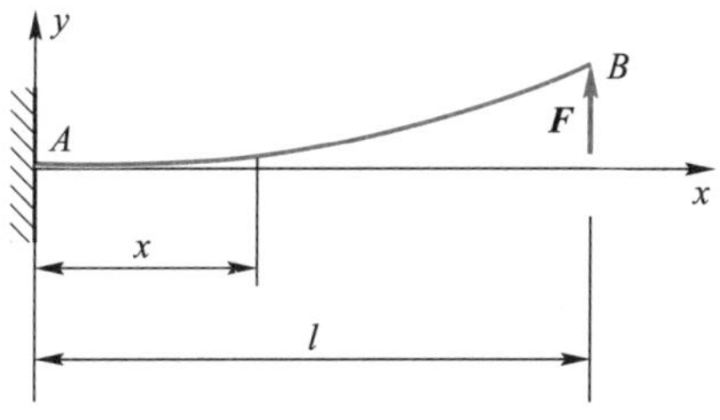

Fig.10-32

Solution (1) Integrate the bending moment equation

Choose the coordinate system shown in the Fig.10−32, the bending moment of any cross section is

$$M(x)=F(l-x)$$

(2) Find the equation of the turning angle and the equation of the deflection curve

The approximate differential equation for the deflection curve is obtained by substituting(10−49) into the equation

$$EIv''(x)=F(l-x)$$

Integrating the above equation twice in succession, we obtain the equation of the angle of rotation and the equation of the deflection curve respectively

$$EI\theta(x)=Flx-\frac{1}{2}Fx^2+C$$

$$EIv(x)=\frac{1}{2}Flx^2-\frac{1}{6}Fx^3+Cx+D$$

Using the boundary conditions $\theta(0)=0$ and $v(0)=0$, we get

$$C=0,\quad D=0$$

So the equation of the angle of rotation and the equation of the deflection curve are

$$EI\theta(x)=Flx-\frac{1}{2}Fx^2$$

$$EIv(x)=\frac{1}{2}Flx^2-\frac{1}{6}Fx^3$$

(3) Determine the maximum deflection and maximum angle of rotation

From the general shape of the deflection curve, it can be determined that the maximum deflection and the maximum angle of rotation are at the free end plane B

$$\theta_{max}=\theta(l)=\frac{Fl^2}{2EI},\quad v_{max}=v(l)=\frac{Fl^3}{3EI} \tag{10-52}$$

The sign of v_{max} and θ_{max} is positive, which indicates that the deflection of section B is in line with the positive direction of y axis and the direction of turning angle is counterclockwise.

Example 10−12 The simply supported beam shown in Fig10−33 is subjected to a concentrated force F. Its length l and the flexural rigidity EI are known. Find the equations of the deflection curve and the angle of rotation of this beam. Determine the maximum deflection and the maximum angle of rotation (set $a>b$).

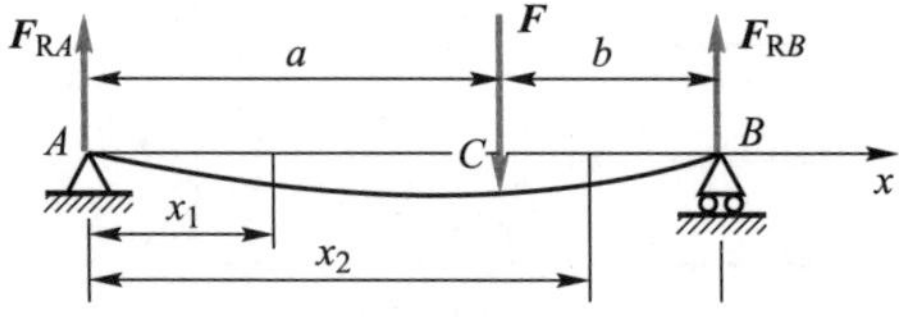

Fig.10−33

Solution (1) Find the bending moment equation

According to the static equilibrium relationship to find the support reaction force

$$F_{RA}=\frac{b}{l}F,\quad F_{RB}=\frac{a}{l}F$$

According to the segmentation law, the bending moment equation is

AC:

$$M_1(x_1)=\frac{b}{l}Fx_1\quad(0\leqslant x_1\leqslant a)$$

CB:

$$M_2(x_2)=\frac{b}{l}Fx_2-F(x_2-a)\quad(a\leqslant x_2\leqslant l)$$

(2) Find the equation of the angle of rotation and the equation of the deflection curve

For the segment AC:

$$EIv_1''(x_1)=\frac{b}{l}Fx_1$$

$$EI\theta_1(x_1)=\frac{b}{2l}Fx_1^2+C_1$$

$$EIv_1(x_1)=\frac{b}{6l}Fx_1^3+C_1x_1+D_1$$

For the segment CB:

$$EIv_2''(x_2)=\frac{b}{l}Fx_2-F(x_2-a)$$

$$EI\theta_2(x_2)=\frac{b}{2l}Fx_2^2-\frac{1}{2}F\ (x_2-a)^2+C_2$$

$$EIv=\frac{b}{6l}Fx_2^3-\frac{1}{6}F\ (x_2-a)^3+C_2x+D_2$$

According to the continuity conditions $\theta_1(a)=\theta_2(a)$ and $v_1(a)=v_2(a)$, we get

$$C_1=C_2,\quad D_1=D_2$$

Then using the boundary conditions $v_1(0)=0$ and $v_2(l)=0$, we can obtain

$$D_1=D_2=0,\quad C_1=C_2=-\frac{Fb}{6l}(l^2-b^2)$$

Therefore, the equation of the angle of the AC section and the equation of the deflection curve are

$$EI\theta_1(x_1)=-\frac{Fb}{6l}(l^2-b^2-3x_1^2)$$

$$EIv_1=-\frac{Fbx_1}{6l}(l^2-b^2-x_1^3)$$

The angle equation of the section CB and the equation of the deflection curve are

$$EI\theta_2(x_2) = -\frac{Fb}{6l}\left[l^2-b^2+\frac{3l}{b}(x_2-a)^2\right]$$

$$EIv_2(x_2) = -\frac{Fb}{6l}\left[(l^2-b^2)x_2-x_2^3+\frac{l}{b}(x_2-a)^3\right]$$

(3) Determine the maximum deflection and maximum angle of rotation

From the general shape of the deflection curve, it can be seen that the maximum corner should be at the two end faces A or B.

$$\theta_A = \theta_1(0) = -\frac{Fab(l+b)}{6EIl}$$

$$\theta_B = \theta_2(l) = \frac{Fab(l+a)}{6EIl}$$

Because $a>b$, θ_B is the maximum angle of rotation.

The maximum deflection of this beam should be at $\theta = v' = 0$. The section AC is studied first. From $\theta_1(x_1^*) = 0$, we get

$$x_1^* = \sqrt{\frac{l^2-b^2}{3}} = \sqrt{\frac{a(a+2b)}{3}} \tag{10-53}$$

When $a>b$, $x_1^*<a$. This indicate: $\theta = v' = 0$ (in the segment AC).

$$v_{\max} = v_1(x_1^*) = \frac{Fb}{9\sqrt{3}EIl}\sqrt{(l^2-b^2)^3} \tag{10-54}$$

Using this example, we discuss the approximate calculation of the maximum deflection of a simply supported beam. First, find the deflection at midpoint D of the above span beam

$$v_D = -\frac{Fb}{48EI}(3l^2-4b^2) \tag{10-55}$$

From equation (10-53), it can be seen that the smaller b is, the larger x_1 is. This indicates that the closer the load is to the right support, the farther the maximum deflection point of the beam is from the midpoint. When the value of b is so small that b^2 is negligible compared to l^2. Two equations (10-64) (10-65) are obtained respectively

$$v_{\max} = -0.064\ 2\times\frac{Fbl^2}{EI}$$

$$v_D = -0.062\ 5\times\frac{Fbl^2}{EI}$$

In this extreme case, the difference between the maximum deflection and the midpoint deflection is very small (less than 3% relative error).If the deflection curve does not appear an inflection point, the maximum deflection can be replaced by the deflection at the midpoint of a simply supported beam. And it doesn't cause a large error.

When the beam is subjected to complex loads, the moment equation must be written in segments. The equations of the deflection curve will be different for each segment. Therefore, when integrating the differential equations from each segment of the beam, two integration constants will appear. To determine these integration constants, both the boundary and continuity conditions are used.

Problems

10-1 Try to find the shear force and bending moment in the cross sections 1—1, 2—2 and 3—3 of each beam shown in Fig.P10-1. These sections are infinitely close to section C or section D. F, q, and a are known.

Answer: (a) Section 1—1 : $F_{Q1}=0$, $M_1=Fa$; Section 2—2 : $F_{Q2}=-F$, $M_2=Fa$; Section 3—3: $F_{Q3}=0$, $M_3=0$

(b) Section 1—1 : $F_{Q1}=1.33$ kN, $M_1=267$ N · m; Section 2—2 : $F_{Q2}=-0.667$ kN, $M_2=333.3$ N · m

(c) Section 1—1: $F_{Q1}=2qa$, $M_1=-\dfrac{3}{2}qa^2$ (clockwise direction); Section 2—2: $F_{Q2}=2qa$, $M_2=-\dfrac{1}{2}qa^2$ (counterclockwise direction)

(d) Section 1—1 : $F_{Q1}=100$ N, $M_1=-20$ N · m; Section 2—2 : $F_{Q2}=-100$ N, $M_2=-40$ N · m; Section 3—3 : $F_{Q3}=200$ N, $M_3=-40$ N · m

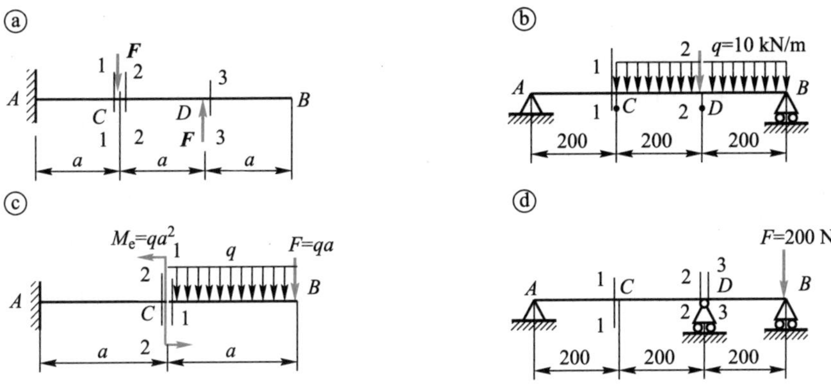

Fig.P10-1

10-2 Establish the equations of shear force and bending moment for each beam shown in the Fig.P10-2. Make shear force and bending moment diagrams. Find the maximum absolute values of shear force and bending moment, $|F_Q|_{max}$ and $|M|_{max}$. Let F, q, M_e, a be known.

Answer: (a) $|F_Q|_{max}=2F$, $|M|_{max}=Fa$

(b) $|F_Q|_{max}=qa$, $|M|_{max}=\dfrac{qa^2}{2}$

(c) $|F_Q|_{max}=30$ kN, $|M|_{max}=15$ kN · m

(d) $|F_Q|_{max}=\dfrac{3M_e}{2a}$, $|M|_{max}=\dfrac{3}{2}M_e$

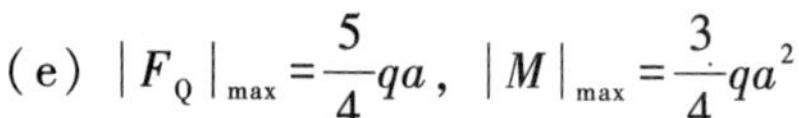

(e) $|F_Q|_{max}=\frac{5}{4}qa$, $|M|_{max}=\frac{3}{4}qa^2$

(f) $|F_Q|_{max}=\frac{9}{4}qa$, $|M|_{max}=\frac{49}{32}qa^2$

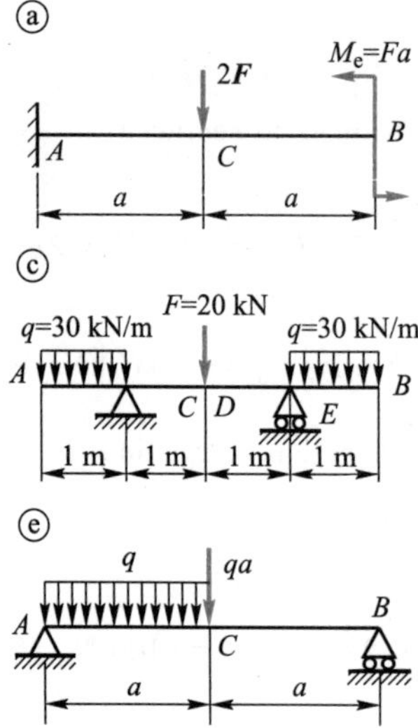

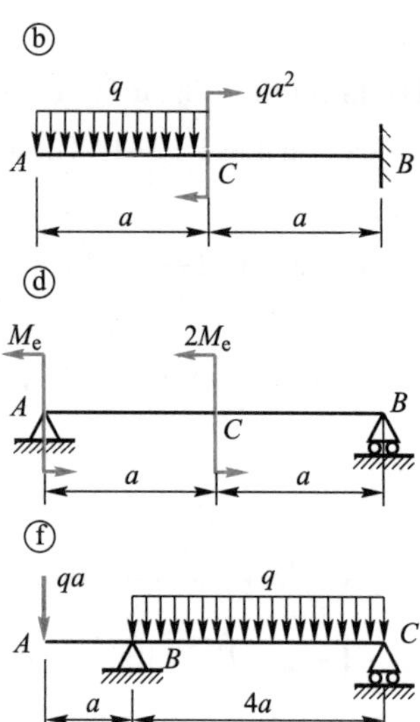

Fig.P10-2

10-3 Try to use the differential relationship between the load, shear force and bending moment to make a shear force and a bending moment diagram (Fig.P10-3).

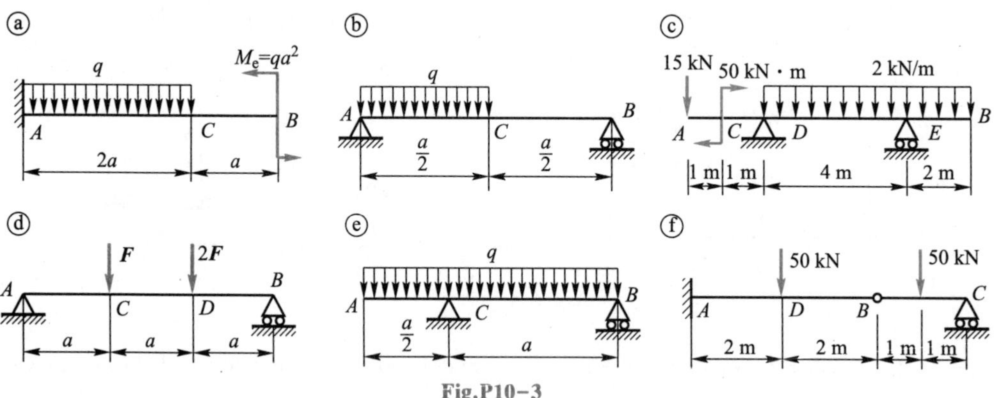

Fig.P10-3

Answer: Omit

10-4 The pressure of each wheel of the trolley on the beam of the overhead crane is F (Fig.P10-4). What is the maximum bending moment in the beam and where does it occur. What is its maximum bending moment? Let the wheelbase of the trolley be d, and let the span length of the beam be l.

Answer: From the left and right $\left(\frac{l}{2}-\frac{d}{4}\right)$, $M_{max}=\frac{F}{2}(l-d)+\frac{Fd^2}{8l}$

10-5 The simply supported beam is subjected to a uniform load as shown in the Fig.P10-5. If

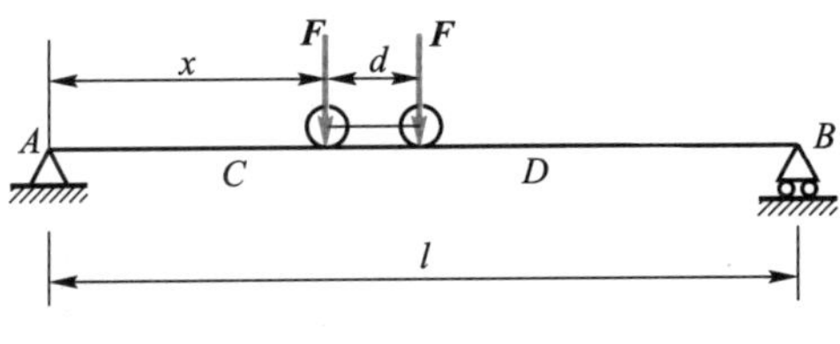

Fig.P10-4

the solid and hollow circular sections of equal area are used respectively, and the diameter of the solid section is $D_1 = 40$ mm, the diameter of the hollow section is related to $\frac{d_2}{D_2} = \frac{3}{5}$. What is the percentage reduction of the maximum normal stress in the hollow circular section compared with that in the solid circular section?

Answer: $\sigma_{max} = \frac{32M_{max}}{\pi(1-\alpha^4)D_2^3}$, 41%

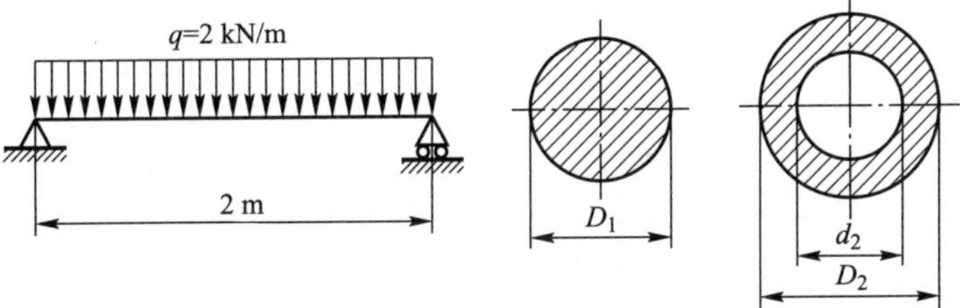

Fig.P10-5

10-6 A cantilever beam of rectangular section is shown in the Fig.P10-6. We have known $l = 4$ m, $\frac{b}{h} = \frac{2}{3}$, $q = 10$ kN/m, $[\sigma] = 10$ MN/m^2. Try to determine the dimensions of the cross section of this beam.

Answer: $h \geqslant 416$ mm, $b \geqslant 277$ mm

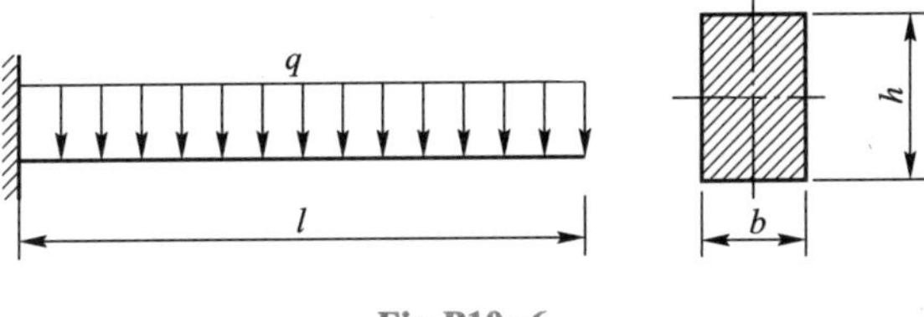

Fig.P10-6

10-7 The load and dimensions of the cast iron beam are shown in the Fig.P10-7. Allowable stress is $[\sigma^+] = 40$ MPa and $[\sigma^-] = 160$ MPa. Try to calibrate the strength of the beam according to the normal stress strength conditions. If the load remains unchanged, but the T-shaped section is inverted(the flange becomes I-shape), is it reasonable and why?

Answer: $\sigma^{+}_{max}=26.2$ MPa, $\sigma^{-}_{max}=52.4$ MPa, reasonable

Unchanged: point B, $\sigma^{+}_{max}=52.4$ MPa>[σ^{+}_{max}], unreasonable

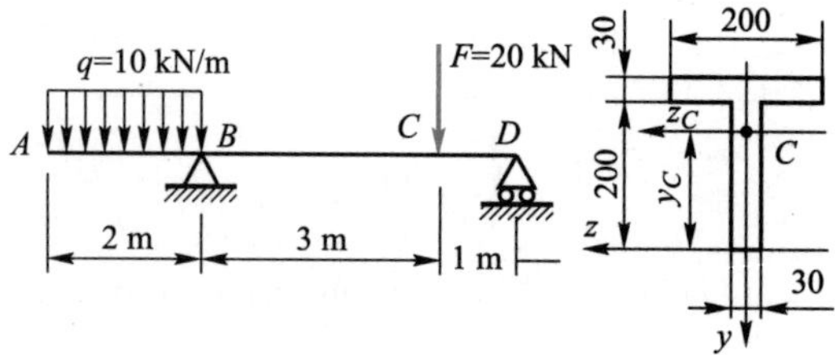

Fig.P10–7

10–8 Try to calculate the normal and shear stresses at points A and B of the section 1—1 of the simply supported rectangular beam (Fig.P10–8).

Answer: point A: $\sigma=-6.04$ MPa, $\tau=0.38$ MPa;

point B: $\sigma=12.9$ MPa, $\tau=0$

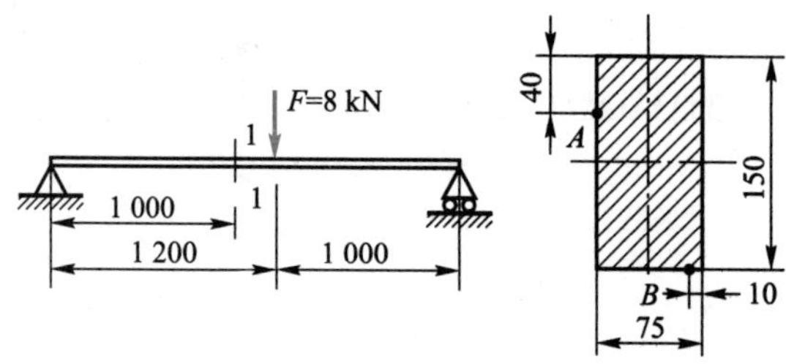

Fig.P10–8

10–9 The beam under the crane consists of two I-shaped beams (Fig.P10–9). The self-weight of the crane is $P_1=50$ kN and the lifting capacity is $P_2=10$ kN. The allowable stress is $[\sigma]=160$ MN/m^2 and $[\tau]=100$ MN/m^2. The self-weight of the beam is not taken into account. Try to select the I-shaped beam type according to the normal stress strength condition, and then calibrate it according to the shear stress strength condition.

Answer: Select I-shaped beam model: $W=438$ cm^3, choose two 28a I-shaped beams, $\tau_{max}=$ 13.9 MPa<[τ], safe

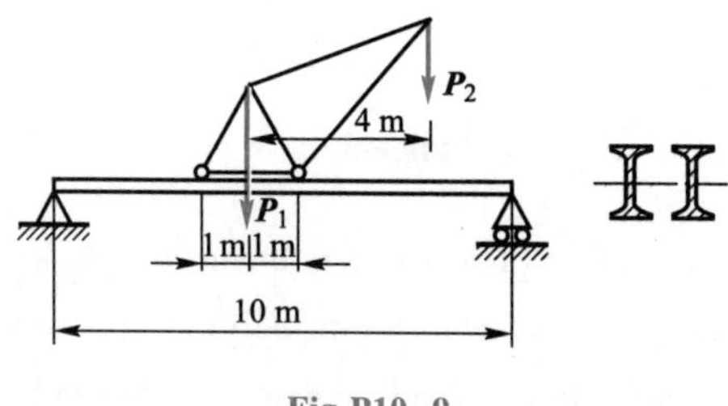

Fig.P10–9

10-10 For each beam shown in the Fig.P10-10, the flexural rigidity EI is constant.

(1) Try to draw the approximate shape of the deflection curve based on the variation of the bending moment along the axis and the support conditions of the beam;

(2) Calculate the maximum deflection and the maximum angle of rotation of the beam by the integration method (find the deflection and angle of rotation of the beam at the outward extension in Fig.P10-10d).

Answer: (a) $f_{max}(x)=-\dfrac{M_e l^2}{9\sqrt{3}EI}$, $\theta_{max}=\dfrac{M_e l}{3EI}$

(b) $f_{max}(x)=-\dfrac{5ql^4}{384EI}$, $\theta_{max}=\dfrac{ql^3}{24EI}$

(c) $f_{max}(x)=-\dfrac{41ql^4}{384EI}$, $\theta_{max}=-\dfrac{7ql^3}{48EI}$

(d) $f_{max}(x)=-\dfrac{Fa^2}{3EI}(l+a)$, $\theta_{max}=\dfrac{Fa}{6EI}(2l+3a)$

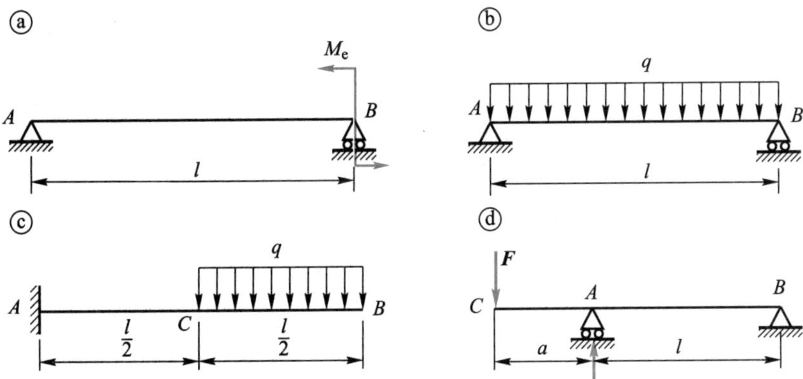

Fig.P10-10

Chapter 11 State of Stress and Theories of Strength

Teaching Scheme of Chapter 11

11.1 State of stress

11.1.1 Concepts of state of stress

Analyzing the stresses in the inclined section of a tensile bar, it can be found that the stresses of different faces through the same point vary with the section orientation. Generally, there are countless sections with different orientations through a point and the stress varies with the orientation of the section. In order to understand the stresses in the member and analyze its strength, it is necessary to study the stresses in each section through a point in the member comprehensively.

The state of stress of a point in various orientations is called the stress state. In order to study the stress state at a point, an element (infinitesimal normal hexahedron) can be taken out around the point. It can be assumed that the stress is uniform on each face of the element and the stresses are the same in the mutually parallel cross sections within the element. Therefore, the stress state of such an element can represent the stress state of a point. The study of the stress state at one point is called stress analysis. The purpose of the stress analysis is to determine the most dangerous place and direction of the stressed member, and to provide a basis for analyzing the strength of the member.

The three mutually perpendicular surface stresses are known element is called the original element. The stress state can be determined by finding the relationship between the stress on any inclined cross section and the known stress on the side of the original element using the section method. For example, in the axially tensioned straight bar, an element is taken out around point A with a pair of cross sections and two pairs of longitudinal sections, as shown in Fig.11−1. The normal stresses on the left and right sides of the monolith are known to be $\sigma_x = \dfrac{F}{A}$ and there is no shear stress. The stress on the rest of the faces is zero. So, the element is the original element of point A. For the torsional circular axis shown in Fig.11−2a, a element is intercepted at point A on the surface with two cross sections, two radial longitudinal sections and two sections perpendicular to the radius, as shown in Fig.11−2b. This element is also the original element. The original element in the direction of $\alpha = 45°$ is shown in Fig.11−2c.

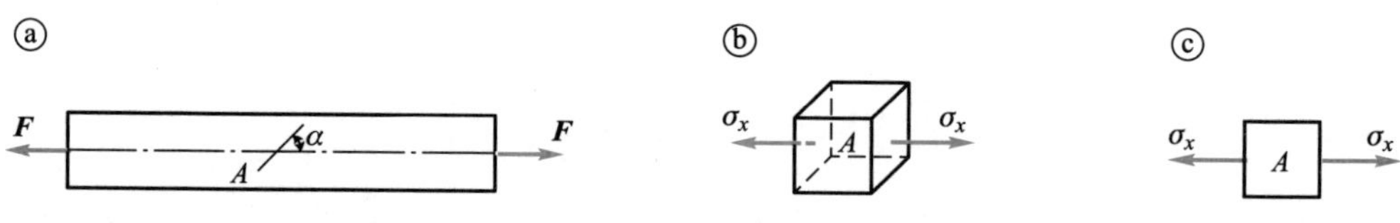

Fig.11−1

In Fig.11−1c and Fig.11−2c, there is no shear stress on all sides of the element. Only normal stress (including zero normal stress) on all sides of the element without shear stress is called the principal element. It can be proved that around any point within the member can always find a principal element. The normal stress on each face of the principal element is called the principal stress. The plane of principal stress is called the principal plane.

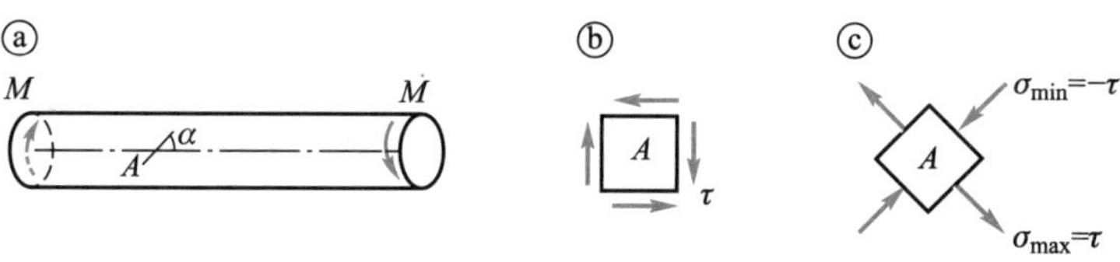

Fig.11−2

There are three pairs of principal stresses on the six faces of the principal element at one point, which are usually denoted as σ_1, σ_2, σ_3 as shown in Fig. 11 − 3. Their algebraic magnitude is $\sigma_1 \geqslant \sigma_2 \geqslant \sigma_3$. According to the number of principal stresses (not zero), the stress states can be classified into three categories.

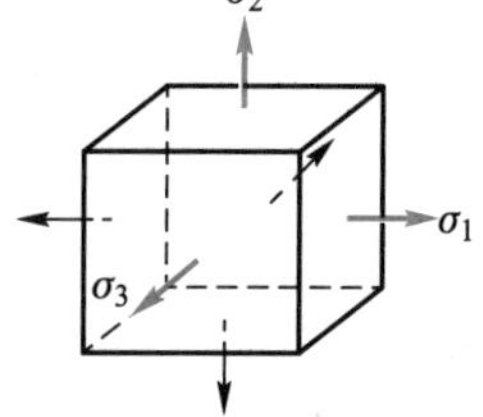

Fig.11−3

1. Uniaxial stress state

Only one principal stress is not zero, the stress state at this point is called the uniaxial stress state. Fig.11−1c shows a uniaxial stress state.

2. Biaxial stress state

If two principal stresses is not zero, the stress state at this point is called biaxial stress state or plane stress state. Fig.11−2c shows an element of $\alpha = 45°$ at point A of torsional circular axis. There are only two normal stresses in the cross section of the element without shear stresses, so the torsional circular shaft at any point is a biaxial stress state.

3. Triaxial stress state

If three principal stresses are not zero, the stress state at that point is called the triaxial stress state or the space stress state. The contact point of the gear is engaged and the stress state of each point near the contact point between the train wheel and the rail are the triaxial stress state.

The uniaxial stress state is also called the simple stress state. The biaxial and triaxial stress states are also called the complex stress state.

11.1.2 Examples for biaxial and triaxial stress

1. Example of biaxial stress state

As an example of a plane stress state, the following is a study of the stress state of a boiler or other cylindrical vessel, as shown in Fig. 11 − 4a. Let the inner diameter of the cylinder be D, the wall thickness is t and $t \ll D$ (for example, $t < D/20$). The internal pressure is p. The pressure at the bottom of the cylinder will stretch the cylinder, while the pressure acting on the inner wall will make the cylinder diameter uniformly larger.

(1) Calculate the axial normal stress.

As in Fig.11-4b, the total pressure acting on the bottom of the cylinder at both ends along the cylinder axis is $F=p\dfrac{\pi D^2}{4}$. Then the axial stress is

$$\sigma'=\frac{F_{\mathrm{N}}}{A}=\frac{F}{\pi Dt}=\frac{pD}{4t} \tag{11-1}$$

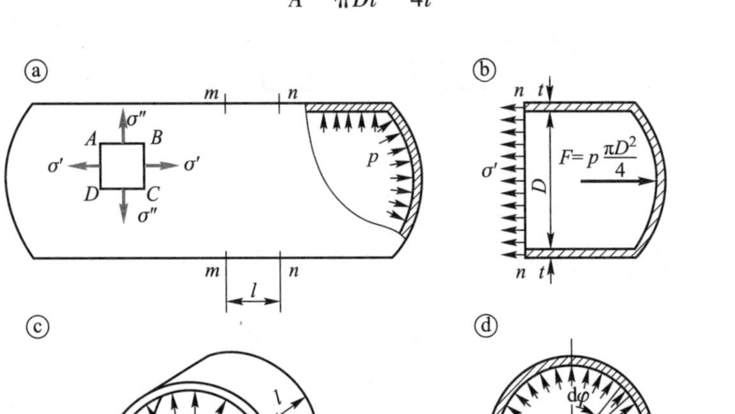

Fig.11-4

(2) Calculate the circumferential normal stress.

Two parallel cross sections are used to intercept a section of length l of the cylinder and a diameter plane is used to cut the cylinder into two parts. Take the upper half as the study object as in Fig.11-4c. If the circumferential stress acting on the longitudinal section of the cylinder is σ'', then the corresponding internal force is

$$F_{\mathrm{N}}=\sigma''tl \tag{11-2}$$

The pressure acting on the micro area $\mathrm{d}A=l\dfrac{D}{2}\mathrm{d}\varphi$ inside the cylinder wall is $\mathrm{d}F=p\mathrm{d}A=pl\dfrac{D}{2}\mathrm{d}\varphi$, whose projection in the vertical direction is

$$\mathrm{d}F_y=pl\frac{D}{2}\mathrm{d}\varphi\sin\varphi \tag{11-3}$$

Then the total pressure in the vertical direction is

$$F_y=\int_0^{\pi}pl\frac{D}{2}\mathrm{d}\varphi\sin\varphi=plD \tag{11-4}$$

From the equilibrium condition $2F_{\mathrm{N}}=F_y$, we have

$$\sigma''=\frac{pD}{2t} \tag{11-5}$$

A element $ABCD$ is taken out in the middle of the cylinder with two longitudinal sections and two transverse sections, as shown in Fig.11−4a. The left and right sections are part of the transverse section with axial stress σ'. The upper and lower sections are part of the longitudinal section and have a circumferential stresses σ''. In the third direction of the element, there is an internal pressure force acting on the inner wall and an atmospheric pressure acting on the outer wall. They are much smaller than σ' and σ'' and can be omitted. It is known that the cylinder is in a biaxial stress state and its principal stresses are

$$\sigma_1=\frac{pD}{2t},\quad \sigma_2=\frac{pD}{4t},\quad \sigma_3=0 \tag{11-6}$$

2. Triaxial stress state

In a ball bearing, the stress state at the contact point between the ball and the outer ring is atriaxial stress state, as shown in Fig.11−5a. A element intercepted in a plane perpendicular and parallel to the pressure F around the contact point A between the outer ring and the ball, as shown in Fig.11−5b. The contact stress σ_3 acting on the contact surface will cause the element at point A to expand around it. As a result, the surrounding materials constrain it with stresses σ_1 and σ_2. From the symmetry conditions know that there is no shear stress on all sides of the element, so the three mutually perpendicular planes of the element are the principal plane. The stress σ_3, σ_1 and σ_2 is the principal stress. Thus, the stress state at point A is a triaxial stress state.

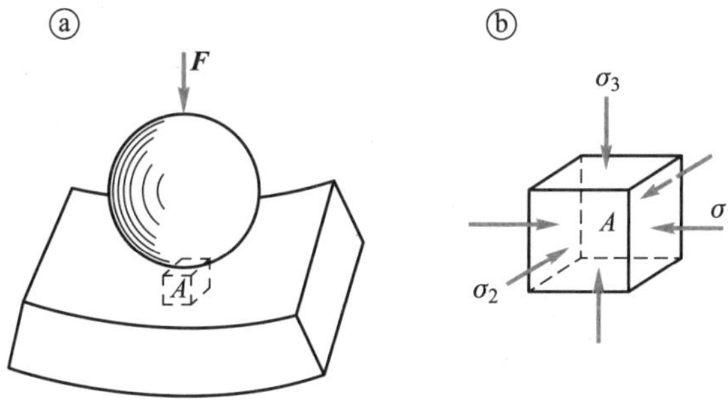

Fig.11−5

11.2 Analysis of plane stress

The general form of the elements inp lane stress is shown in Fig. 11 − 6a. Let all the stresses on the element be known. Fig.11−6b shows the orthogonal projection of the element in the x−y plane. On the plane with the x axis as the normal direction, the normal stress is denoted by σ_x and the shear stress is denoted by τ_{xy}. With the y axis as the normal direction, the normal stress is σ_y and the shear stress is τ_{yx}. There are two subscripts for the shear stress. For example, the subscript x of τ_{xy} indicates that the direction of the shear stress is normal to the x axis, and the subscript y indicates that the direction of the shear stress is parallel to the y axis. The sign of the normal stress: tension is positive and compression is negative. Shear stress to make the element produce a clockwise rotation trend is positive and counterclockwise rotation trend is negative. In Fig.11-6a, σ_x, σ_y and τ_{xy} are positive.

11.2.1 Stresses on the inclined sections

Study the inclined section *ef* where the normal of the section is at an angle of α with the x axis, as in Fig.11-6b. It is specified that when the x axis is counterclockwise turned to the normal n, α angle is positive; the opposite is negative. Cross section *ef* divide the element into two parts and take *aef* part for analysis. Inclined section *ef* with normal stress σ_α and shear stress τ_α is shown in Fig.11-6c. If the area of *ef* is dA, the area of *af* and *ae* are dAsinα and dAcosα, respectively, as shown in Fig.11-6d. According to the equilibrium of *aef*, the equilibrium equation of the normal n direction and the tangent t direction can be written as

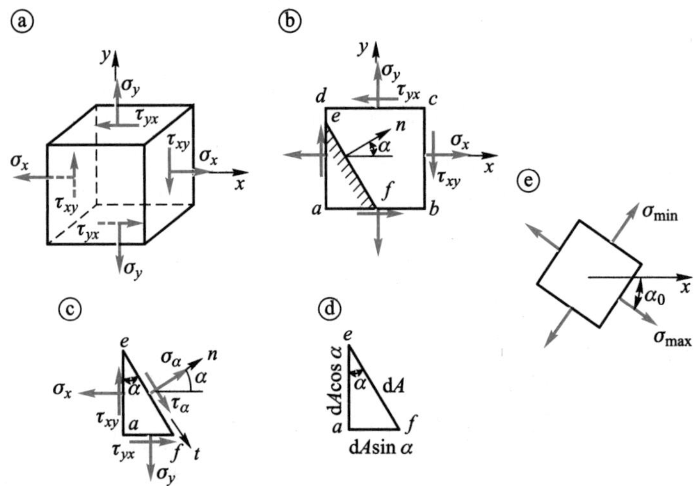

Fig.11-6

$$\sum F_n = 0,\quad \sigma_\alpha \mathrm{d}A + (\tau_{xy}\mathrm{d}A\cos\alpha)\sin\alpha - (\sigma_x\mathrm{d}A\cos\alpha)\cos\alpha + (\tau_{yx}\mathrm{d}A\sin\alpha)\cos\alpha - (\sigma_y\mathrm{d}A\sin\alpha)\sin\alpha = 0 \tag{11-7}$$

$$\sum F_t = 0,\quad \tau_\alpha \mathrm{d}A - (\tau_{xy}\mathrm{d}A\cos\alpha)\cos\alpha - (\sigma_x\mathrm{d}A\cos\alpha)\sin\alpha + (\sigma_y\mathrm{d}A\sin\alpha)\cos\alpha + (\tau_{yx}\mathrm{d}A\sin\alpha)\sin\alpha = 0 \tag{11-8}$$

According to the theorem of complementary shearing stresses, we have $\tau_{xy}=-\tau_{yx}$ and substituting it into the above two equations and simplifying to get

$$\sigma_\alpha = \sigma_x\cos^2\alpha+\sigma_y\sin^2\alpha-2\tau_{xy}\sin\alpha\cos\alpha = \frac{\sigma_x+\sigma_y}{2}+\frac{\sigma_x-\sigma_y}{2}\cos 2\alpha-\tau_{xy}\sin 2\alpha \tag{11-9}$$

$$\tau_\alpha = \frac{\sigma_x-\sigma_y}{2}\sin 2\alpha+\tau_{xy}\cos 2\alpha \tag{11-10}$$

Equations (11-9) and (11-10) are the equations for calculating the stresses in any inclined section of the element. It can be seen that the normal and shear stresses in the inclined section are a function of α angle and vary with α angle.

11.2.2 Principal stress and principal plane

The maximum value of stress is always of interest in engineering analysis. The maximum and minimum values of normal stress can be determined by using equation (11-9). The derivative of equation (11-9) for α derivative, we get

$$\frac{d\sigma_\alpha}{d\alpha}=-(\sigma_x-\sigma_y)\sin 2\alpha-2\tau_{xy}\cos 2\alpha$$

When $\left.\frac{d\sigma_\alpha}{d\alpha}\right|_{\alpha=\alpha_0}$, there is

$$(\sigma_x-\sigma_y)\sin 2\alpha_0+2\tau_{xy}\cos 2\alpha_0=0 \tag{11-11}$$

The result is

$$\tan 2\alpha_0=-\frac{2\tau_{xy}}{\sigma_x-\sigma_y} \tag{11-12}$$

By using equation (11-12) to find $\sin2\alpha_0$ and $\cos2\alpha_0$, substituting into equation (11-9), the maximum normal stress and minimum normal stress can be obtained

$$\left.\begin{matrix}\sigma_{max}\\ \sigma_{min}\end{matrix}\right\}=\frac{\sigma_x+\sigma_y}{2}\pm\sqrt{\left(\frac{\sigma_x-\sigma_y}{2}\right)^2+\tau_{xy}^2} \tag{11-13}$$

From equation (11-12), two α_0 angles with a difference of 90° can be found, which corresponds to two perpendicular planes. There is a maximum normal stress in one of the planes and a minimum normal stress in the other plane. It can be proved that if $\sigma_x \geqslant \sigma_y$, the smaller absolute value of α_0 determine the plane of maximum normal stress.

Comparing equation (11-10) and equation (11-11), we can see that α_0 of equation (11-11) makes τ_α equal to zero. That is, the shear stress in the plane where the maximum and minimum normal stresses is zero. Thus, the principal stress is the maximum or minimum normal stress.

11.2.3 Maximum (minimum) shear stress and its plane

Taking the derivative of equation (11-10) to α, we get

$$\frac{d\tau_\alpha}{d\alpha}=(\sigma_x-\sigma_y)\cos 2\alpha-2\tau_{xy}\sin 2\alpha \tag{11-14}$$

When $\left.\frac{d\tau_\alpha}{d\alpha}\right|_{\alpha=\alpha_1}$, there is

$$(\sigma_x-\sigma_y)\cos 2\alpha_1-2\tau_{xy}\sin 2\alpha_1=0$$

The result is

$$\tan 2\alpha_1=\frac{\sigma_x-\sigma_y}{2\tau_{xy}} \tag{11-15}$$

From equation (11-15), two angles α_1 with 90° difference can also be determined, corresponding to the maximum shear stress and the minimum shear stress plane, respectively. The maximum shear stress and the minimum shear stress planes are perpendicular to each other.

Solving from equation (11-15) to find $\sin 2\alpha_1$ and $\cos 2\alpha_2$, substitute into equation (11-10), the maximum shear stress and minimum shear stress can be obtained

$$\left.\begin{matrix}\tau_{max}\\ \tau_{min}\end{matrix}\right\} = \pm\sqrt{\left(\frac{\sigma_x-\sigma_y}{2}\right)^2+\tau_{xy}^2} \tag{11-16}$$

Comparing equation (11-12) and equation (11-15), we can see that

$$\tan 2\alpha_1 = -\cot 2\alpha_0 = \tan(2\alpha_0+90°)$$

So there was

$$\alpha_1 = \alpha_0 + 45° \tag{11-17}$$

The plane where the maximum and minimum shear stresses are located is at an angle of 45° to the principal plane.

Substituting $\sin 2\alpha_1$ and $\cos 2\alpha_2$ into equation (11-9), we get the normal stress in the plane of the maximum and minimum shear stress

$$\sigma_{\alpha_1} = \frac{\sigma_x+\sigma_y}{2}$$

Example 11-1 The original element at a point is shown in Fig.11-7a. Try to calculate: (1) the principal stress and the principal plane orientation; (2) the maximum and minimum shear stresses, and the orientation of the planes in which they are located.

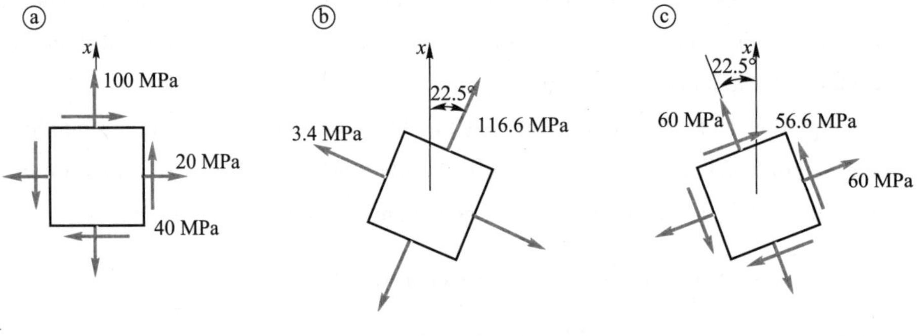

Fig.11-7

Solution From Fig.11-7a, we can see that $\sigma_x = 100$ MPa, $\sigma_y = 20$ MPa, $\tau_{xy} = 40$ MPa.

(1) Find the principal stress and the principal plane orientation

Let the angle of the principal plane be α_0, then

$$\tan 2\alpha_0 = -\frac{2\tau_{xy}}{\sigma_x-\sigma_y} = -\frac{2\times 40\ \text{MPa}}{(100-20)\ \text{MPa}} = -1$$

$$2\alpha_0 = -45° \quad \text{or} \quad 2\alpha_0 = -225°$$

$$\alpha_0 = -22.5° \quad \text{or} \quad \alpha_0 = -112.5°$$

Here $\sigma_x > \sigma_y$. The principal stress on the principal plane of $\alpha_0 = -22.5°$ (clockwise direction) from the x axis is σ_{max}, while the principal stress on the plane of $\alpha_0 = -122.5°$ is σ_{min}. The maximum and minimum normal stresses are

$$\left.\begin{matrix}\sigma_{max}\\ \sigma_{min}\end{matrix}\right\} = \left[\frac{100+20}{2} \pm \sqrt{\left(\frac{100-20}{2}\right)^2 + 40^2}\right] \text{MPa} = (60 \pm 56.6) \text{ MPa} = \begin{cases}116.6 \text{ MPa}\\ 3.4 \text{ MPa}\end{cases}$$

Therefore, the principal stress is $\sigma_1 = 116.6$ MPa, $\sigma_2 = 3.4$ MPa, and $\sigma_3 = 0$. The principal plane orientation (principal element) is shown in Fig.11-7b.

(2) Find the maximum and minimum shear stress and its orientation in the plane

The angle of the plane where the maximum shear stress is located is

$$\alpha_1 = \alpha_0 + 45° = -22.5° + 45° = 22.5°$$

From equation (11-16), the maximum and minimum shear stresses are

$$\left.\begin{matrix}\tau_{max}\\ \tau_{min}\end{matrix}\right\} = \sqrt{\left(\frac{100-20}{2}\right)^2 + 40^2} \text{ MPa} = \pm 56.6 \text{ MPa}$$

The normal stress in the maximum and minimum shear stresses plane is $\sigma = \frac{116.6+3.4}{2}$ MPa $= 60$ MPa, as shown in Fig.11-7c.

Example 11-2 Fig.11-8 gives the normal stresses and shear stresses on the plane AB and the shear stresses on the plane AC. Find: (1) the normal stress σ_y in the plane perpendicular to AB; (2) the shear stress τ_α in the plane AC; (3) principal stress and principal plane orientation.

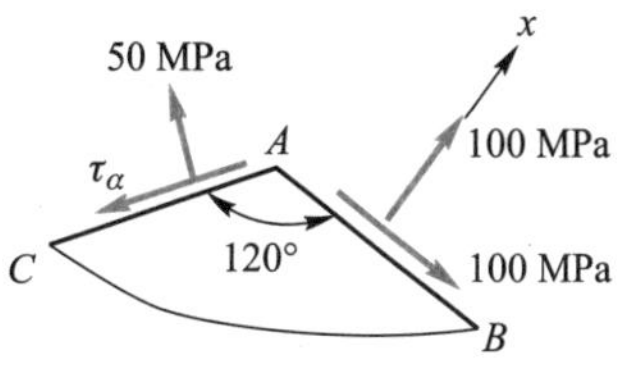

Fig.11-8

Solution (1) Find the normal stress σ_y in the plane perpendicular to the plane AB

Given $\sigma_x = 100$ MPa, $\tau_{xy} = 100$ MPa, $\alpha = 60°$, $\sigma_\alpha = 50$ MPa, obtained from equation (11-9)

$$\sigma_\alpha = \frac{\sigma_x + \sigma_y}{2} + \frac{\sigma_x - \sigma_y}{2}\cos 2\alpha - \tau_{xy}\sin 2\alpha$$

$$= \frac{100 \text{ MPa} + \sigma_y}{2} + \frac{100 \text{ MPa} - \sigma_y}{2} \times \cos 120° - 100 \text{ MPa} \times \sin 120° = 50 \text{ MPa}$$

The solution is

$$\sigma_y = 148.8 \text{ MPa}$$

(2) Find the shear stress τ_α on the surface AC

From equation (11-10) we get

$$\tau_\alpha = \frac{\sigma_x - \sigma_y}{2}\sin 2\alpha + \tau_{xy}\cos 2\alpha$$

$$= \frac{100-148.8}{2} \text{ MPa } \sin 120° + 100 \text{ MPa} \cdot \cos 120° = -71.1 \text{ MPa}$$

(3) Find the principal stress and the principal plane orientation

From equation (11−12) we get

$$\tan 2\alpha_0=\frac{-2\tau_{xy}}{\sigma_x-\sigma_y}=\frac{-2\times 100\ \text{MPa}}{(100-148.8)\ \text{MPa}}\approx 4.098$$

$$2\alpha_0=256.3° \quad \text{or} \quad 2\alpha_0=76.3°$$

$$\alpha_0=128.15° \quad \text{or} \quad \alpha_0=38.15°$$

From equation (11−13), the maximum and minimum normal stresses are

$$\left.\begin{matrix}\sigma_{\max}\\ \sigma_{\min}\end{matrix}\right\}=\frac{100+148.8}{2}\ \text{MPa}\pm\sqrt{\left(\frac{100-148.8}{2}\right)^2+100^2}\ \text{MPa}=\begin{cases}227.3\ \text{MPa}\\ 21.5\ \text{MPa}\end{cases}$$

Therefore, the principal stress is $\sigma_1=2\ 212.3$ MPa, $\sigma_2=21.5$ MPa and $\sigma_3=0$.

Example 11−3 A beam under transverse-force bending is shown in Fig.11−9a. After finding the bending moment M and shear force F_Q in the cross section m—n, the normal and shear stresses at point A of the cross section are $\sigma=-70$ MPa and $\tau=50$ MPa, respectively. Try to determine the principal stress at point A and the orientation of the principal plane, and discuss the stress state at other points on the same section.

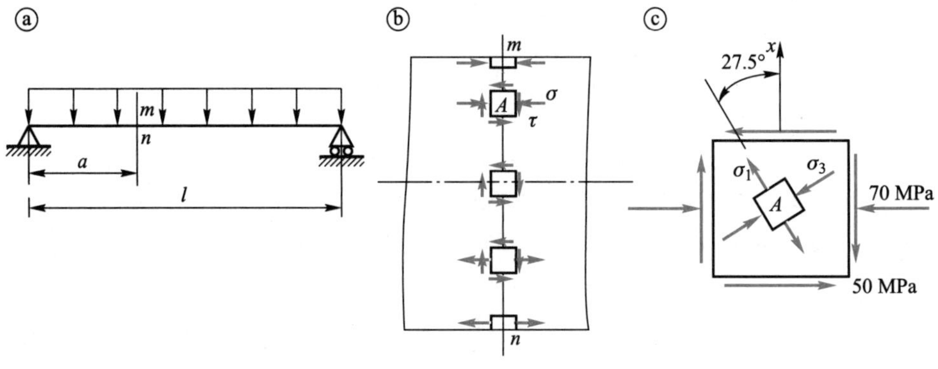

Fig.11−9

Solution The element of point A is enlarged and shown in Fig.11−9c. We get $\sigma_x=0$, $\sigma_y=-70$ MPa, $\tau_{xy}=-50$ MPa. From equation (11−12), we have

$$\tan 2\alpha_0=-\frac{2\tau_{xy}}{\sigma_x-\sigma_y}=-\frac{2\times(-50)\ \text{MPa}}{[0-(-70)]\ \text{MPa}}=1.429$$

$$2\alpha_0=55° \quad \text{or} \quad 2\alpha_0=235°$$

$$\alpha_0=27.5° \quad \text{or} \quad \alpha_0=117.5°$$

From equation (11−13), the maximum and minimum normal stresses are

$$\left.\begin{matrix}\sigma_{\max}\\ \sigma_{\min}\end{matrix}\right\}=\frac{0+(-70)}{2}\ \text{MPa}\pm\sqrt{\left(\frac{0-(-70)}{2}\right)^2+(-50)^2}\ \text{MPa}=\begin{cases}26\ \text{MPa}\\ -96\ \text{MPa}\end{cases}$$

Therefore, the principal stress is $\sigma_1=26$ MPa, $\sigma_2=0$ and $\sigma_3=-96$ MPa. Because $\sigma_x>\sigma_y$, then the stress in the $\alpha_0=27.5°$ plane is $\sigma_{\max}$, while the stress in the $\alpha_0=117.5°$ plane is $\sigma_{\min}$. The principal element and the principal stress are shown in Fig.11−9c.

The stress state at other points can be analyzed in the same way. The points at the upper and lower edges are in uniaxial compression or tension, and the cross section is their principal plane. On the neutral axis, the stress state at each point is in pure shear, and the principal plane is at an angle of 45° to the beam axis. From the upper edge to the lower edge, the stress state at each point is shown in Fig.11-9b.

11.3 Mohr circle for plane stress

11.3.1 Mohr circle

From equation (11-9) and equation (11-10), it can be seen that the stresses σ_α and τ_α in any inclined section are as a function of 2α. In the $\sigma-\tau$ cartesian coordinate system, the curve given by $\sigma_\alpha = f(\tau_\alpha)$ can reflect the stresses on all inclined sections of the element. The equation (11-9) is rewritten as

$$\sigma_\alpha - \frac{\sigma_x+\sigma_y}{2} = \frac{\sigma_x-\sigma_y}{2}\cos 2\alpha - \tau_{xy}\sin 2\alpha$$

The above equation is squared with the left and right sides of equation (11-10) and then added together to obtain

$$\left(\sigma_\alpha - \frac{\sigma_x+\sigma_y}{2}\right)^2 + \tau_\alpha^2 = \left(\frac{\sigma_x-\sigma_y}{2}\right)^2 + \tau_{xy}^2 \qquad (11\text{-}18)$$

If the σ_x, σ_y, τ_{xy} are known, equation (11-18) is an equation of a circle with σ_α, τ_α as variables. In the $\sigma-\tau$ cartesian coordinate system, the coordinates of the center C of this circle are $\left(\frac{\sigma_x+\sigma_y}{2},\ 0\right)$ and the radius is $\sqrt{\left(\frac{\sigma_x-\sigma_y}{2}\right)^2 + \tau_{xy}^2}$. The horizontal and vertical coordinates of any point on the circle represent the normal and shear stresses in the corresponding section, respectively. This circle is called the Mohr circle.

11.3.2 Draw Mohr circle

The following is an example of how to draw the Mohr circle in the plane stress state, as shown in Fig. 11-10a. First, we establish the $\sigma-\tau$ cartesian coordinate system as shown in Fig.11-10b. Measure $\overline{OA} = \sigma_x$ on the σ axis according to the selected scale, make a vertical line through point A and let $\overline{AD} = \tau_{xy}$. By this way, point D is determined. The coordinates of point D represent the stress in the plane with the x axis as normal line. Measure $\overline{OB} = \sigma_y$, make a vertical line through point B, and let $\overline{BD'} = \tau_{yx}$. Since τ_{yx} is negative, the vertical coordinate of D' is also negative. By this way, point D' is determined. The coordinates of point D' represent the stress in the plane with the y axis as normal line.

After connecting points D and D', DD' intersects with the horizontal axis at point C. With point C as the center of the circle, we can make a circle with $\overline{CD}$ or $\overline{CD'}$ as the radius, as shown in Fig.11-10b

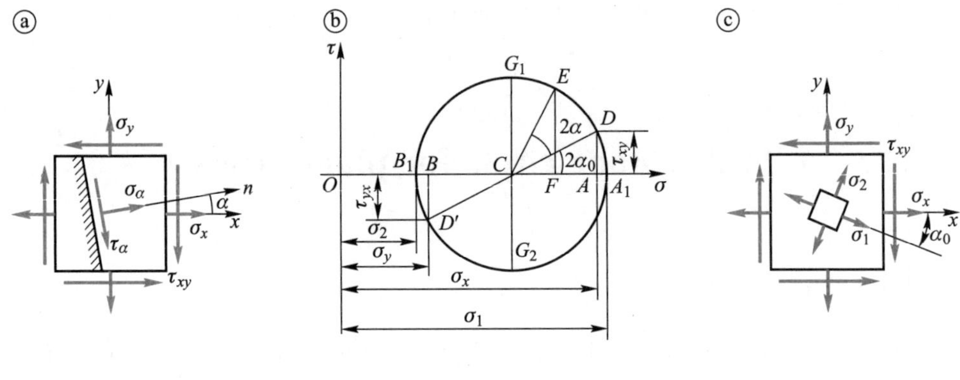

Fig.11-10

The center C of this circle is on the σ axis and its coordinate is 0, horizontal coordinate is

$$\overline{OC}=\frac{1}{2}(\overline{OA}+\overline{OB})=\frac{\sigma_x+\sigma_y}{2} \tag{11-19}$$

The radius of this circle is

$$\overline{CD}=\sqrt{\overline{CA}^2+\overline{AD}^2}=\sqrt{\left(\frac{\sigma_x-\sigma_y}{2}\right)^2+\tau_{xy}^2} \tag{11-20}$$

It can be seen that the circle is the Mohr circle corresponding to equation (11-18).

11.3.3 Application of Mohr circle

A Mohr circle graphically describes the stress at a point in all directions. The Mohr circle can be used to determine the stress in any inclined section of the element by using the following method. In Fig.11-10a, let the angle from the x axis to the normal n of any inclined section be the counterclockwise α angle. On the Mohr circle, from point D turn along the circle in the counterclockwise direction to point E, and make the angle of the center of the circle corresponding to the DE arc is 2α. The coordinates of point E represent the stress on the inclined section normal to n. The proof is as follows.

The horizontal coordinate of point E is

$$\begin{aligned}\overline{OF}&=\overline{OC}+\overline{CE}\cos(2\alpha_0+2\alpha)\\&=\overline{OC}+\overline{CE}\cos 2\alpha_0\cos 2\alpha-\overline{CE}\sin 2\alpha_0\sin 2\alpha\\&=\overline{OC}+\overline{CA}\cos 2\alpha-\overline{AD}\sin 2\alpha\\&=\frac{\sigma_x+\sigma_y}{2}+\frac{\sigma_x-\sigma_y}{2}\cos 2\alpha-\tau_{xy}\sin 2\alpha\end{aligned}$$

The vertical coordinate of point E is

$$\overline{EF}=\overline{CE}\sin(2\alpha_0+2\alpha)$$
$$=\overline{CD}\sin 2\alpha_0\cos 2\alpha+\overline{CD}\cos 2\alpha_0\sin 2\alpha$$
$$=\overline{AD}\cos 2\alpha+\overline{CA}\sin 2\alpha$$
$$=\frac{\sigma_x-\sigma_y}{2}\sin 2\alpha+\tau_{xy}\cos 2\alpha$$

Comparing with equations (11-9) and (11-10), it is known that $\overline{OF}=\sigma_\alpha$ and $\overline{FE}=\tau_\alpha$. The coordinates of point E on the Mohr circle represent the stress in the inclined section.

A_1 and B_1 on the stress circle have maximum and minimum transverse coordinates, respectively. Since the vertical coordinates of both points are zero, the transverse coordinates of these two points are the principal stress in the principal plane

$$\sigma_{max}=\overline{OA_1}=\overline{OC}+\overline{CA_1}=\overline{OC}+\overline{CD}=\frac{\sigma_x+\sigma_y}{2}+\sqrt{\left(\frac{\sigma_x-\sigma_y}{2}\right)^2+\tau_{xy}^2}=\sigma_1$$

$$\sigma_{min}=\overline{OB_1}=\overline{OC}-\overline{CB_1}=\overline{OC}-\overline{CD}=\frac{\sigma_x+\sigma_y}{2}-\sqrt{\left(\frac{\sigma_x-\sigma_y}{2}\right)^2+\tau_{xy}^2}=\sigma_2$$

The location of the principal plane is also easily determined by the Mohr circle. Points D and A_1 on the Mohr circle correspond to the plane normal to the x axis and the principal plane where the σ_1 are located, respectively. Because the angle of the circle between point D and point A_1 is clockwise. In the element, the normal of principal plane of σ_1 is obtained by rotating the x axis α_0 clockwise, as shown in Fig. 11-10b. The normal of the principal plane of σ_2 is perpendicular to it. As seen in Fig.11-10c

$$\tan 2\alpha_0=-\frac{\overline{AD}}{\overline{AC}}=-\frac{2\sigma_{xy}}{\sigma_x-\sigma_y}$$

Make a vertical radius $\overline{CG_1}$ and $\overline{CG_2}$ through point C, the vertical coordinates of points G_1 and G_2 are the maximum and minimum shear stresses respectively. So we have

$$\tau_{max}=\overline{CG_1}=\overline{CD}=\sqrt{\left(\frac{\sigma_x-\sigma_y}{2}\right)^2+\tau_{xy}^2}$$

$$\tau_{min}=-|\overline{CG_2}|=-|\overline{CD}|=-\sqrt{\left(\frac{\sigma_x-\sigma_y}{2}\right)^2+\tau_{xy}^2}$$

Because the absolute values of τ_{max} and τ_{min} are equal to the radius of the Mohr circle, we get

$$\left.\begin{matrix}\tau_{max}\\ \tau_{min}\end{matrix}\right\}=\pm\frac{\sigma_1-\sigma_2}{2} \qquad (11\text{-}21)$$

Point A_1 and G_1 corresponds to σ_1 and τ_{max}, respectively. Since the angle from point A_1 to the point G_1 is 90° counterclockwise, the normal of the σ_1 plane to the normal of τ_{max} plane is 45° counterclockwise.

Example 11-4 Analyze normal stress σ_α and the shear stress τ_α of each element shown in

Fig.11−11a,b using the Mohr circle method.

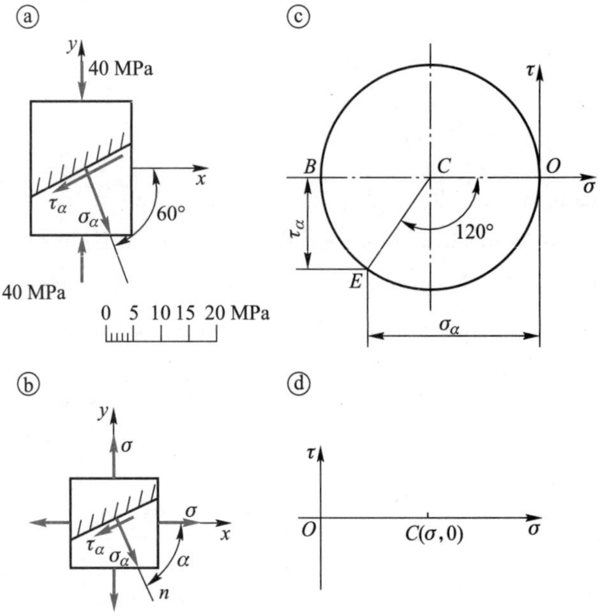

Fig.11−11

Solution For the element shown in Fig.11−11a, draw the point with coordinates ($\sigma_x=0, \tau_{xy}=0$), which is the origin of the coordinates O. Then, plot point B coordinates ($\sigma_y=-40$ MPa, $\tau_{xy}=0$). Plot a Mohr circle with OB as the diameter as shown in Fig.11−11c.

In the element, the angle from the x axis to the normal of inclined section is 60° clockwise. The stress circle should be measured clockwise from point O along the circumference of the circle at an angle of 120° to determine point E. The coordinate of point E is the stress in the inclined section. It is possible to measure $\sigma_\alpha=-30$ MPa, $\tau_\alpha=-112.4$ MPa.

For the element shown in Fig.11−11b, the point of the section normal to the x axis and the point of the section normal to the y axis are the same point $C(\sigma,0)$. The Mohr circle metamorphoses into a point circle, as shown in Fig.11−11d. In this case, the stress on any inclined section is $(\sigma_\alpha, \tau_\alpha)=(\sigma, 0)$.

Example 11−5 In the original element shown in Fig.11−12a, it have the $\sigma_x=80$ MPa, $\sigma_y=-40$ MPa and $\tau_{xy}=-60$ MPa. Try to find the principal stresses and the location of the principal plane using the Mohr circle method.

Solution At the selected scale, point D is plotted in the $\sigma-\tau$ cartesian coordinate system with $\sigma_x=80$ MPa and $\tau_{xy}=-60$ MPa. Point D' is plotted with $\sigma_y=-40$ MPa and $\tau_{yx}=60$ MPa. DD' intersects with the horizontal axis at point C and plot a stress circle with radius $\overline{CD}$. In Fig.11−12b, we have

$$\sigma_1=\overline{OA_1}=105\ \text{MPa}$$

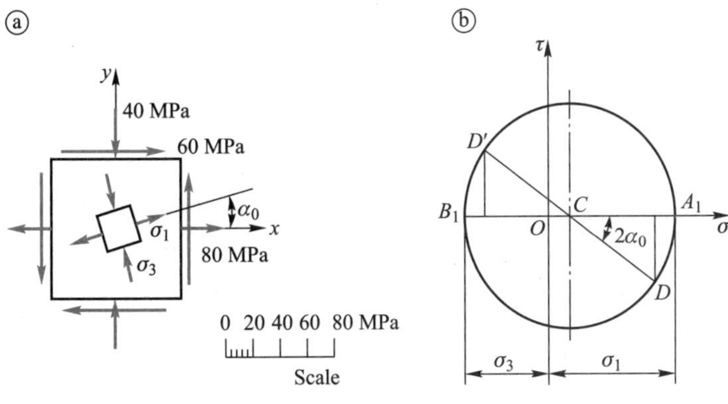

Fig.11-12

$$\sigma_3=-\overline{OB_1}=-65\ \text{MPa}$$

Here, another principal stress $\sigma_2=0$. The angle of the circle measured from point D to point A_1 is equal to twice the angle between the σ_1 plane and the x axis. The measured angle is

$$\angle DCA_1=2\alpha_0=45°$$

Therefore $\alpha_0=22.5°$.

11.4 Triaxial stress state

In this section, only the special case of the triaxial stress state is briefly analyzed. The triaxial stress state with three principal stresses ($\sigma_1 \geqslant \sigma_2 \geqslant \sigma_3 \neq 0$) acting on each of the three opposite sides of the element is studied, as shown in Fig.11-13a.

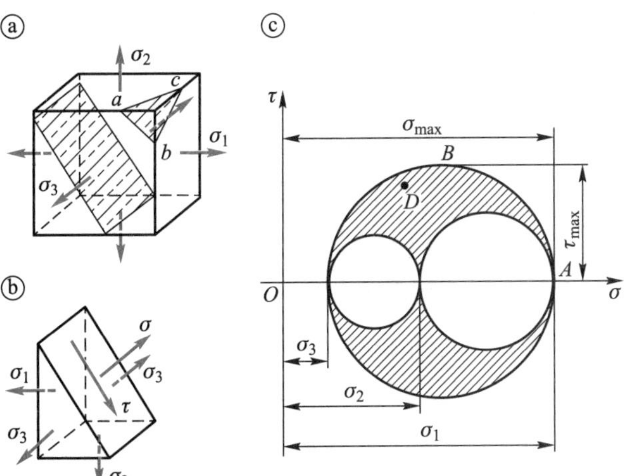

Fig.11-13

Let us first study the stresses on an arbitrary inclined section parallel to the principal stress σ_3. Imagine using an arbitrary section parallel to σ_3 to cut the element and taking any part of it to study, as shown in Fig.11-13b.

Both the normal stress σ and the shear stress τ in the inclined section are independent of σ_3 and are determined by the principal stresses σ_1 and σ_2. Thus, when the Mohr circle method is used for the analysis, the points corresponding to the inclined sections of this class are located on the stress circle determined by σ_1 and σ_2. Similarly, stress circle made by σ_2 and σ_3 represents the stress on each inclined section parallel to the σ_1, and the stress circle made by σ_3 and σ_1 represents the stress on each inclined section parallel to σ_2. Further study proves that point D, which represents the stress on any inclined section (such as *abc* in Fig. 11 - 13a) that intersects obliquely with the three principal stresses, must be located within the shaded area enclosed by the above three stress circles. It follows that the points representing the stresses in any cross section of the element in the $\sigma - \tau$ cartesian coordinate system must lie on the circumference of the three stress circles and within the shaded area enclosed by them, as shown in Fig.11-13c.

According to the above analysis, the maximum normal stress in the triaxial stress state should be equal to the transverse coordinate of point A on the maximum Mohr circle, i.e.

$$\sigma_{max}=\sigma_1 \tag{11-22}$$

The maximum shear stress is equal to the vertical coordinate of point B on the maximum Mohr circle, that is

$$\tau_{max}=\frac{\sigma_1-\sigma_3}{2} \tag{11-23}$$

From the position of point B, it is known that the section where the maximum shear stress is located is perpendicular to σ_2 plane. Moreover, this plane is at an angle of 45° to both σ_1 plane and σ_3 plane.

The above conclusions also apply to plane stress states. If a plane stress state of $\sigma_1>0$, $\sigma_2=0$, $\sigma_3<0$, the maximum normal and maximum shear stresses at that point are expressed as in equations (11-22) and (11-23). If the principal stress is $\sigma_1 \geqslant \sigma_2>0$, $\sigma_3=0$, the expression of normal stress remains unchanged. According to the equation (11-23), the expression of maximum shear stress should be

$$\tau_{max}=\frac{\sigma_1}{2} \tag{11-24}$$

This maximum shear stress is clearly larger than $\tau=\frac{\sigma_1-\sigma_2}{2}$ which is obtained from equation (11-21) in section 11.3. Because only planes parallel to σ_3 planes are investigated. The maximum value of shear stress in such planes is also $\frac{\sigma_1-\sigma_2}{2}$. If we also consider those planes parallel to σ_2 planes, we will obtain the maximum shear stress expressed by equation (11-24).

Example 11-6 A point is in a triaxial stress state, and its element is shown in the Fig.11-14. Find its principal stress and maximum shear stress.

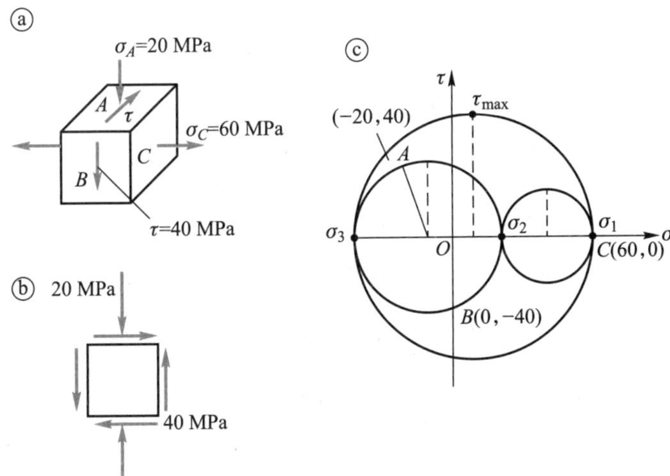

Fig.11−14

Solution The element is in a triaxial stress state shown in Fig.11−14a. One principal plane and the principal stress on that face are known. The other two principal stresses can be found in a similar way to the plane stress state shown in Fig.11−14b.

Take two points $A(-20,40)$, $B(0,-40)$ in the $\sigma-\tau$ cartesian coordinate system, corresponding to the stresses on both sides of the element A, B respectively, as shown in Fig.11−14c. With AB as the diameter of the Mohr circle, the two intersection points of the Mohr circle on the σ axis are the other two principal stresses, which are 31 MPa and −51 MPa, respectively. Since the known principal stress is 60 MPa, we have

$$\sigma_1 = 60\ \text{MPa},\quad \sigma_2 = 31\ \text{MPa},\quad \sigma_3 = -51\ \text{MPa}$$

Then make two other Mohr circles. The maximum shear stress is on the Mohr circle of the largest diameter, we get

$$\tau_{\max} = 55.5\ \text{MPa}$$

11.5 Generalized Hooke law and volumetric strain

11.5.1 Generalized Hooke law in its general form

In uniaxial tensile or compressive deformation, the relationship between stress and strain in the linear elastic range is

$$\sigma = E\varepsilon \quad \text{or} \quad \varepsilon = \frac{\sigma}{E} \tag{11-25}$$

The equation (11−25) is the tension (pressure) Hooke law. The transverse deformation caused by the axial force is also an elastic deformation that cannot be ignored. Transverse strain ε' can be expressed as

$$\varepsilon' = -\mu\varepsilon = -\mu \frac{\sigma}{E} \tag{11-26}$$

The experimental results show that the relationship between shear stress and shear strain obeys Hooke law in shear when the shear stress does not exceed the shear proportional limit

$$\gamma = \frac{\tau}{G} \tag{11-27}$$

In general, nine stress components are required to describe the stress state at a point, as shown in Fig.11-15. Considering the theorem of complementary shearing stresses, the τ_{xy} and τ_{yx}, τ_{yz} and τ_{zy}, τ_{zx} and τ_{xz} are equal in value, respectively. Thus, only 6 of the 9 stress components are independent. The normal strain at a point for an isotropic material is only related to the normal stress at that point, but not to the shear stress. At the same time, the shear strain at that point is only related to the shear stress. These two types of relationships are investigated separately. First discuss the relationship between the normal strain ε_x in the x-direction and the normal stress σ_x, σ_y and σ_z. The normal strain in the x-direction caused by σ_x alone is

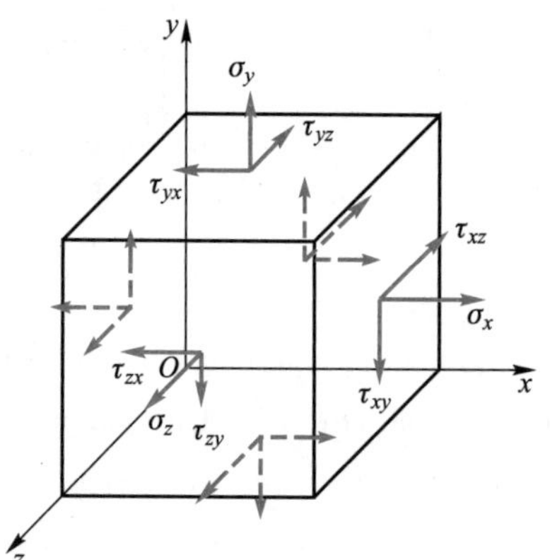

Fig.11-15

$$\varepsilon_x' = \frac{\sigma_x}{E}$$

The normal strains caused by σ_y and σ_z in the x-direction are

$$\varepsilon_x'' = -\mu \frac{\sigma_y}{E}, \quad \varepsilon_x''' = -\mu \frac{\sigma_z}{E}$$

The normal strain in the x-direction can be obtained by summing the above 3 equations

$$\varepsilon_x = \varepsilon_x' + \varepsilon_x'' + \sigma_x''' = \frac{\sigma_x}{E} - \mu \frac{\sigma_y}{E} - \mu \frac{\sigma_z}{E} = \frac{1}{E}[\sigma_x - \mu(\sigma_y + \sigma_z)] \tag{11-28}$$

Similarly, we can get the normal strain in y and z direction. Finally, we get

$$\left.\begin{aligned} \varepsilon_x &= \frac{1}{E}[\sigma_x - \mu(\sigma_y + \sigma_z)] \\ \varepsilon_y &= \frac{1}{E}[\sigma_y - \mu(\sigma_z + \sigma_x)] \\ \varepsilon_z &= \frac{1}{E}[\sigma_z - \mu(\sigma_x + \sigma_y)] \end{aligned}\right\} \tag{11-29}$$

As for the relationship between shear stress and shear strain, it still obeys the Hooke law in shear

$$\gamma_{xy} = \frac{\tau_{xy}}{G}, \quad \gamma_{yz} = \frac{\tau_{yz}}{G}, \quad \gamma_{zx} = \frac{\tau_{zx}}{G} \tag{11-30}$$

Equation (11-29) and equation (11-30) are called the generalized Hooke law in its general form.

When all faces of the element are in the principal plane, the stress on the element is only the principal stress. Let the axis direction be the same as $\sigma_1, \sigma_2, \sigma_3$ direction, then $\sigma_x = \sigma_1$, $\sigma_y = \sigma_2$, $\sigma_z = \sigma_3$, $\tau_{xy} = \tau_{yz} = \tau_{zx} = 0$, the generalized Hooke law becomes

$$\left.\begin{aligned}\varepsilon_1 &= \frac{1}{E}[\sigma_1 - \mu(\sigma_2 + \sigma_3)] \\ \varepsilon_2 &= \frac{1}{E}[\sigma_2 - \mu(\sigma_3 + \sigma_1)] \\ \varepsilon_3 &= \frac{1}{E}[\sigma_3 - \mu(\sigma_1 + \sigma_2)]\end{aligned}\right\} \tag{11-31}$$

Here, $\varepsilon_1, \varepsilon_2, \varepsilon_3$ represent the strains along the three principal stress directions and are called principal strains. They satisfy the relationship $\varepsilon_1 \geqslant \varepsilon_2 \geqslant \varepsilon_3$. Equation (11-31) is the generalized Hooke law expressed in terms of principal stresses.

In the plane stress state shown in Fig.11-6a, substituting $\sigma_z = \tau_{yz} = \tau_{zx} = 0$ into equation (11-29) and equation (11-30), the non-zero strain component is obtained

$$\left.\begin{aligned}\varepsilon_x &= \frac{1}{E}(\sigma_x - \mu\sigma_y) \\ \varepsilon_y &= \frac{1}{E}(\sigma_y - \mu\sigma_x) \\ \varepsilon_z &= -\frac{\mu}{E}(\sigma_x + \sigma_y) \\ \gamma_{xy} &= \frac{\tau_{xy}}{G}\end{aligned}\right\} \tag{11-32}$$

Using strain to express stress, equation (11-32) becomes

$$\left.\begin{aligned}\sigma_x &= \frac{E}{1-\mu^2}(\varepsilon_x + \mu\varepsilon_y) \\ \sigma_y &= \frac{E}{1-\mu^2}(\varepsilon_y + \mu\varepsilon_x) \\ \tau_{xy} &= G\gamma_{xy}\end{aligned}\right\} \tag{11-33}$$

Equation (11-32) and equation (11-33) are Hooke law for the plane stress state.

Example 11-7 The slotted rigid body shown in Fig.11-16a has a square steel block placed inside it. The modulus of elasticity of the steel is known to be $E = 200$ GPa, Poisson ratio is $\mu = 0.3$, and the top surface of the steel block is subjected to a combined force $F = 8$ kN of uniform pressure. Try to find the three principal stresses in the block.

Solution Under the action of pressure F, the steel block is directly compressed in addition to the top surface. The lateral compressive stress σ_x is simultaneously induced due to its lateral deformation obstruction. The steel block is in a plane stress state, as shown in Fig.11-16b. It satisfy the condition

$$\varepsilon_x = 0$$

The compressive stress on the top surface of the steel block is

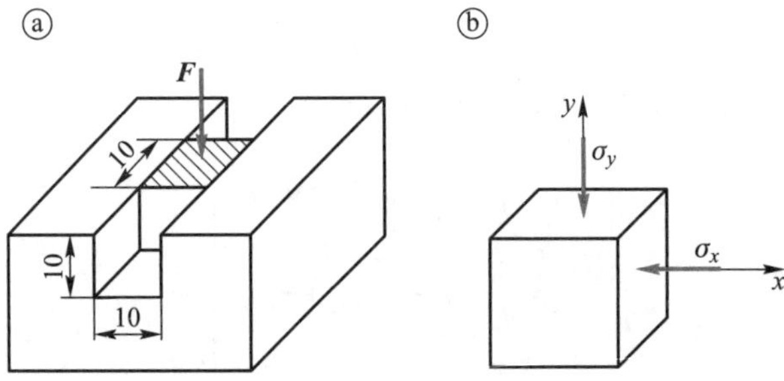

Fig.11-16

$$\sigma_y = -\frac{F}{A} = -\frac{8\times10^3}{0.01^2}\ \text{Pa} = -80\ \text{MPa}$$

From equation (11-32), we know that

$$\varepsilon_x = \frac{\sigma_x}{E} - \mu\frac{\sigma_y}{E} = 0$$

We get

$$\sigma_x = \mu\sigma_y = 0.3\times(-80)\ \text{MPa} = -24\ \text{MPa}$$

It can be seen that the three principal stresses of the steel block are

$$\sigma_1 = 0,\quad \sigma_2 = -24\ \text{MPa},\quad \sigma_3 = -80\ \text{MPa}$$

11.5.2 Volumetric strain

As shown in Fig.11-17, let the lengths of the three edges of a, b, c before deformation be $\mathrm{d}x, \mathrm{d}y, \mathrm{d}z$, then the volume of the element is

$$V_0 = \mathrm{d}x\mathrm{d}y\mathrm{d}z$$

If the normal strains of the three prismatic edges of a, b, c after deformation are ε_1, ε_2, ε_3, the volume of the element becomes

$$V = \mathrm{d}x(1+\varepsilon_1)\cdot\mathrm{d}y(1+\varepsilon_2)\cdot\mathrm{d}z(1+\varepsilon_3)$$

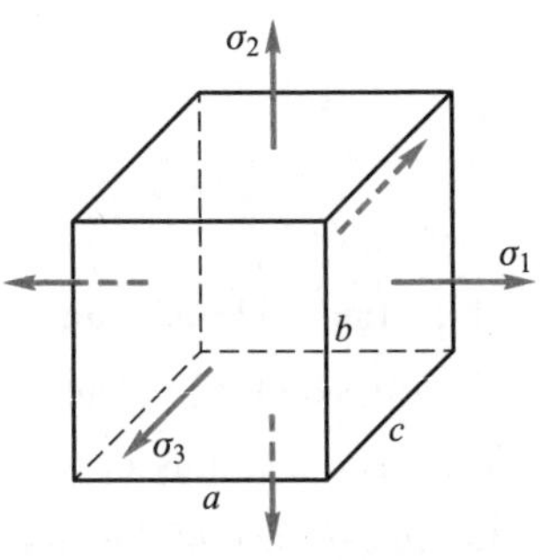

Fig.11-17

Expanding the above equation and noting that the higher-order differential is negligible in the case of small deformations, we have

$$V = V_0(1+\varepsilon_1+\varepsilon_2+\varepsilon_3)$$

Therefore, the change per unit volume is

$$\Theta = \frac{V-V_0}{V_0} = \varepsilon_1+\varepsilon_2+\varepsilon_3$$

Θ is called the volumetric strain. Substituting equation (11-31) into the above equation, we get

$$\Theta = \frac{1-2\mu}{E}(\sigma_1+\sigma_2+\sigma_3) \tag{11-34}$$

Introduction of symbols

$$K=\frac{E}{3(1-2\mu)} \tag{11-35}$$

$$\sigma_m=\frac{1}{3}(\sigma_1+\sigma_2+\sigma_3) \tag{11-36}$$

Then equation (11−34) becomes

$$\Theta=\frac{\sigma_m}{K} \tag{11-37}$$

where K is called the bulk modulus of elasticity; σ_m is called the average principal stress. From equation (11−37), it can be seen that the volumetric strain Θ is proportional to the average principal stress σ_m which is the volume Hooke law. Equation (11−37) also shows that the change per unit volume Θ is only related to the sum of the three principal stresses, but not to the ratio between the three principal stresses. Therefore, the unit volume change is the same whether the element is acting on three unequal principal stresses or their average principal stresses.

11.6 Strain energy density of triaxial stress

11.6.1 Strain energy density

The elastic strain energy will be accumulated inside the object which is deformed by external forces. The elastic strain energy within each unit volume of the object is called elastic strain energy density. In uniaxial tension or compression, the strain energy is numerically equal to the work done by the external force. By using the Hooke law, the formula of elastic strain energy is

$$\mu=\frac{1}{2}\sigma\varepsilon \tag{11-38}$$

In a complex stress state, the strain energy of the object is still numerically equal to the work done by the external force. When its deformation is small, the strain energy accumulated in the object depends on the final value of the external force and is independent of the order of force application.It is assumed that all the external forces increase from zero to the final value in the same proportion, so that the stresses on each face of element also increase from zero to the final value in the same proportion. Now we study a triaxial stress element. In the linear elastic case, the relationship between each principal stress and its corresponding principal strain is still linear, which is the same as the relationship between σ and ε in the uniaxial stress state. Thus, the strain energy density corresponding to each principal stress can be calculated according to equation (11−38). The strain energy density in the triaxial stress state is calculated as

$$u=\frac{1}{2}\sigma_1\varepsilon_1+\frac{1}{2}\sigma_2\varepsilon_2+\frac{1}{2}\sigma_3\varepsilon_3 \tag{11-39}$$

Substituting equation (11−31) into the above equation, we get

$$u=\frac{1}{2E}[\sigma_1^2+\sigma_2^2+\sigma_3^2-2\mu(\sigma_1\sigma_2+\sigma_2\sigma_3+\sigma_3\sigma_1)] \tag{11-40}$$

11.6.2 Volume change energy density and distortion energy density

Generally speaking, when an object is deformed, it contains both volume change and shape change. Therefore, the total strain energy density should be composed of two parts

$$u=u_t+u_x \tag{11-41}$$

where u_t is the strain energy density corresponding to the volume change; u_x is the strain energy density corresponding to the shape change, and is called distortion energy density or distortion specific energy.

For the principal element of the triaxial stress state shown in Fig. 11-18a, the three principal stresses on the element are not equal. The principal strains corresponding to σ_1, σ_2 and σ_3 are ε_1, ε_2, ε_3, and the volumetric strain is Θ.

The average stress $\sigma_m=\dfrac{\sigma_1+\sigma_2+\sigma_3}{3}$ is introduced. The principal stresses on the element is decomposed into two parts: $(\sigma_m, \sigma_m, \sigma_m)$ and $(\sigma_1-\sigma_m, \sigma_2-\sigma_m, \sigma_3-\sigma_m)$. Thus, the stress state shown in Fig.11-18a is decomposed into the superposition of the stress states shown Fig.11-18b and Fig.11-18c.

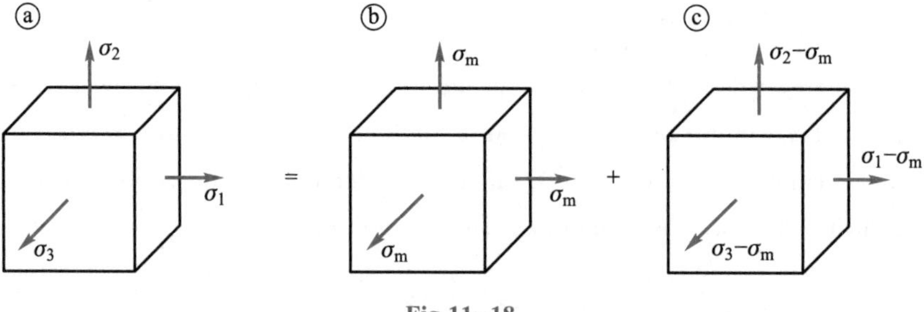

Fig.11-18

The stress state shown in Fig. 11-18b is a triaxial isotropic stress state in which the element produces only volume change without shape change. It is known from Section 11.5 that the volume strain at this time is equal to the volume strain of the original element Θ. Thus, the strain energy density of this case is also u_t of the original element, so

$$u_t=\frac{1}{2}\sigma_m\varepsilon_m+\frac{1}{2}\sigma_m\varepsilon_m+\frac{1}{2}\sigma_m\varepsilon_m=\frac{3}{2}\sigma_m\varepsilon_m \tag{11-42}$$

By the Hooke law

$$\varepsilon_m=\frac{1}{E}[\sigma_m-\mu(\sigma_m+\sigma_m)]=\frac{(1-2\mu)}{E}\sigma_m$$

Substituting it into equation (11-42), we get

$$u_t=\frac{3(1-2\mu)}{2E}\sigma_m^2=\frac{3(1-2\mu)}{2E}\left(\frac{\sigma_1+\sigma_2+\sigma_3}{3}\right)^2=\frac{1-2\mu}{6E}(\sigma_1+\sigma_2+\sigma_3)^2 \tag{11-43}$$

In the stress state shown in Fig.11−18c, there is no volume change in the element and only shape change. The strain energy density in this case is the distortion energy density. Substituting equation (11−43) and equation (11−40) into equation (11−41), we get

$$u_x = u - u_t = \frac{1}{2E}[\sigma_1^2+\sigma_2^2+\sigma_3^2-2\mu(\sigma_1\sigma_2+\sigma_2\sigma_3+\sigma_3\sigma_1)] - \frac{1-2\mu}{6E}(\sigma_1+\sigma_2+\sigma_3)^2$$

After simplification, the distortion energy density is

$$u_x = \frac{1+\mu}{6E}[(\sigma_1-\sigma_2)^2+(\sigma_2-\sigma_3)^2+(\sigma_3-\sigma_1)^2] \tag{11-44}$$

Example 11−8 Fig.11−19 shows a element in a pure shear stress state with isotropic material. Try to prove that the following relationship exists between the three material constants E, G and μ. The following relationship exists between

$$G = \frac{E}{2(1+\mu)} \tag{11-45}$$

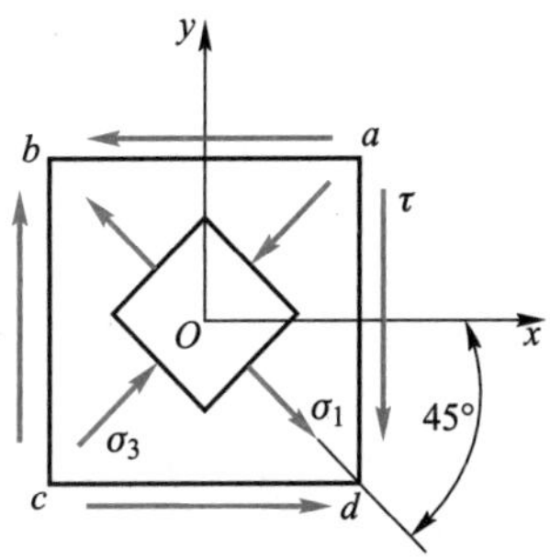

Fig.11−19

Solution Let the length of each side of the element be dx, dy and t. γ is the shear strain corresponding to the shear stress τ. The shear force of the left and right faces of the element are $\tau t dy$. The misalignment of the left and right sides of the element due to shear is γdx. The stored strain energy dU is equal to the microwork done by the shear force $\tau t dy$ on the misalignment γdx. In the elastic range, there is

$$dU = \frac{1}{2}\tau t dy \cdot \gamma dx \tag{11-46}$$

The above equation is removed by the volume $tdxdy$ to obtain the strain energy density of the element

$$\mu_1 = \frac{1}{2}\tau\gamma = \frac{\tau^2}{2G} \tag{11-47}$$

The principal stresses on the element in the pure shear stress state are

$$\sigma_1 = \tau, \quad \sigma_2 = 0, \quad \sigma_3 = -\tau \tag{11-48}$$

The principal element is also shown in Fig.11−19. Substituting equation (11−48) into equation (11−40), we get

$$u_2 = \frac{1}{2E}(\tau^2+\tau^2+2\mu\tau^2) = \frac{1+\mu}{E}\tau^2 \tag{11-49}$$

From $u_1 = u_2$ yields

$$\frac{1}{2G}\tau^2 = \frac{1+\mu}{E}\tau^2 \tag{11-50}$$

So

$$G = \frac{E}{2(1+\mu)}$$

11.7 Theories of strength

11.7.1 Forms of material failure

The failure behavior under room temperature and static load due to insufficient strength includes the following two forms.

(1) Plastic yielding (plastic flow)

Mild steel is in a uniaxial stress state when the material is in simple tension. The yielding phenomenon occurs when the normal stress reaches the yield stress σ_s, The stress remains essentially constant while the tensile deformation continues to increase. Strain at the end of yield can sometimes reach 2%. In addition, when a thin-walled cylinder of mild steel is twisted, the material is in a plane stress state (pure shear), and yielding occurs when the shear stress reaches the shear yield stress. Once yielding begins, the material proceeds to the plastic deformation stage. At this point, the member can no longer work properly because the deformation is too large and most of the deformation is unrecoverable plastic deformation. This situation is a form of material failure.

(2) Brittle fracture

Cast iron breaks along the cross section (surface of maximum tensile stress) in uniaxial tension. Cast iron circular shafts break in torsion at an inclination of 45° to the axis of the bar (surface of maximum tensile stress). In these cases, there is no significant plastic deformation before fracture. Another form of material failure is the brittle fracture of a plastic material such as mild steel, which can occur without significant plastic deformation under a triaxial equivalent tensile stress.

11.7.2 Theorise of strength

The strength conditions in the uniaxial stress state are established directly from the results of the tensile test. The ultimate stress of the material is determined by the tensile test. The bar will yield or fracture when the stress reaches the ultimate stress of the material.

In complex stress states, it is not possible to know when a material fails and how to establish the failure criterion by experiment alone. The failure of a material at a defined state of stress is related not only to the magnitude of the principal stresses σ_1, σ_2, σ_3, but also with their ratio. It was thought to make some hypotheses about the theories of failure based on limited experimental data. It is believed that no matter what stress state, as long as the same form of failure occurs, the causes of failure and the factors leading to failure are the same. In this way, it is possible to use the results of simple experiments to predict the failure behavior of materials under various stresses and to establish failure criteria. Such a hypothesis is called a theories of strength. Based on the theories of strength, the corresponding strength conditions can be established.

The quantities that measure the deformation are mainly stress, strain and strain energy. They are the main factors considered by various strength theories. Since strength theory is a hypothesis, it must be tested by practice.

This chapter presents only four commonly used strength theories.There are far more than these four strength theories, but the various existing strength theories do not yet satisfactorily solve all strength problems. This area still needs to be developed.

11.7.3 Four commonly used strength theories

Since there are two forms of material damage, brittle fracture and plastic flow, there are different damage mechanisms. Therefore, strength theories are also divided into two categories. One is to explain the brittle fracture of materials, commonly used are the maximum tensile stress theory and maximum elongation normal strain theory. One is to explain the plastic yield damage of materials, commonly used are the maximum shear stress theory and maximum distortion energy theory. They are described separately below.

(1) Maximum tensile stress criterion (first theories of strength)

The maximum tensile stress σ_1 is the main factor causing the brittle fracture of the material. Whether it is a uniaxial stress state or a complex stress state, the material will fracture if the maximum tensile stress σ_1 reaches the limit value of the material σ^0. The limit value of the maximum tensile stress is intrinsic properties of material and independent of the stress state. It can be determined by simple experiments. The uniaxial tensile test is the simplest and most basic test, which is usually used to determine the limit value of such materials. Fracture of the material occurs when the maximum tensile stress σ_1 reaches the strength limit σ_b of the material in the case of axial tension. Therefore, according to the first theories of strength, the fracture criterion of the material is

$$\sigma_1 = \sigma_b \tag{11-51}$$

In equation (11-51), we have $\sigma_1 > 0$. The allowable stress $[\sigma]$ is obtained by dividing σ_b by the factor of safety. The strength condition established by the first strength theory is

$$\sigma_1 \leqslant [\sigma] \tag{11-52}$$

Experiments have shown that the maximum tensile stress theory is very close to the experimental results when brittle materials are subjected to uniaxial, biaxial or triaxial tension. For the presence of compressive stresses, the theory is also applicable if the maximum compressive stress value does not exceed the maximum tensile stress value or does not exceed it by much. This theory does not consider the effects of the other two principal stresses and cannot be applied to states without tensile stresses.

(2) Maximum tensile strain criterion (second theories of strength)

This theory believes that the maximum elongation normal strain is the main factor causing brittle fracture of the material. Whether it is a uniaxial stress state or a complex stress state, the material will fracture if the elongation normal strain ε_1 reaches the limit value of the material ε^0. The limit value of the maximum elongation normal strain is intrinsic properties of material and independent of the stress state. It is determined by the uniaxial tensile test. In the case of an axially stretched material, it is assumed that the ultimate normal strain of the material can still be calculated by Hooke law until fracture occurs, then we have

$$\varepsilon^0=\frac{\sigma_b}{E} \tag{11-53}$$

From the Hooke law, the maximum elongation normal strain of the member under any stress state is

$$\varepsilon_1=\frac{1}{E}[\sigma_1-\mu(\sigma_2+\sigma_3)]$$

According to the second theories of strength, the fracture criterion of the material is

$$\sigma_1-\mu(\sigma_2+\sigma_3)=\sigma_b \tag{11-54}$$

Divide σ_b by the factor of safety to obtain the allowable stress $[\sigma]$ and the strength condition established by the second theories of strength is

$$\sigma_1-\mu(\sigma_2+\sigma_3)\leqslant[\sigma] \tag{11-55}$$

When brittle materials are mainly subjected to compressive stress, this theory is basically consistent with the experimental results. When brittle materials such as stone or concrete are subjected to axial compression, if lubricant is added to the contact surface of the experimental machine and the specimen to reduce the effect of friction, the specimen will fracture and damage along the direction perpendicular to the pressure. This direction is the same as the maximum elongation normal strain. However, in uniaxial compression, the stress value in the direction of the maximum elongation normal strain is zero. The cross section is not subject to the action of the force and the material fractures along this cross section does not seem to be in line with the mechanics of common sense. The second strength theory is only suitable for formal explanation or supplementation of brittle fracture damage when tensile stresses do not exist (first strength theory is not available).

(3) Maximum shear stress theory (third theories of strength)

This theory believes that the maximum shear stress τ is the main factor causing plastic yielding damage of the material. Whether it is a uniaxial stress state or a complex stress state, the material will undergo plastic yielding damage if the maximum shear stress τ_{max} reaches the limit value of the material τ^0. The Limit shear stress is intrinsic properties of material and independent of the stress state. It can be obtained by uniaxial tensile experiments. In the case of axial tension, when the tensile stress in the cross section reaches the yield stress σ_s of the material, the maximum shear stress occurs in the inclined section at an angle of 45° to the axis is

$$\tau^0=\frac{\sigma_s}{2} \tag{11-56}$$

In the complex stress state, from equation (11-23) we have

$$\tau_{max}=\frac{\sigma_1-\sigma_3}{2}$$

Thus the plastic yield criterion corresponding to the third theories of strength is

$$\sigma_1-\sigma_3=\sigma_s \tag{11-57}$$

Equation (11-57) is also known as the yielding criterion of Tresca. σ is divided by the factor of safety to obtain the allowable stress $[\sigma]$. The strength condition established by the third theories of strength is

$$\sigma_1-\sigma_3\leqslant[\sigma] \tag{11-58}$$

This theory can better explain the phenomenon of plastic yielding in plastic materials. For example, when mild steel is stretched, slip lines appear in an inclined section at an angle of 45° to the axis. The maximum shear stress occurs in this section. The experiments show that for plastic materials, the maximum shear stress theory is in better agreement with the experimental results and is more safe. The shortcoming of this theory is that it does not take into account the principal stress σ_2 (or rather, the effect of other principal shear stresses) and is only applicable to materials with the same tensile and compressive yield stresss.

(4) Maximum distortion energy theory (fourth theories of strength)

This theory believes that the distortion energy density is the main factor causing yield damage. Whether it is a uniaxial stress state or a complex stress state, the material will undergo plastic yielding damage if the distortion energy density u_x reaches the limit value of u_x^0. The ultimate distortion energy density is intrinsic properties of material and independent of the stress state. It is given by equation (11-44)

$$u_x=\frac{1+\mu}{6E}[(\sigma_1-\sigma_2)^2+(\sigma_2-\sigma_3)^2+(\sigma_3-\sigma_1)^2]$$

In the case of axial tension, when the tensile stress in the cross section reaches the yield stress of the material σ_s, there is $\sigma_1=\sigma_s$, $\sigma_2=\sigma_3=0$. The corresponding ultimate distortion energy density is

$$u_x^0=\frac{1+\mu}{6E}(2\sigma_s^2)$$

Thus, the fourth theories of strength is

$$\sqrt{\frac{1}{2}[(\sigma_1-\sigma_2)^2+(\sigma_2-\sigma_3)^2+(\sigma_3-\sigma_1)^2]}=\sigma_s \tag{11-59}$$

Equation (11-59) is also known as the yielding criterion of Mises. The allowable stress $[\sigma]$ is obtained by dividing σ_s by the factor of safety. The strength condition established by the fourth strength theory is

$$\sqrt{\frac{1}{2}[(\sigma_1-\sigma_2)^2+(\sigma_2-\sigma_3)^2+(\sigma_3-\sigma_1)^2]}\leqslant[\sigma] \tag{11-60}$$

This theory is in agreement with the experimental results that plastic materials do not break even under very high hydrostatic pressure. It takes into account the effect of each principal stress on the yield strength in a more comprehensive way than the third theories of strength.

The yield stresses predicted by the third and fourth strength theories for the uniaxial tensile and compressive stress states are the same (this is related to the fact that each theories of strength is calibrated by tensile experiments). The yield stress results for the pure shear stress state have the largest deviation of about 15%. This deviation cannot be considered to be significant in terms of the accuracy of the strength theories. Both theories are also consistent in their scope of application and can explain the usual yield damage phenomena. Experiments show that the fourth theories of strength is closer to the measured results than the third theories of strength in most cases. But the third theories of strength is safer than the fourth theories of strength.

The strength conditions for the four strength theories can be written in the following unified form

$$\sigma_{xdi} \leqslant [\sigma] \tag{11-61}$$

where σ_{xdi} is called the equivalent stress. The equivalent stresses corresponding to the four strength theories are:

First theories of strength

$$\sigma_{xd1} = \sigma_1 \tag{11-62}$$

Second theories of strength

$$\sigma_{xd2} = \sigma_1 - \mu(\sigma_2 + \sigma_3) \tag{11-63}$$

Third theories of strength

$$\sigma_{xd3} = \sigma_1 - \sigma_3 \tag{11-64}$$

Fourth theories of strength

$$\sigma_{xd4} = \sqrt{\frac{1}{2}[(\sigma_1-\sigma_2)^2+(\sigma_2-\sigma_3)^2+(\sigma_3-\sigma_1)^2]} \tag{11-65}$$

In summary, when establishing the strength conditions of a member in a complex stress state according to theories of strength, the form is to compare a certain integrated value of the three principal stresses with the allowable stress uniaxial tension. The strength problem of the complex stress state is expressed as the strength problem of the uniaxial stress state. The equivalent stress σ_{xdi} is the stress of a uniaxial tensile stress state equivalent to the complex stress state in terms of material damage or failure.

Example 11-9 The thin-armed cylindrical container shown in Fig. 11-4 is known to have a maximum internal pressure of p. The inner diameter of the cylinder is D, the thickness is $t(t \ll D)$ and the allowable stress of the material is $[\sigma]$. Try to establish the strength conditions of the container according to the fourth theories of strength.

Solution From the calculation in 11.1 section, the three principal stresses in the cylinder wall are

$$\sigma_1 = \sigma'' = \frac{pD}{2t}, \quad \sigma_2 = \sigma' = \frac{pD}{4t}, \quad \sigma_3 = 0 \tag{11-66}$$

Substituting into equation (11-65), we get

$$\sigma_{xd4} = \sqrt{\frac{1}{2}[(\sigma_1-\sigma_2)^2+(\sigma_2-\sigma_3)^2+(\sigma_3-\sigma_1)^2]} = \sqrt{3}\frac{pD}{4t}$$

The corresponding strength condition is

$$\sqrt{3}\frac{pD}{4t} \leqslant [\sigma]$$

If the thin-walled cylindrical container of $p = 1.5$ MPa, $D = 1$ m, $t = 10\times10^{-3}$ m, $[\sigma] = 100$ MPa, then

$$\sqrt{3}\frac{pD}{4t} = \sqrt{3}\times\frac{1.5\times1}{4\times10\times10^{-3}} \text{ MPa} = 64.9 \text{ MPa} \leqslant [\sigma]$$

It can be seen that the vessel satisfies the strength condition.

Example 11-10 Try to establish the strength conditions for the pure shear stress state according to the third and fourth strength theories. Find the relationship between the allowable shear stress $[\tau]$ and the allowable tensile stress $[\sigma]$ of plastic materials.

Solution According to the discussion of the example 11-8, the pure shear stress state is a tensile-compressive biaxial stress state, and

$$\sigma_1=\tau, \quad \sigma_2=0, \quad \sigma_3=-\tau$$

According to the third theories of strength, the strength condition of the pure shear stress state is

$$\sigma_1-\sigma_3=\tau-(-\tau)=2\tau\leqslant[\sigma]$$

$$\tau\leqslant\frac{[\sigma]}{2} \tag{11-67}$$

It is also known that the shear strength condition is

$$\tau\leqslant[\tau] \tag{11-68}$$

Comparing equation (11-67) with equation (11-68), we get

$$[\tau]=\frac{[\sigma]}{2}=0.5[\sigma] \tag{11-69}$$

Eqution (11-69) is the relationship between $[\tau]$ and $[\sigma]$ obtained according to the third theories of strength.

According to the fourth theories of strength, the strength condition of the pure shear stress state is

$$\sqrt{\frac{1}{2}[(\sigma_1-\sigma_2)^2+(\sigma_2-\sigma_3)^2+(\sigma_3-\sigma_1)^2]}=$$

$$\sqrt{\frac{1}{2}[(\tau-0)^2+(0+\tau)^2+(-\tau-\tau)^2]}=\sqrt{3}\tau\leqslant[\sigma]$$

$$\tau\leqslant\frac{[\sigma]}{\sqrt{3}} \tag{11-70}$$

Comparing equation (11-70) with equation (11-68), we get

$$[\tau]=\frac{[\sigma]}{\sqrt{3}}\approx0.577[\sigma] \tag{11-71}$$

Equation (11-71) is the relationship between $[\tau]$ and $[\sigma]$ obtained according to the fourth theories of strength, which is closer to the experimental results.

Example 11-11 A simply supported beam subjected to three concentrated forces is shown in Fig. 11-20. It is known that $F=32$ kN, $a=1$ m, the allowable tensile stress of the material $[\sigma]=160$ MPa, the allowable shear stress $[\tau]=100$ MPa. If the beam is made of I-shaped beam, try to choose the type.

Solution (1) The I-shaped beam type is selected by the normal stress strength

Make the shear and bending moment diagrams of the beam as shown in Fig.11-20a. Obviously there is a maximum normal stress in section E, so

$$\sigma_{\max}=\frac{M_{\max}}{W_z}=\frac{48\times10^3\ \text{N}\cdot\text{m}}{W_z}\leqslant160\times10^6\ \text{Pa}$$

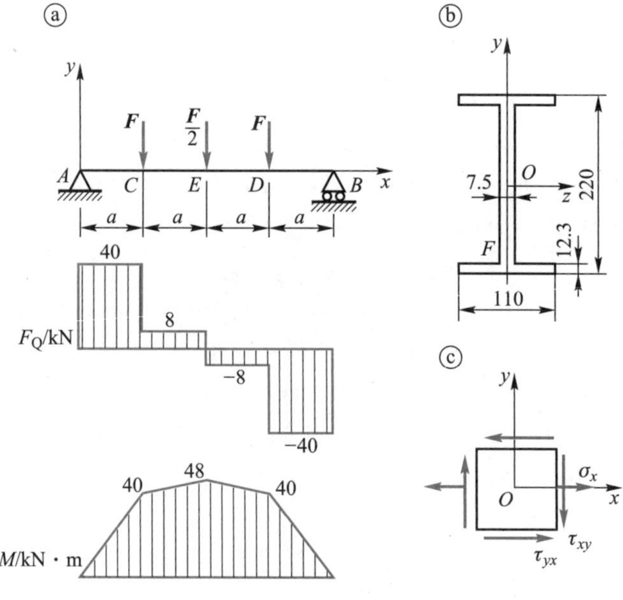

Fig.11−20

$$W_z \geqslant 3\times10^{-4}\ \text{m}^3 = 300\ \text{cm}^3$$

The I−shaped beam type 22a is selected from the section table. Its $W_z = 309\ \text{cm}^3$, $I_z = 3\ 400\ \text{cm}^4$, $\dfrac{I_z}{S_{max}} = 18.9$ cm. The dimensions are shown in Fig.11−20b.

(2) Calibrate the shear stress strength of the beam

The maximum shear stress on the beam occurs in the section *AC* or section *DB*. The maximum shear stress is

$$\tau_{max} = \frac{F_{Qmax}}{b\dfrac{I_z}{S_{max}}} = \frac{40\times10^3}{7.5\times18.9\times10^{-5}}\ \text{Pa} = 28.2\ \text{MPa} < [\tau]$$

(3) Using theories of strength to calibrate the strength of other dangerous points of the beam

The stress state of point *F* at the junction of the flange and web on the cross section slightly left of section *E* and slightly left of section *C* is shown in Fig.11−20c. The normal and shear stresses at this point are relatively large, so the strength should also be calibrated.

For point *F* on the slightly left cross section of section *E*:

$$\sigma_x = \frac{M_E y_F}{I_z} = \frac{48\times10^3\times(110-12.3)\times10^{-3}}{3\ 400\times10^{-8}}\ \text{Pa} = 138\ \text{MPa}$$

$$\tau_{xy} = \frac{F_{QE}S_z^*}{I_z b} = \frac{8\times10^3\times(110\times12.3\times103.9)\times10^{-9}}{3\ 400\times10^{-8}\times7.5\times10^{-3}}\ \text{Pa} = 4.4\ \text{MPa}$$

From equation (11−13), and note that $\sigma_y = 0$, the principal stress is obtained as

$$\sigma_1 = \frac{\sigma_x}{2} + \sqrt{\left(\frac{\sigma_x}{2}\right)^2 + \tau_{xy}^2}$$

$$\sigma_3=\frac{\sigma_x}{2}-\sqrt{\left(\frac{\sigma_x}{2}\right)^2+\tau_{xy}^2}$$

The strength of this point is calibrated according to the third theories of strength

$$\sigma_{xd3}=\sigma_1-\sigma_3=\sqrt{\sigma_x^2+4\tau_{xy}^2}=\sqrt{138^2+4\times4.4^2}\ \text{MPa}=138.3\ \text{MPa}<[\sigma]$$

Calibrate the strength of this point according to the fourth theories of strength

$$\sigma_{xd4}=\sqrt{\frac{1}{2}[(\sigma_1-\sigma_2)^2+(\sigma_2-\sigma_3)^2+(\sigma_3-\sigma_1)^2]}$$

$$=\sqrt{\sigma_x^2+3\tau_{xy}^2}=\sqrt{138^2+3\times4.4^2}\ \text{MPa}=138.2\ \text{MPa}<[\sigma]$$

Similarly, strength calibration is performed for point F on the slightly left cross section of section C. Since

$$\sigma_x=\frac{138}{48}\times40\ \text{MPa}=115\ \text{MPa}$$

$$\tau_{xy}=\frac{4.4}{8}\times40\ \text{MPa}=22\ \text{MPa}$$

Calibrate the strength of this point according to the third and fourth theories of strength respectively

$$\sigma_{xd3}=\sqrt{\sigma_x^2+4\tau_{xy}^2}=\sqrt{115^2+4\times22^2}\ \text{MPa}=123\ \text{MPa}<[\sigma]$$

$$\sigma_{xd4}=\sqrt{\sigma_x^2+3\tau_{xy}^2}=\sqrt{115^2+3\times22^2}\ \text{MPa}=121\ \text{MPa}<[\sigma]$$

The strength of the beam is satisfied.

Problems

11-1 The member is loaded as shown in the Fig.P11-1, try to find the danger point and use the element to indicate the stress state of the danger point.

Answer: Omit

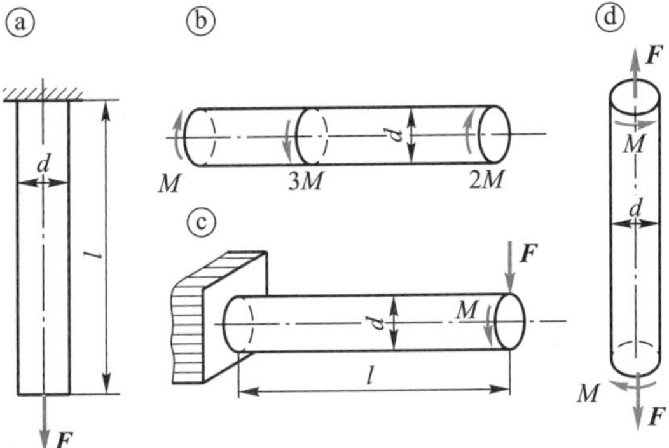

Fig.P11-1

11-2 The plane stress state as shown in the Fig.P11-2 (stress unit is MPa). Try to calculate by analytical method: (1) the principal direction and the principal stress value; (2) the maximum shear stress.

Answer: (a) $\sigma_1=57$ MPa, $\sigma_3=-7$ MPa; $\alpha_0=19°20'$, $\tau_{max}=32$ MPa

(b) $\sigma_1=25$ MPa, $\sigma_3=-25$ MPa; $\alpha_0=-45°$, $\tau_{max}=25$ MPa

(c) $\sigma_1=11.2$ MPa, $\sigma_3=-71.2$ MPa; $\alpha_0=-37°59'$, $\tau_{max}=41.2$ MPa

(d) $\sigma_1=4.7$ MPa, $\sigma_3=-84.7$ MPa; $\alpha_0=-13°17'$, $\tau_{max}=44.7$ MPa

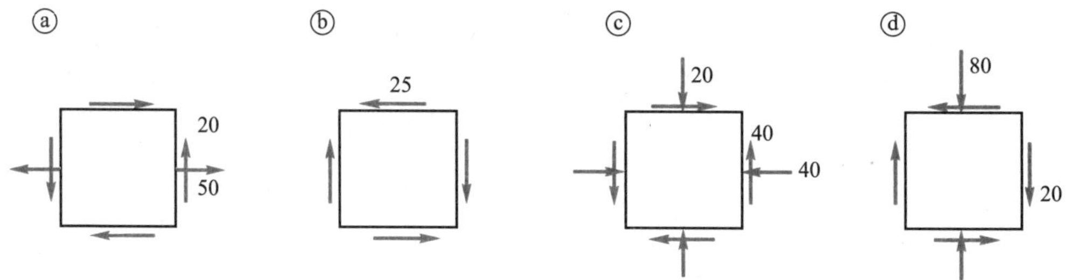

Fig.P11-2

11-3 The plane stress state is shown in the Fig.P11-3 (stress unit is MPa), try to calculate the stress in the specified inclined section by the analytical method.

Answer: (a) $\sigma_\alpha=-27.3$ MPa, $\tau_\alpha=-27.3$ MPa

(b) $\sigma_\alpha=52.3$ MPa, $\tau_\alpha=18.7$ MPa

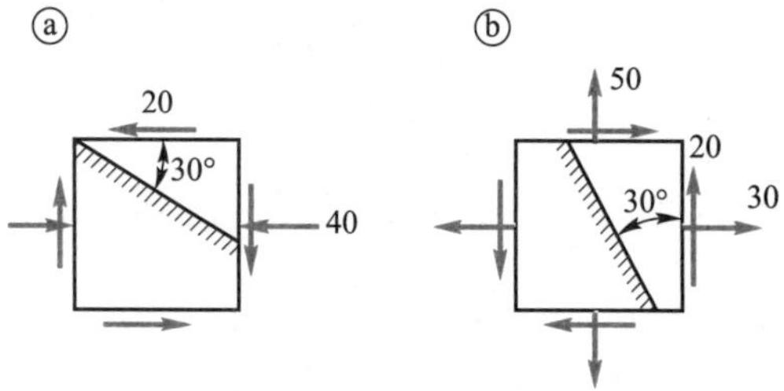

Fig.P11-3

11-4 The forces in the laminated panel element are shown in the Fig.P11-4. The layers are glued to each other, and the joint direction angle is shown. If the allowable stress of the adhesive layer is known $[\tau]=2$ MPa, try to calibrate the safety of this element.

Answer: $\tau_{30°}=1.55\ \text{MPa}<[\tau]$

11-5 The bending moment and shear force in a section of a rectangular section beam (Fig.P11-5) are $M=10$ kN·m and $F_Q=120$ kN, respectively. Try to draw the stress state of point 1,2,3,4 and find their principal stress.

Answer: Point 1: $\sigma_1=\sigma_2=0$, $\sigma_3=-120$ MPa;

Point 2: $\sigma_1=36$ MPa, $\sigma_2=0$, $\sigma_3=-36$ MPa;

Point 3: $\sigma_1 = 70.3$ MPa, $\sigma_2 = 0$, $\sigma_3 = -10.3$ MPa;

Point 4: $\sigma_1 = 120$ MPa, $\sigma_2 = \sigma_3 = 0$

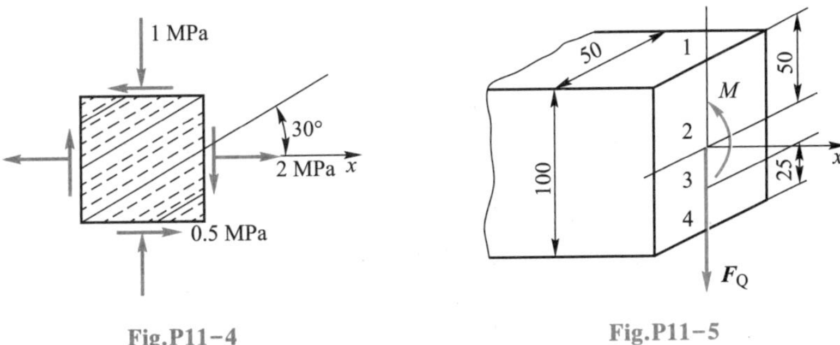

Fig.P11-4　　Fig.P11-5

11-6　The stress state at a point is a superposition of two simple stress states (Fig.P11-6). Try to find the principal stress, the maximum in-plane shear stress and the maximum shear stress of that point.

Answer: (a) $\sigma_1 = \sigma_0(1+\cos\theta)$, $\sigma_2 = 0$, $\sigma_3 = \sigma_0(1-\cos\theta)$, $\tau'_{max} = \tau_{max} = \sigma_0 \cos\theta$;

(b) $\sigma_1 = 10$ MPa, $\sigma_2 = \sigma_3 = 0$, $\tau'_{max} = \tau_{max} = 50$ MPa

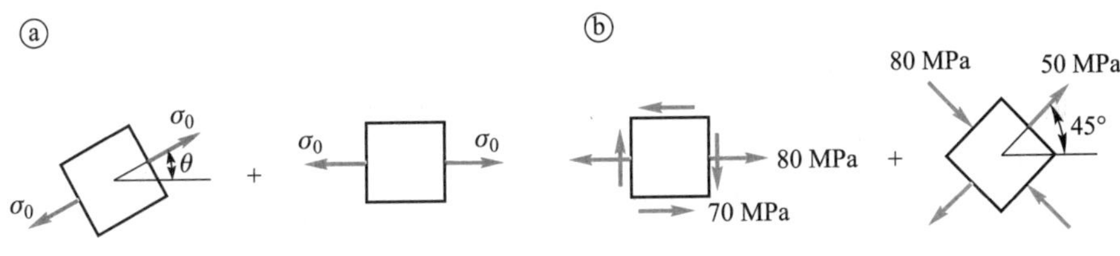

Fig.P11-6

11-7　As shown in the Fig.P11-7, if the maximum shear stress in the face is required to be less than or equal to 85 MPa, try to determine the value of τ_{xy}.

Answer: $\tau_{xy} \leqslant 40$ MPa

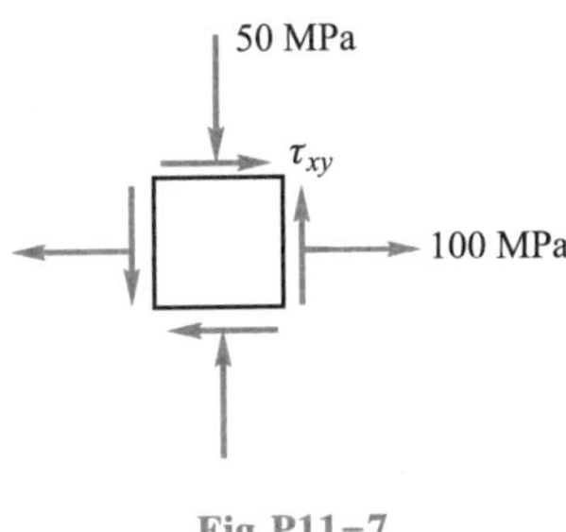

Fig.P11-7

11-8　The schematic diagram of a thin-walled cylinder torsion and stretching experiment is shown in Fig.P11-8. If we know $F = 20$ kN, $M_e = 600$ N · m, $d = 5$ cm and $\delta = 2$ mm. Try to find: (1) the stress at point A in the specified inclined section; (2) the principal stress at point A and the orientation of the principal plane (expressed in terms of the element).

Answer: (1) $\sigma_\alpha = -45.8$ MPa, $\tau_\alpha = 8.79$ MPa;

(2) $\sigma_1 = 108$ MPa, $\sigma_3 = -42.3$ MPa; $\alpha_0 = 33°17'$

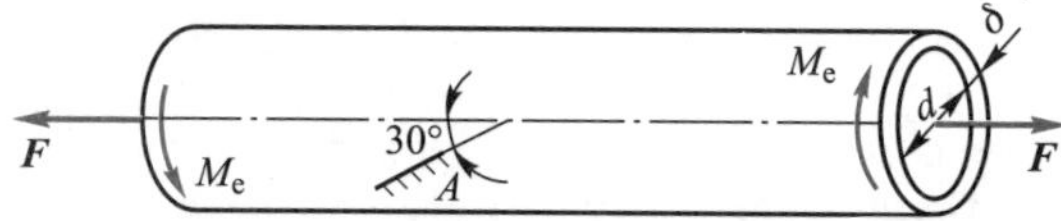

Fig.P11-8

11-9 The simply supported beam is made of 36a I-beam (Fig.P11-9). We have $F = 140$ kN, $l = 4$ m. The section where point A is located is at the left of the concentrated force F. Find: (1) the stress at point A in the inclined section; (2) the principal stress and the rincipal plane of point A (expressed as an element).

Answer: (1) $\sigma_\alpha = 2.08$ MPa, $\tau_\alpha = 24.3$ MPa;

(2) $\sigma_1 = 84.7$ MPa, $\sigma_2 = 0$, $\sigma_3 = -5$ MPa; $\alpha_0 = -13°36'$

11-10 In two planes passing through a point, the stresses are shown in the Fig. P11-10 (stresses in MPa). Try to find the principal stress and the principal plane (expressed in elements).

Answer: $\sigma_1 = 120$ MPa, $\sigma_2 = 20$ MPa, $\sigma_3 = 0$; $\alpha_0 = 45°$

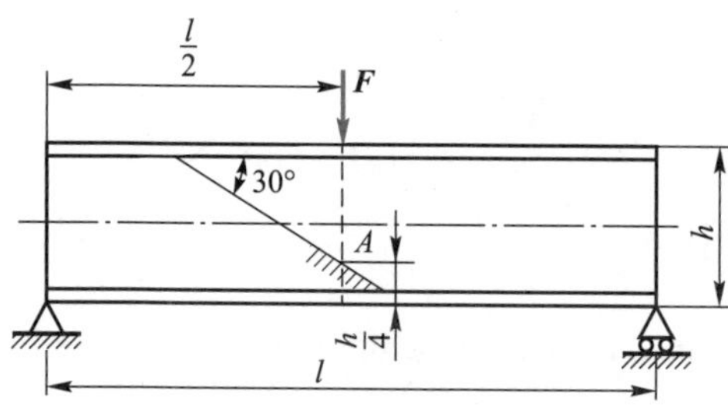

Fig.P11-9

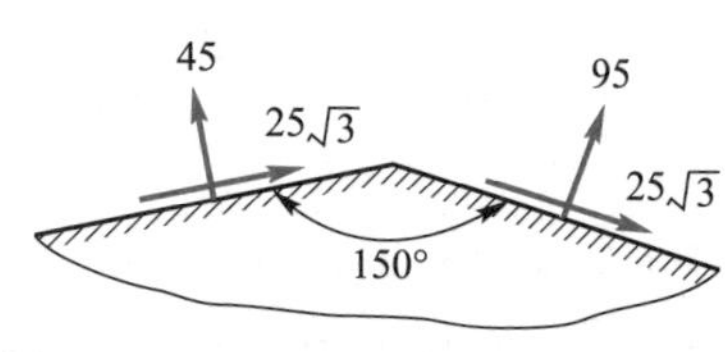

Fig.P11-10

11-11 The triaxial stress state is shown in Fig. P11-11 (stress in MPa), try to find the principal stress and the maximum shear stress.

Answer: (a) $\sigma_1 = 50$ MPa, $\sigma_2 = 50$ MPa, $\sigma_3 = -50$ MPa, $\tau_{max} = 50$ MPa;

(b) $\sigma_1 = 52.2$ MPa, $\sigma_2 = 50$ MPa, $\sigma_3 = -42.2$ MPa, $\tau_{max} = 47.2$ MPa;

(c) $\sigma_1 = 130$ MPa, $\sigma_2 = 30$ MPa, $\sigma_3 = -30$ MPa, $\tau_{max} = 80$ MPa

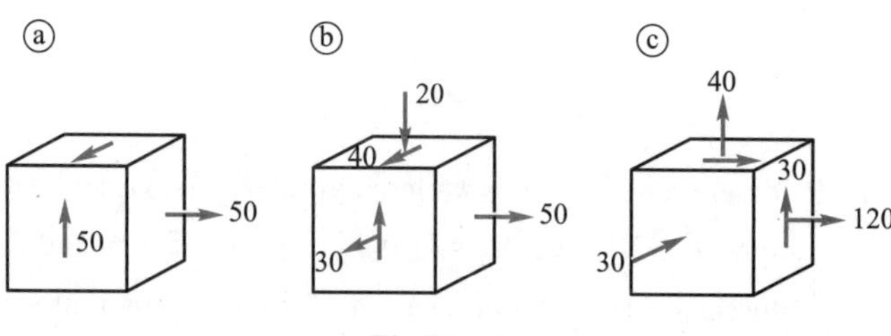

Fig.P11-11

11-12 The inner diameter of the long transmission pipe is 0.75 m, subject to internal pressure 2.0 MPa action. We have $\mu = 0.3$ and allowable stress $[\sigma] = 50$ MPa. Try to use the fourth theories of strength to calculate the wall thickness. (Hint: the axial strain of the pipe can be set to zero.)

Answer: 1.33 cm

11-13 The gun barrel cross section is shown in the Fig.P11-13. At the danger point, we have $\sigma_t = 550$ MPa, $\sigma_r = -350$ MPa. The third principal stress is perpendicular to the cross section and is a tensile stress. Its magnitude is 420 MPa. Try to calculate the equivalent stresses at the danger point according to the third and the fourth theories of strength, respectively.

Answer: $\sigma_{xd3} = 900$ MPa, $\sigma_{xd4} = 842$ MPa

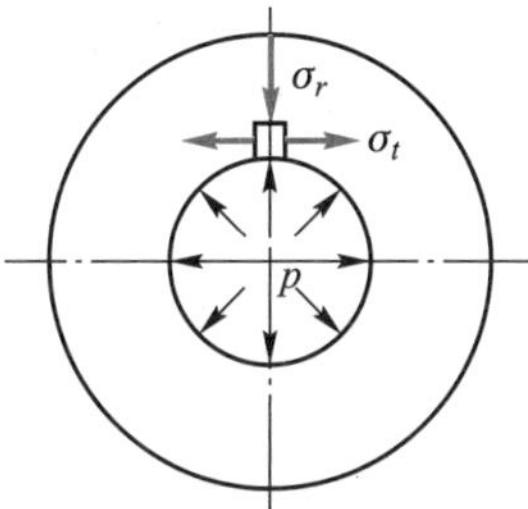

Fig.P11-13

Chapter 12 Combined Deformation

Teaching Scheme of Chapter 12

12.1 Introduction

Axial tension and compression, shear, torsion and bending are the four basic forms of deformation of bar. However, many components in engineering structures and machinery, the deformation often includes more than two basic deformations. When the stress (deformation) corresponding to various deformations belongs to the same order of magnitude, it is called combined deformation or composite deformation. Take the frame of a press (Fig.12-1) for example. The frame column is subjected to both tensile and bending deformations under the external force F. The gear transmission shaft (Fig.12-2) is subjected to both torsional and bending deformations. The ship spindles (Fig.12-3) is subjected to compression, bending and torsional deformations at the same time.

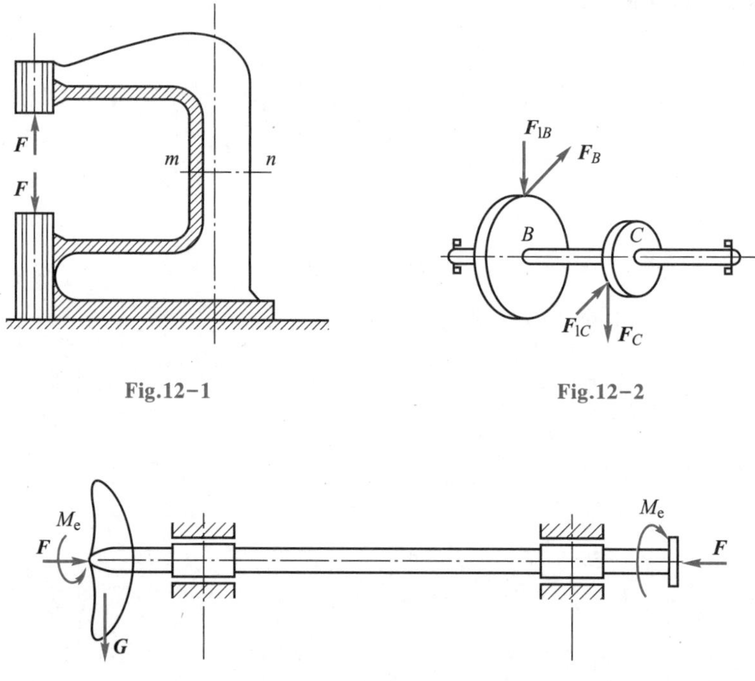

Fig.12-1

Fig.12-2

Fig.12-3

The common combinations of deformations in engineering are summarised in the following three deformations.

(1) Combination of two plane bending (skew bending).

(2) A combination of tension (or compression) and bending. Eccentric tension (or compression).

(3) A combination of torsion with bending or tension (or compression).

When analysing combined deformation problems, the external forces can first be decomposed or simplified. And then, the external forces on the member are transformed into several groups of statically equivalent loads. Each group of loads corresponding to a basic deformation. For small deformations and linear elasticity, the principle of superposition can be used. Calculate the stress at a point under each deformation and then superimpose all, the stress at that point in the combined deformation can be obtained.

In general, there are six internal force components in the cross section of the bar: F_{Nx}, F_{Qy}, F_{Qz}, M_x, M_y, M_z. It is rare that six internal force components exist at the same time in engineering. In addition, the normal stress in bending are more important than the shear forces F_{Qy} and F_{Qz}. So the these shear forces are usually not considered in the strength calculation of the combined deformation.

The approximate steps for strength calculation in case of combined deformation are

(1) Simplify or decompose the external force into several simple forms.

(2) Draw the internal forces for each basic deformation and find the location of possible dangerous sections. Then give the distribution of stresses in the dangerous sections.

(3) Find the location of the danger point based on the stress distribution in the danger section. Use the principle of superposition to determine the stress state at the danger point.

(4) Establish the corresponding strength conditions and perform strength calculations.

12.2 Skew bending

The previously discussed bending problem refers to plane bending. The condition for plane bending is that the external bending force (transverse force or in-plane force couple) is over the "center of bending" and parallel to the principal plane of inertia of the centroid. In plane bending, the bending plane of the beam (the plane where the deflection curve is located) coincides with the plane of action of the external force. In engineering practice, the bending forces acting on a beam sometimes may not satify these conditions. For example, in the case of a rectangular cross-sectional standard strip on a roof frame (Fig.12-4), the vertical downward load F passes through the centroid, but does not coincide with the two main inertia axis y and z (principal axis) of the centroids. After deformation, the deflection curve of the purlin will no longer be in the plane of action and this bending will not be the plane bending. When the load passes through the center of the bend, but the plane of action is not parallel to either of the principal centroid planes of inertia, the resulting bend becomes an skew bending, also known as a bi-directional bend.

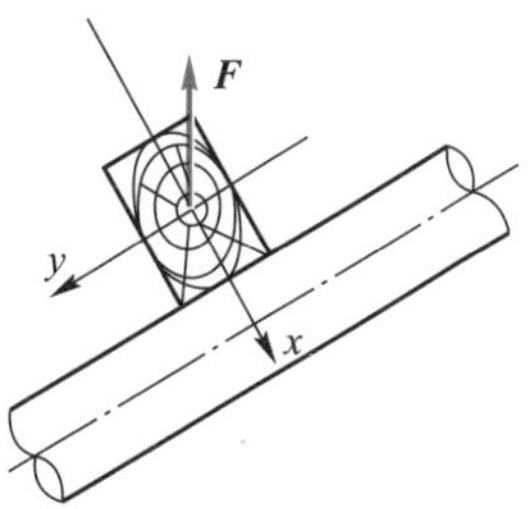

Fig.12-4

The external force F is decomposed into F_y and F_z along the principal axis in the directions y and z, respectively. F_y and F_z will cause the standard strip in the two planes perpendicular to each other to produce plane bending.

The cantilever beam of rectangular section shown in Fig.12-5a is used as an example to illustrate the calculation of stresses and deformations in skew bending.

12.2.1 Internal forces and stresses

In Fig.12−5a, a transverse concentrated force F is applied at the centroid of the section at the free end of the beam, and the angle between the line of action and the z axis is φ. The force F along the principal axes y and z are

$$F_y = F\sin\varphi,\quad F_z = F\cos\varphi \tag{12-1}$$

The beam under F_y and F_z action will bend in plane with the z and y axes as neutral axes respectively, as shwon in Fig.12−5b, c. The bending moments around the z and y axes in the section m—m at a distance x from the fixed end are

$$\left.\begin{aligned} M_z &= F_y(l-x) = F(l-x)\sin\varphi = M\sin\varphi \\ M_y &= F_z(l-x) = F(l-x)\cos\varphi = M\cos\varphi \end{aligned}\right\} \tag{12-2}$$

where $M = F(l-x)$ is the total bending moment of the force F to the cross section m—m. In section m—m, as mentioned earlier, the shear stresses caused by shear force F_{Qy} and F_{Qz} are not considered.

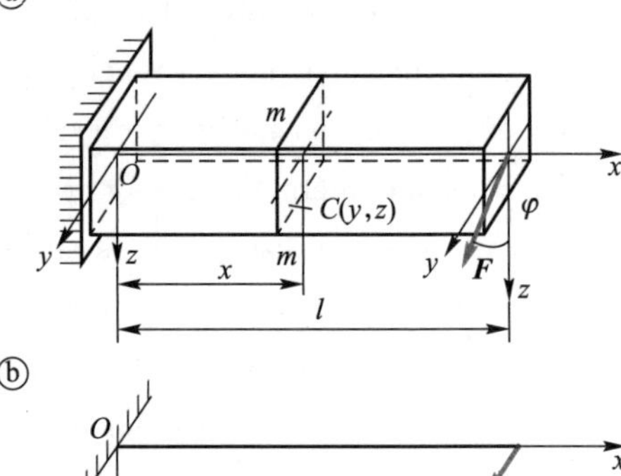

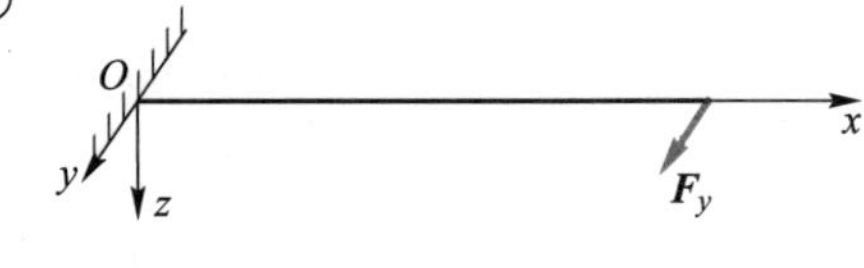

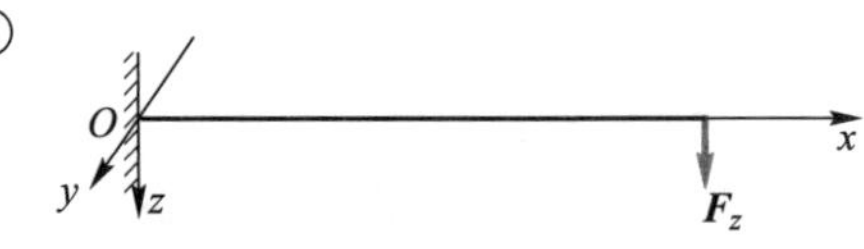

Fig.12−5

The distribution of the normal stresses in the section m—m is shown in Fig. 12 − 6a, b, which is generated by the bending moment M_y and M_z. Since both plane bendings cause linear stress distributions, the stresses remain linear after superposition as shown in Fig.12−6c. For cross sections with angle such as rectangles, the location of the danger point is easy to determine. As can be seen from Fig.12−6c, the two vertices D_1 and D_2 of the section are the danger points. After superposition, the normal stresses of the points D_1 and D_2 are the maximum tensile stress and the maximum compressive stress, respectively. Their values are equal and can be calculated according to the following formula

$$\sigma_{\max} = \frac{M_{z\max}}{W_z} + \frac{M_{y\max}}{W_y}$$

In order to calculate the position of the neutral axis on the section, the normal stress at any point $C(y,z)$ on the cross section m—m is discussed. The normal stress in bending at this point caused by M_z is

$$\sigma' = \frac{M_z y}{I_z} = \frac{M\sin\varphi}{I_z} y$$

The normal stress in bending at this point caused by M_y is

$$\sigma'' = \frac{M_y z}{I_y} = \frac{M\cos\varphi}{I_z} z$$

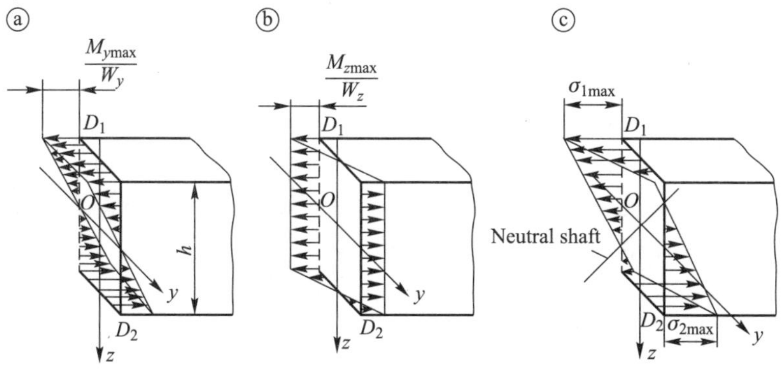

Fig.12-6

Thus, we have

$$\sigma=\sigma'+\sigma''=M\left(\frac{\sin\varphi}{I_z}y+\frac{\cos\varphi}{I_y}z\right) \tag{12-3}$$

Whether σ' and σ'' are tensile or compressive stress can be determined by the deformation of the bar in the specific problem.

If the point on the neutral axis is marked as (y_0, z_0), the normal stress on neutral axis is zero. The equation for the neutral axis can be obtained using equation (12-3) as

$$\frac{y_0\sin\varphi}{I_z}+\frac{z_0\cos\varphi}{I_y}=0 \tag{12-4}$$

It can be seen that the neutral axis is a line passing through the centroid of the section, and its angle θ with the z axis can be determined by the following equation

$$\tan\theta=\frac{y_0}{z_0}=-\frac{I_z}{I_y}\cot\varphi \tag{12-5}$$

From the above equation, we know that when $I_y\neq I_z$, $\theta+\varphi\neq 90°$, the direction of external force is not perpendicular to the neutral axis.

The point furthest from the neutral axis has the highest normal stress and is the danger point. For the rectangular section shown in Fig.12-6c, points D_1 and D_2 are the danger point. For a section without sharp corners, two tangents parallel to the neutral axis are made on the perimeter of the section. The tangency point furthest from the neutral axis is the danger point.

From the above analysis it can be seen that the danger points of the inclined bending beam are in a uniaxial stress state. If the coordinates of the danger points are (y_1, z_1) and (y_2, z_2), and the tensile and compressive strengths of the material are equal, the strength condition for diagonal bending can be expressed as

$$|\sigma_{max}|=\left|M\left(\frac{\sin\varphi}{I_z}y_{1,2}+\frac{\cos\varphi}{I_y}z_{1,2}\right)\right|\leqslant[\sigma] \tag{12-6}$$

For sections such as rectangles, the strength condition can be simply expressed as

$$\sigma_{max}=\frac{M_{zmax}}{W_z}+\frac{M_{ymax}}{W_y}\leqslant[\sigma] \tag{12-7}$$

Using the above strength conditions, the same strength calculations can be carried out for three common types of engineering problems. However, it should be noted that, if the tensile and compressive strengths of the materials are different, strength checks should be carried out for the maximum tensile and maximum compressive stresses, respectively. When selecting sections, since W_y, W_z(or I_y, I_z and y_1, z_1) in the intensity conditions are unknown, the intensity conditions cannot be used to determine the values of W_y, W_z at the same time. So we need to set a value of W_y/W_z according to experience and then to solve by trial calculation.

12.2.2 Deflection of skew bending

The deflection of a beam in skew bending can also be calculated using the method of superposition. For example, to find the deflection at the free end of the cantilever beam shown in Fig.12-5a, the force F is first decomposed into two components F_y and F_z. Then the deflection f_y caused by F_y and the deflection f_z caused by F_z are calculated according to the of plane bending. Finall, the magnitude and direction of the total deflection f are derived by vector synthesis.

In plane bending, there is

$$f_y=\frac{F_yl^3}{3EI_z},\quad f_z=\frac{F_zl^3}{3EI_y}$$

The total deflection at the free end of the beam is

$$f=\sqrt{f_y^2+f_z^2} \tag{12-8}$$

Let the angle of the total deflection with respect to the z axis be ψ, we have

$$\tan\psi=\frac{f_y}{f_z}=\frac{I_y}{I_z}\tan\varphi \tag{12-9}$$

From equation (12-9) we get that $\psi\neq\varphi$ for a rectangular section with $I_y\neq I_z$. Thus, the total deflection f is not in the same direction of the external force $\boldsymbol{F}$ for skew bending. Considering equation (12-5) and equation (12-9), we know that $\psi+\theta=90°$. The total deflection f is always perpendicular to the neutral axis. Therefore, the direction of the external force is not perpendicular to the neutral axis, which is the main difference between skew bending and plane bending. It is only when $I_y=I_z$(e.g. circular, square or other normal polygonal cross section), that $\psi\neq\varphi$. Moreover, the direction of f coincides with the direction of $\boldsymbol{F}$, forming a plane bend. For a cross-sectional of $I_y=I_z$, transverse forces act in any longitudinal plane past the section center, plane bending always occurs.

Example 12-1 Bridge crane girder made from 32a I-beam. The material is steel Q235 and the allowable stress is $[\sigma]=160\ \text{MN/m}^2$, the girder length is $l=4$ m. As shown in Fig.12-7, the direction of the load F deviates from the longitudinal symmetry plane by an angle $\varphi=15°$. If the load is $F=30$ kN, try to check the strength of the beam.

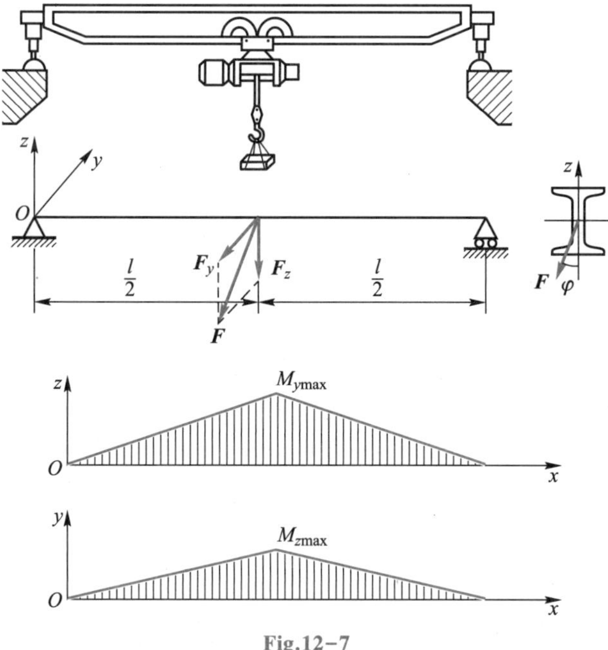

Fig.12-7

Solution When the lifting trolley goes to the midpoint of the beam, it is in the most unfavorable state of stress. At this time, the bending moment of the mid-span section is the largest.

In the $x-y$ plane, the maximum bending moment caused by F_y is

$$M_{zmax}=\frac{F_y l}{4}=\frac{Fl\sin\varphi}{4}=\frac{30\times10^3\times4\times\sin 15°}{4}\ \text{N}\cdot\text{m}\approx7.76\times10^3\ \text{N}\cdot\text{m}$$

In the $x-z$ plane, the maximum bending moment caused by F_z is

$$M_{ymax}=\frac{F_z l}{4}=\frac{Fl\cos\varphi}{4}=\frac{30\times10^3\times4\times\cos 15°}{4}\ \text{N}\cdot\text{m}\approx29\times10^3\ \text{N}\cdot\text{m}$$

The section modulus of bending of 32a I-beam is found from the table of sections

$$W_y=692.2\ \text{cm}^3,\quad W_z=70.8\ \text{cm}^3$$

So

$$\sigma_{max}=\frac{M_{zmax}}{W_z}+\frac{M_{ymax}}{W_y}=\left(\frac{7.76\times10^3}{70.8\times10^{-6}}+\frac{29\times10^3}{692.2\times10^{-6}}\right)\times10^{-6}\ \text{MPa}=151\ \text{MPa}<[\sigma]$$

In this example, if the load F is always undeviated along the vertical direction, i.e. $\varphi=0$, the maximum normal stress is

$$\sigma_{max}=\frac{M_{max}}{W_y}=\frac{Fl/4}{W_y}=\frac{30\times10^3}{692.2\times10^{-6}}\times10^{-6}\ \text{MPa}=43.4\ \text{MPa}$$

For I-beams, the maximum normal stress increases considerably when the direction of the load deviates by a modest angle. The reason for this result is that W_z of I-beam is much smaller than W_y. So it is unfavorable to use a beam with largely different W_z, W_y to resist skew bending or to withstand external forces with less fixed directions. Box-section beams are superior to I-beams in this respect.

12.3 Combined axial load and bending

The bar will be deformed in a combination of tension (compression) and bending under two types of loading: (1) combined axial and transverse loads; (2) eccentric axial loads.

12.3.1 Strength checks

Under the above two loading methods, the bending moment M_y or M_z (or both) and the axial force F_N will be generated in the cross section of the bar. In the case of transverse loads, there are also shear forces in the cross section. These forces are omitted since the shear stresses caused by them are small. Therefore, only the normal stresses caused by M_y, M_z and F_N are considered here.

Firstly, according to M_y, M_z and F_N, determine the location of the dangerous section of the bar. Then based on the actual direction of M_y, M_z and F_N, the location of the dangerous point in the dangerous section is determined. By superimposing the stresses corresponding to each internal force component at the danger point, the stress value at the danger point is obtained.

When a combination of tensile (compression) and bending deformation occurs, the dangerous point on the member is in a uniaxial stress state, and its strength condition is

$$[\sigma_{max}] \leqslant [\sigma] \tag{12-10}$$

The following specific examples are used to illustrate the method of strength calculation in combined deformation of tensile (compression) and bending.

Example 12-2 Fig.12-8a shows a schematic diagram of a small press. The frame material is cast iron and the allowable tensile stress of cast iron is known to be $[\sigma^+] = 30$ MPa. The allowable compressive stress is $[\sigma^-] = 120$ MPa. The cross-sectional sizes of the frame column are shown in Fig.12-8b. Try to determine the maximum allowable pressure F_{max} according to the strength of the column.

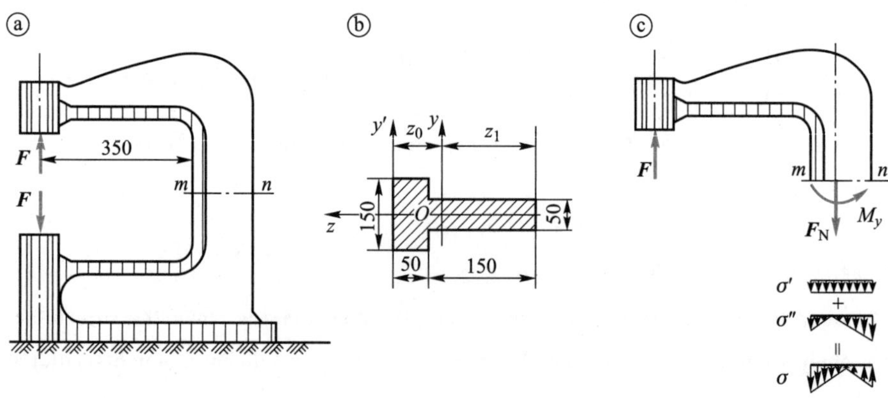

Fig.12-8

Solution (1) Calculate the geometric parameters of the section

The cross-sectional area of the column is

$$A=15\times5\ \mathrm{cm}^2+15\times5\ \mathrm{cm}^2=150\ \mathrm{cm}^2$$

The coordinates are chosen as shown. The coordinates of the position of the shape center O are

$$z_0=\frac{\sum A_i z_i}{\sum A_i}=\frac{15\times5\times2.5+15\times5\times(5+7.5)}{15\times5+15\times5}\ \mathrm{cm}=7.5\ \mathrm{cm}$$

From the parallel shift formula, we get

$$I_y=\left(\frac{15\times5^3}{12}+15\times5\times5^2+\frac{5\times15^3}{12}+15\times5\times5^2\right)\ \mathrm{cm}^4=5\ 310\ \mathrm{cm}^4$$

(2) Calculate the internal forces and stresses in the column

In the section m—n, the column is cut and the upper half is taken for study. From the equilibrium conditions, the internal force of the section m—n is

$$F_N=F$$

$$M_y=F\times(35+7.5)\times10^{-2}\ \mathrm{m}=0.425\ \mathrm{m}\cdot F$$

Axial forces F_N produce uniformly distributed tensile stresses σ' and bending moment M_y produces linearly distributed normal stresses σ''. The inner side is subject to tensile and the outer side to compressive. After superposition, the tensile stress is greatest at the point on the inner edge and the compressive stress is greatest at the point on the outer edge. So the points on both the inner and outer edges are dangerous.

(3) Find the maximum allowable pressure F

The allowable pressure F is calculated separately for tensile and compressive strength conditions. Tensile strength condition gives

$$\sigma^+_{\max}=\frac{F_N}{A}+\frac{M_y z_0}{I_y}\leqslant[\sigma^+]$$

i.e.

$$\frac{F}{150\times10^{-4}\ \mathrm{m}^2}+\frac{0.425\ \mathrm{m}\cdot F\times7.5\times10^{-2}\ \mathrm{m}}{5\ 310\times10^{-8}\ \mathrm{m}^4}\leqslant30\times10^6\ \mathrm{Pa}$$

So $F\leqslant45.1\times10^3$ N.

Compressive strength condition gives

$$\sigma^-_{\max}=\left|\frac{F_N}{A}-\frac{M_y z_1}{I_y}\right|\leqslant[\sigma^-]$$

Therefore, the maximum allowable pressure $F_{\max}$ can be taken as 45.1 kN.

Example 12-3 A bar of rectangular cross section has two resistive strain gauges attached to its upper and lower surfaces. The ends of the bar are subjected to an axial linear distributed force, as shown in Fig.12-9a. The modulus of elasticity E of the material is known, try to find the reading of two resistance strain gauges.

Solution (1) Calculate the internal force of the bar

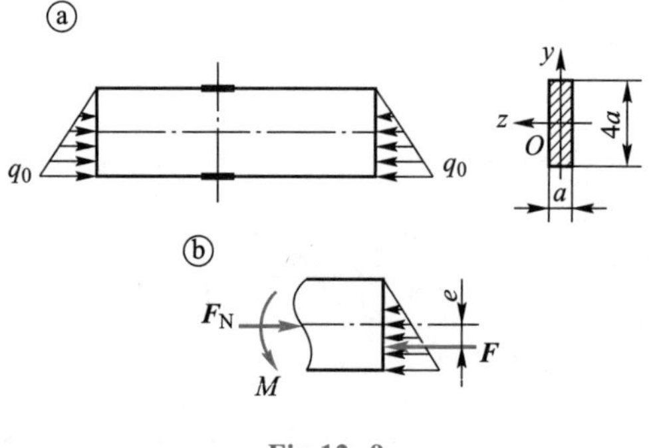

Fig.12-9

The combined force of the line distribution forces is

$$F=\frac{1}{2}\cdot 4a\cdot q_0=2q_0a$$

The distance from its action point to the axis is

$$e=\frac{2}{3}\cdot 4a-\frac{4a}{2}=\frac{2}{3}a$$

The internal forces on any cross section of the bar are thus shown in Fig.12-9b, and there are

$$F_N=F=2q_0a$$

$$M=Fe=\frac{4}{3}q_0a^2$$

(2) Calculate the reading of the resistance strain gauge

The stress near the upper surface of the bar is

$$\sigma=\frac{M}{W_z}-\frac{F_N}{A}=\frac{\frac{4}{3}q_0a^2}{\frac{1}{6}a\cdot(4a)^2}-\frac{2q_0a}{a\cdot 4a}=0$$

The stress near the lower surface of the bar is

$$\sigma=-\frac{M}{W_z}-\frac{F_N}{A}=-\frac{\frac{4}{3}q_0a^2}{\frac{1}{6}a\cdot(4a)^2}-\frac{2q_0a}{a\cdot 4a}=-\frac{q_0}{a}$$

According to Hooke law for uniaxial stress states, the readings of the resistance strain gauges on the upper and lower surfaces are 0 and $-\frac{q_0}{Ea}$ respectively.

12.3.2 Core of Section

Fig.12-10a shows a short column subject to eccentric compression. y, z axes are the principal centroidal axes of inertia in the cross section. The coordinates of point A where the pressure F acts is (y_F, z_F). By simplifying F towards the axis and decomposing the resulting force couple moment into the two principal planes of inertia, we obtain the pressure F and bending moment M_y, M_z, as shown in Fig. 12-10b. There is $M_y = Fz_F$ and $M_z = Fy_F$. In the cross section shown in Fig.12-11, at point $B(y, z)$, the stress components corresponding to the three deformations are

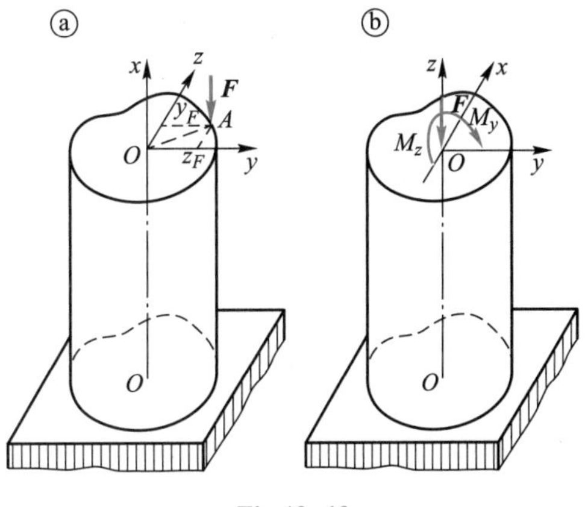

Fig.12-10

$$\sigma' = -\frac{F}{A}$$

$$\sigma'' = \frac{M_z y}{I_z} = -\frac{F_{y_F} y}{I_z}$$

$$\sigma''' = \frac{M_y z}{I_y} = -\frac{F_{z_F} z}{I_y}$$

By the method of superposition and taking into account the relationship $I_z = Ai_z^2$ and $I_y = Ai_y^2$, the normal stress at point B is

$$\sigma = -\frac{F}{A}\left(1+\frac{y_F y}{i_z^2}+\frac{z_F z}{i_y^2}\right)$$

If we see (y_0, z_0) as the coordinates of any point on the neutral axis, the equation of the neutral axis can be obtained as

$$\frac{y_F y_0}{i_z^2}+\frac{z_F z_0}{i_y^2}+1=0 \tag{12-11}$$

As can be seen, the neutral axis in eccentric compression is a straight line that does not pass through the section's centroid, as shown in Fig.12-11. The neutral axis divides the section into two parts. The part with the shaded line is in tension and the other part in compression. Points D_1 and D_2 furthest away from the neutral axis have the highest stresses.

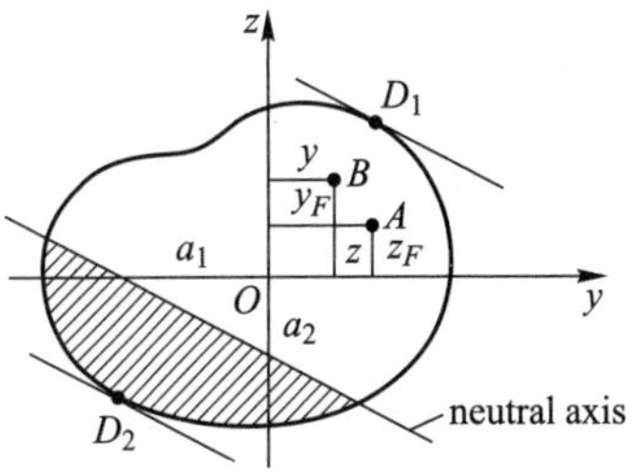

Fig.12-11

In equation (12-11), let $z_0 = 0$ and $y_0 = 0$, the intercepts of the neutral axis on the y axis and z axis can be obtained as

$$a_y = -\frac{i_z^2}{y_F}, \quad a_z = -\frac{i_y^2}{z_F} \tag{12-12}$$

It can be seen from equation (12-12) that, a_y and y_F, a_z and z_F are of opposite sign. So the neutral axis and the action point A of the external force F are on each side of the section centroid. In addition, if the action point of the pressure F gradually approaches the centroid, the intercept will gradually increase. The neutral axis will gradually move away from the centroid of the section. When the pressure is applied within this area, the neutral axis will not cross the cross section, and there will be no tensile stresses in the cross section. This area is called core of section. When the pressure acts on the boundary of the core of section, the neutral axis is tangential to the perimeter of the section. Therefore, equation (12-12) can be used to determine the location of the core boundary of the section.

To determine the boundary of the core of an arbitrary section, any line ① tangent to the perimeter of the section can be regarded as the neutral axis. Let its intercepts on the main inertia axes y, z of the centroid be a_{y1}, a_{z1}. From these two values, point 1 of external force corresponding to this neutral axis (a point on the boundary of the core of section) can be calculated from equation (12-12). Its coordinates are

$$y_{F_1}=-\frac{i_z^2}{a_{y_1}},\quad z_{F_1}=-\frac{i_y^2}{a_{z_1}} \tag{12-13}$$

Similarly, the other lines ② ③… tangent to the perimeter of the section can be regarded as neutral axes. The coordinates of points 2, 3, … on the boundary of the corresponding core of section can be obtained in the above method in turn. Connect these points to obtain a closed curve, which is the boundary of the core of section.

Taking rectangular section as an example. The specific method to determine the core of section is described below.

Fig.12-12 shows a rectangular section with side lengths b and h. Axes y, z are the principal centroidial axes of inertia. Firstly, the line ① tangent to the side AB is regarded as the neutral axis, and its intercepts on the axes y, z are

$$a_{y_1}=\frac{h}{2},\quad a_{z_1}=\infty$$

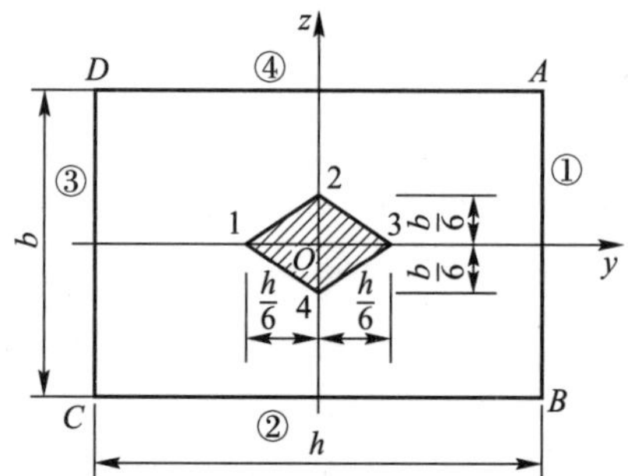

Fig.12-12

The square of the radius of gyration of this rectangular section is

$$i_y^2=\frac{I_y}{A}=\frac{b^2}{12},\quad i_z^2=\frac{I_z}{A}=\frac{h^2}{12}$$

Substituting the above geometric quantities into equation (12-13), the coordinates of point 1 on the boundary of the core of section corresponding to the neutral axis ① can be obtained as

$$y_{F_1}=-\frac{h}{6},\quad z_{F_1}=0$$

Similarly, the lines②③④ tangent to the sides of BC, CD and DA respectively are regarded as neutral axes, and the coordinates of points 2, 3 and 4 on the boundary of the corresponding core of section can be found in the above method as

$$y_{F_2}=0,\quad z_{F_2}=\frac{b}{6};\quad y_{F_3}=\frac{h}{6},\quad z_{F_3}=0;\quad y_{F_4}=0,\quad z_{F_4}=-\frac{b}{6}$$

This identifies the four points on the boundary of the core of section.

In order to use these four points to determine the boundary of core of section, it is necessary to solve the problem: when the neutral axis is rotated from position ① to position ② around the vertex B of the section, what kind of curve is the trajectory of the corresponding point of external force? During the rotation of the neutral axis around B, a series of neutral axes through point B with different slopes will be obtained. Point B is the common point of the series of neutral axes. Substitute the coordinates of point $B(y_B, z_B)$ into the equation (12–12) for the neutral axis, we get

$$\frac{y_F y_B}{i_z^2}+\frac{z_F z_B}{i_y^2}+1=0$$

Since y_F, z_F are constant, the equation is the linear equation of the coordinate of the action point y_F, z_F. The corresponding straight line is the trajectory of the corresponding point of external force as the neutral axis rotates around B. Since points 1 and 2 are on this line, which correspond to the rotation start point ① and end point ②, the boundary of the core of section past points 1 and 2 should be the line connecting these two points. By analogy, the core of section of a rectangular cross section is the shaded diamond area in Fig.12–12.

A similar cross-sectional core for a circular section can be obtained as shown in Fig.12–13.

It must be noted that the coordinate axis used in determining the location of the boundary of the core of section must be the main inertia axis of the section's centroid.

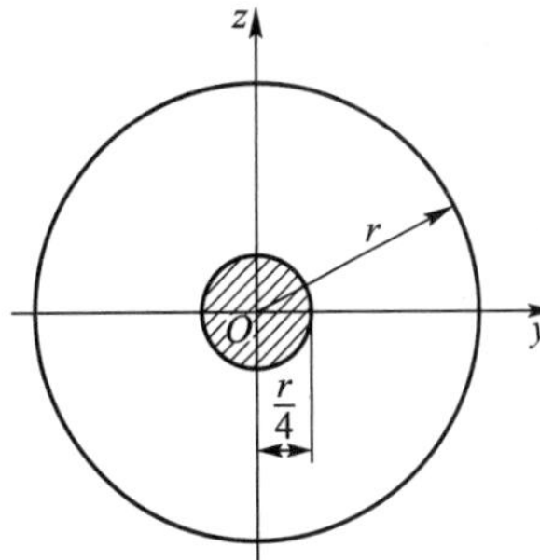

Fig.12–13

12.4 Combined bending and torsion

Most of the common shafts in engineering (e.g. gear shafts, pulley shafts, motor shafts, crank shafts, etc.) are subjected to torsion along with bending deformation. When the bending deformation is small, the calculation can be based on the torsional deformation only. But when the bending deformation can not be ignored, We should deal with it according to the combined deformation of bending and torsion. The internal force components of the cross section are bending moment and torque. Due to the shear stress corresponding to the shear force is small, its influence can be ignored. Most of the shaft members in the project are circular cross sections (solid or hollow), so the combined bending and torsion for the circular shaft is discussed first.

12.4.1 Deformation of circular shafts subjected to bending and torsion

When calculating the strength of a circular shaft with bending and torsional deformation, it is first necessary to draw the bending moment (M_y and M_z) diagrams and torque diagrams. Since the two centroids of a circular section have equal principal moments of inertia, we have $I_y = I_z$. So the circular section bar will not undergo skew bending, only plane bending is possible (section 12.2). The M_y and M_z can be vector synthesized, the magnitude of the synthetic bending moment is

$$M_W=\sqrt{M_y^2+M_z^2} \tag{12–14}$$

The action plane of the synthetic bending moment M_W is perpendicular to the vector M_W, as shown in Fig.12-14. From the synthetic bending moment M_W and the torque diagram, the position of the dangerous section of the circular shaft can be determined.

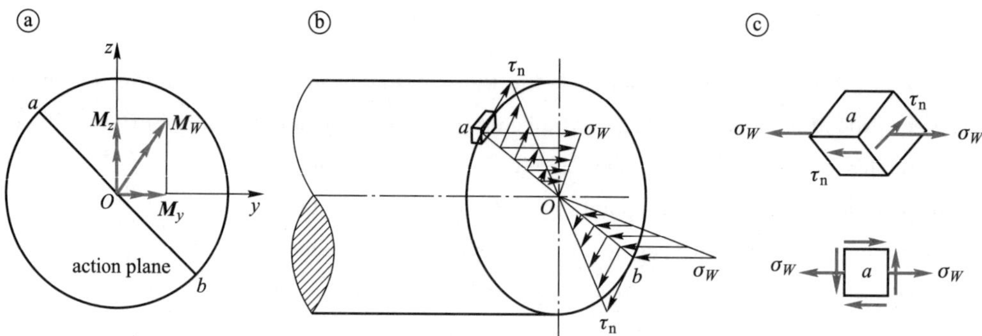

Fig.12-14

The maximum value of the torsional shear stress corresponding to the torque M_n in the dangerous section occurs at the edge of the cross section with the value

$$\tau_n = \frac{M_n}{W_n} \tag{12-15}$$

The normal stress in bending corresponding to the synthetic bending moment M_W is maximum at points a and b, and its value is

$$\sigma_W = \frac{M_W}{W} \tag{12-16}$$

The distribution of shear stresses and normal stresses are along the diameter ab of the section shown in Fig.12-14b. There are maximum torsional shear stress and maximum normal bending stress at points a and b. These two points are dangerous points. A element is cut at point a with a cross section, a radial longitudinal section and a tangential longitudinal section. The stresses in each section of the element are shown in Fig.12-14c.

The danger point is in a plane stress state, so strength conditions should be established according to theories of strength. The principal stress at point a is

$$\left\{\begin{matrix}\sigma_1 \\ \sigma_3\end{matrix}\right. = \frac{\sigma_W}{2} \pm \frac{1}{2}\sqrt{\sigma_W^2 + 4\tau_n^2}, \quad \sigma_2 = 0 \tag{12-17}$$

If the shaft is made of a plastic material, the third or fourth theories of strength should be used for strength calculations. If the third theories of strength is used, substitute the above mentioned principal stresses σ_1 and σ_3 into the corresponding strength condition expressions

$$\sigma_{xd3} = \sigma_1 - \sigma_3 \leqslant [\sigma]$$

After simplification, we get

$$\sigma_{xd3} = \sqrt{\sigma_W^2 + 4\tau_n^2} \leqslant [\sigma] \tag{12-18}$$

If the fourth theories of strength is used, substitutethe principal stresses given in equation (12-17) into the corresponding expressions for the strength conditions

$$\sigma_{xd4}=\sqrt{\frac{1}{2}[(\sigma_1-\sigma_2)^2+(\sigma_2-\sigma_3)^2+(\sigma_3-\sigma_1)]}\leqslant[\sigma]$$

After simplification, we get

$$\sigma_{xd4}=\sqrt{\sigma_W^2+3\tau_n^2}\leqslant[\sigma] \tag{12-19}$$

Substitute equations (12-15) and (12-16) into equations (12-18) and (12-19), respectively. For circular sections, there is $W_n=2W$. The alternative formulation of the strength condition for a circular shaft in combined deformation of bending and torsional can be obtained. For the third theories of strength, we get

$$\sigma_{xd3}=\frac{\sqrt{M_W^2+M_n^2}}{W}=\frac{\sqrt{M_y^2+M_z^2+M_n^2}}{W}\leqslant[\sigma] \tag{12-20}$$

For fourth theories of strength, we get

$$\sigma_{xd4}=\frac{\sqrt{M_W^2+0.75M_n^2}}{W}=\frac{\sqrt{M_y^2+M_z^2+0.75M_n^2}}{W}\leqslant[\sigma] \tag{12-21}$$

In equations (12-20) and (12-21), $\sqrt{M_y^2+M_z^2+M_n^2}$ and $\sqrt{M_y^2+M_z^2+0.75M_n^2}$ are called the "calculated bending moments" corresponding to the third and fourth strength theories, respectively. Obviously, after the introducing the concept of "calculated bending moment", when checking the strength of a circular shaft of the combined deformation, it is not necessary to find out the value of the stress component at the danger point. We just need to calculate the value of the bending moment and torque at the danger section. And then the strength check can be carried out directly using equation (12-20) or (12-21). At the same time, when applying the strength conditions to select the section, it is very convenient to apply equations (12-20) and (12-21) to the selection of section sizes.

The two expressions for the intensity conditions above have different applications.

(1) Equations (12-20) and (12-21) apply only to combined deformations of bending and torsion of circular or hollow circular section.

(2) Equations (12-18) and (12-19) are more widely applicable, It is applicable as long as the dangerous point is in the plane stress state shown in Figure.12-14c.

There are often bars in engineering, such as ship propulsion shafts, which usually undergo the bending, torsion and axial compression (tension) deformations at the same time. The danger point of these bars is also in the stress state shown in Fig.12-14c. The normal stress on the element should be the sum of the normal stress in axial compression (tensile) and bending normal stress, $\sigma=\frac{F_N}{A}+\frac{M_W}{W}$. For this type of bar, the equation (12-18) or (12-19) can be used for strength calculation, but not the equation (12-20) or (12-21).

Example 12-4 A steel circular shaft is fitted with tape wheels A and B, as shown in Fig.12-15a. Both wheels have the same diameter $D=1$ m and weight $P=5$ kN. The tension in the tape on wheel A is horizontal and the tension in the tape on wheel B is in the plumb direction. The allowable stress for the circular shaft is $[\sigma]=80$ MPa. Try to find the required diameter of the shaft according to the third theories of strength.

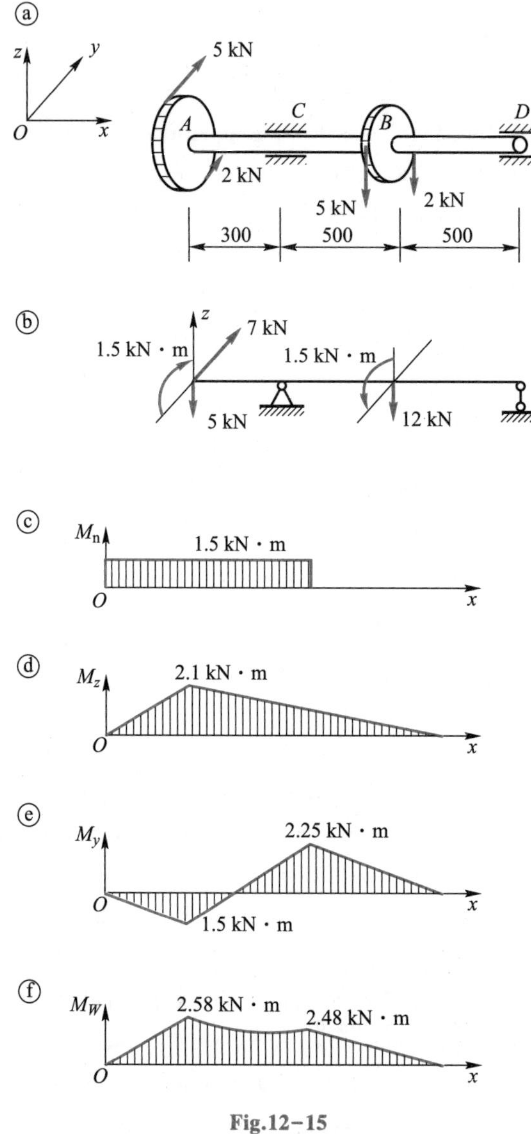

Fig.12−15

Solution (1) A sketch of the calculation is given in Fig.12−15b

(2) Make a diagram of the internal forces of the shaft

The torque diagram for the shaft is shown in Fig.12−15c and the bending moment diagram is shown in Fig.12−15d,e. Then the synthetic bending moments of section C and section B are

$$M_C=\sqrt{M_y^2+M_z^2}=\sqrt{1.5^2+2.1^2}\ \text{kN}\cdot\text{m}\approx 2.58\ \text{kN}\cdot\text{m}$$

$$M_B=\sqrt{M_y^2+M_z^2}=\sqrt{2.25^2+2.1^2}\ \text{kN}\cdot\text{m}\approx 2.48\ \text{kN}\cdot\text{m}$$

Synthetic bending moment diagram M_W is shown in Fig.12-15f. It can be proved that the curve of diagram M_W of section BC is concave. Obviously, section C is a dangerous section.

(3) Calibrated strength

By third theories of strength, from equation (12-20)

$$\frac{\sqrt{M_W^2+M_n^2}}{W}\leqslant[\sigma]$$

By substituting the corresponding data, there are

$$\frac{32\times\sqrt{(2.58\times10^3)^2+(1.5\times10^3)^2}}{\pi d^3}\leqslant 80\times10^6$$

This gives the required diameter $d=72$ mm.

Example 12-5 Fig.12-16a shows the propulsion shaft of a cargo ship. The power of the main engine is known as $P=7\ 277$ kW, speed $n=119$ r/min, effective propulsive force $F=767$ kN, the weight of the propeller blades $W_1=180$ kN, the total weight of the shaft outreach $W_2=45$ kN, diameter of the shaft $d=51.5$ cm. And there is $a_1=1.9$ m, $a_2=1.2$ m. The material is high quality carbon steel with a yield stress of $\sigma_s=250$ MPa, allowable factor of safety $[n]=4$. Check the strength of section A of the propulsion shaft according to the fourth theories of strength.

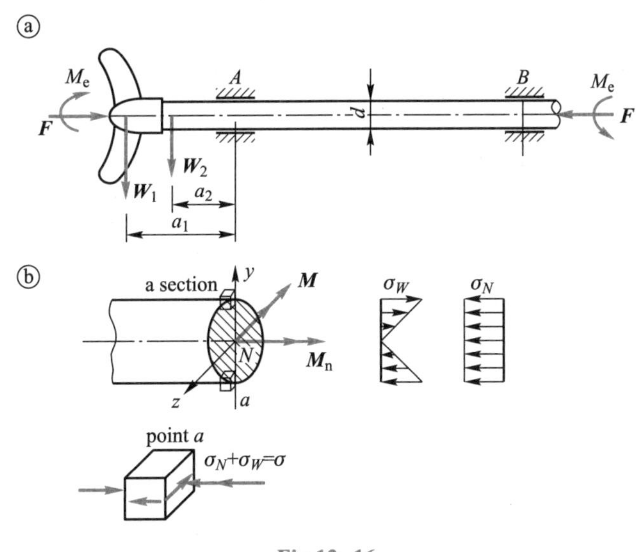

Fig.12-16

Solution (1) Calculate the internal force of the propulsion shaft

Torsional torque

$$M_e=9\ 549\times\frac{7\ 277}{119}\ \text{N}\cdot\text{m}=584\times10^3\ \text{N}\cdot\text{m}$$

The axial force, torque and bending moment on section A are

$$F_N=F=-767\ \text{kN}$$

$$M_n = M_e = 584 \text{ kN} \cdot \text{m}$$

$$M = W_1 a_1 + W_2 a_2 = (180\times1.9+4.5\times1.2) \text{ kN} \cdot \text{m} = 396 \text{ kN} \cdot \text{m}$$

The propulsion bearing is therefore subjected to a combination of compression, torsion and bending deformation.

(2) Calculate the stresses at the point of danger

Normal stresses σ_W and σ_N due to bending moments and axial forces are shown in Fig. 12-16b. The compressive stress is greatest at point a of the lower section edge, with a value of

$$\sigma = \frac{F_N}{A} + \frac{M}{W} = \left(\frac{767\times10^3\times4}{\pi\times0.515^2} + \frac{396\times10^3\times32}{\pi\times0.515^3}\right)\times10^{-6} \text{ MPa} = 33.2 \text{ MPa}$$

The maximum torsional shear stress occurs at the edge of the circular section and the value of point a is

$$\tau = \frac{M_n}{W_n} = \frac{584\times10^3\times16}{\pi\times0.515^3}\times10^{-6} \text{ MPa} = 21.8 \text{ MPa}$$

By fourth theories of strength, equation (12-21) gives

$$\sigma_{xd4} = \sqrt{\sigma^2+3\tau^2} = \sqrt{33.2^2+3\times21.8^2} \text{ MPa} = 50.3 \text{ MPa}$$

(3) Calibrated strength

The factor of safety of a shaft represents the safety reserve of the shaft in operation and is equal to the breaking stress divided by the working stress. The operating factor of safety for propulsion shafts is

$$n = \frac{\sigma_s}{\sigma_{xd4}} = \frac{250}{50.3} = 4.97 > 4$$

Therefore the shaft is safe.

Note that the strength condition in this example cannot be written as

$$\sigma_{xd3} = \frac{\sqrt{M_y^2+M_z^2+0.75M_n^2}}{W} + \frac{F_N}{A} \leqslant [\sigma]$$

12.4.2 Deformation of rectangular bars subjected to bending and torsion

For rectangular section bars, when there are both bending moments M_y and M_z acting on the section, they cannot be synthesized into a total bending moment M_W as in the case of circular axes. Because of $I_y \neq I_z$. So skew bending usually occurs for rectangular section. Plane bending of the beam occurs only when M_y or M_z acts alone. When M_y and M_z act simultaneously, we should calculate the normal stresses caused by M_y and M_z separately, and then superimpose them at the same point. Under the action of torque M_n, the shear stress at the midpoint of the long side of the rectangular section is the maximum. The shear stress at the midpoint of the short side is the local extreme value, and the shear stress at the corner point is zero. When a rectangular section bar undergoes bending and torsional deformation simultaneously, there will be bending moments M_y, M_z and torque M_n at the same time. At this time, the stress state of the dangerous point can have two types: (1) rectangular corner point, only normal stress, no shear stress, in a uniaxial stress state and the same strength conditions as bending; (2) the midpoint of the long or short side of the rectangular section, both bending stress and torsional shear stress, the same stress state as combined bending and torsional deformation of the circular axis. Therefore, the strength conditions (12-18) and (12-19) is still applicable, but the expression "calculated bending moment" cannot be used.

Example 12-6 A half-ring fixed at one end is subjected to forces as shown in Fig.12-17a. The circular bar has square section of 30 mm×30 mm. If the allowable stress in the material is known as $[\sigma]=150$ MPa. Check strength of sections 1—1 and 2—2 according to the third theories of strength.

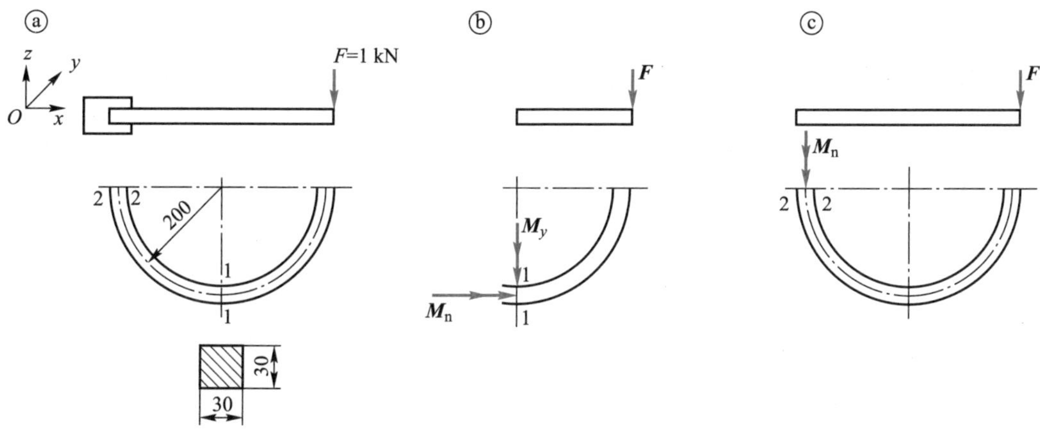

Fig.12-17

Solution Establish the coordinate system. The bar is truncated along the section 1—1 and 2—2 with a section perpendicular to the bar axis as shown in Fig.12-17b,c, respectively. From the section method, the internal force components on the section 1—1 are

$$M_n=10^3\times0.2\ \text{N}\cdot\text{m}=200\ \text{N}\cdot\text{m},\quad M_y=10^3\times0.2\ \text{N}\cdot\text{m}=200\ \text{N}\cdot\text{m},\quad M_z=0$$

The internal force components on the cross section 2—2 are

$$M_n=10^3\times0.4\ \text{N}\cdot\text{m}=400\ \text{N}\cdot\text{m},\quad M_y=M_z=0$$

Bending-torsional combination deformation occurs on the section 1—1. Under the action of torque, the maximum shear stress is at the midpoint of each side of the square section. And the largest normal stress is at two points farthest from the neutral axis. These two points are dangerous points. The normal and shear stresses at the two points are

$$\sigma=\frac{M_y}{W_y}=\frac{200\times6}{30^3\times10^{-9}}\times10^{-6}\ \text{MPa}=44.4\ \text{MPa}$$

$$\tau=\frac{M_n}{ahb^2}=\frac{200}{0.208\times30^3\times10^{-9}}\times10^{-6}\ \text{MPa}=35.5\ \text{MPa}$$

By third theories of strength, we have

$$\sigma_{xd3}=\sqrt{\sigma^2+4\tau^2}=\sqrt{44.4^2+4\times35.5^2}\ \text{MPa}=86.9\ \text{MPa}<[\sigma]$$

Torsional deformation occurs on section 2—2 only, the maximum torsional shear stress is

$$\tau=\frac{M_n}{ahb^2}=\frac{400}{0.208\times30^3\times10^{-9}}\times10^{-6}\ \text{MPa}=71.3\ \text{MPa}$$

Substituting the above strength condition equation gives

$$\sigma_{xd3}=2\tau=142.6\ \text{MPa}<[\sigma]$$

Therefore, both sections are safe.

Problems

12-1 A cantilever beam with a box section is subjected to the forces shown in Fig. P12-1. Calculate the normal stresses at points A, B, C of the fixed end.

Answer: $\sigma_A=2.60$ MPa, $\sigma_B=3.63$ MPa, $\sigma_C=27.54$ MPa

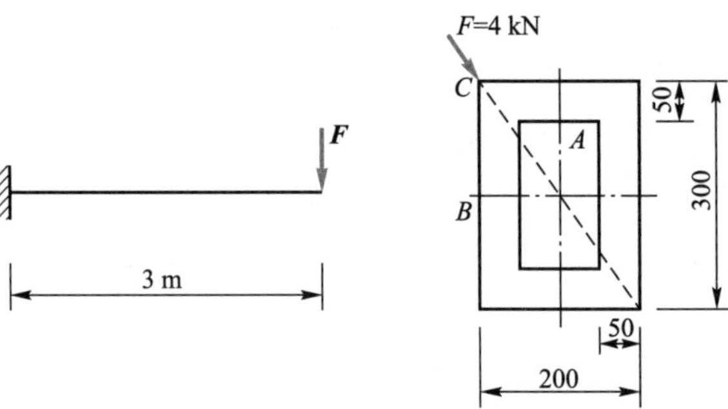

Fig.P12-1

12-2 The cantilever beam shown in the Fig. P12-2 is subjected to horizontal force F_1 and a perpendicular force F_2 at two different sections. $F_1=800$ N, $F_2=1\ 650$ N. Try to find the maximum normal stress in the beam and its location for the following two cases.

(1) The beam has a rectangular cross section with a width and height of $b=9$ cm, $h=18$ cm.

(2) The beam has circular section of diameter $d=13$ cm.

Answer: (1) $\sigma_{max}=9.98$ MPa; (2) $\sigma_{max}=10.7$ MPa

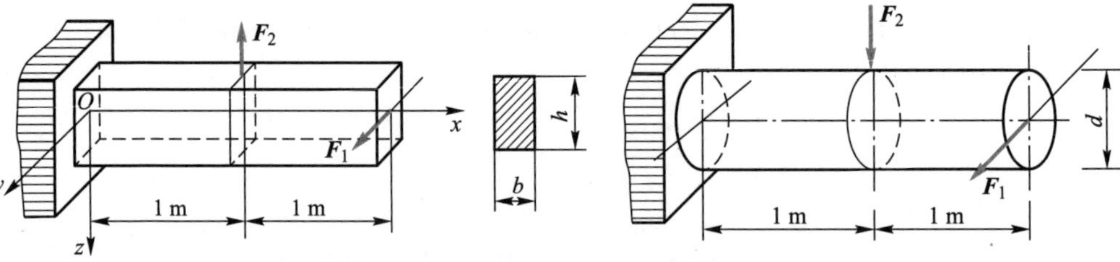

Fig.P12-2

12-3 The maximum lifting weight of the crane shown (including travelling trolley, etc.) is $P=40$ kN. The crossbeam AC consists of two No.18 channels of steel Q235 with allowable stress $[\sigma]=120$ MPa. Try to check the strength of the crossbeam (Fig.P12-3).

Answer: $\sigma_{max}=121$ MPa, exceeding the allowable stress by 0.75%, it can still be used

12-4 The column of the drilling machine shown in Fig.P12-4 is made of cast iron with a allowable tensile stress $[\sigma^+]=35$ MPa and $F=15$ kN. Try to determine the required diameter d of the column.

Answer: $d=122$ mm

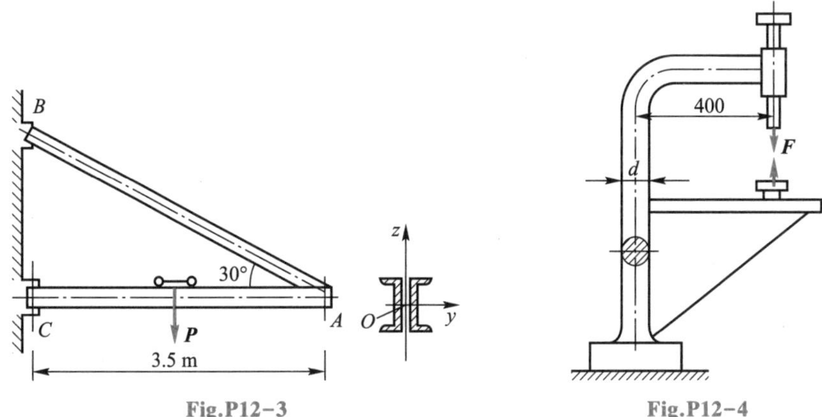

Fig.P12-3　　Fig.P12-4

12-5 The hand crank hoist is shown in the Fig.P12-5. Diameter of the shaft is $d=30$ mm. The material is steel Q235 and the allowable stress is $[\sigma]=80$ MPa. Try to find the maximum lifting weight of the hoist according to the third theories of strength.

Answer: $P=788$ N

12-6 Fig.P12-6 shows the main shaft of a certain type of water turbine. The output power of the turbine is $P=37\ 500$ kW, speed $n=150$ r/min. The axial thrust is known as $F_z=4\ 800$ kN, the weight of the runner $W_1=390$ kN, inner diameter of the main shaft $d=34$ cm, outer diameter $D=75$ cm, self weight $W=285$ kN. The spindle material is No.45 steel and the allowable stress is $[\sigma]=80$ MPa. Try to check the strength of this spindle according to the fourth theories of strength.

Answer: $\sigma_{xd4}=54.4$ MPa $<[\sigma]$

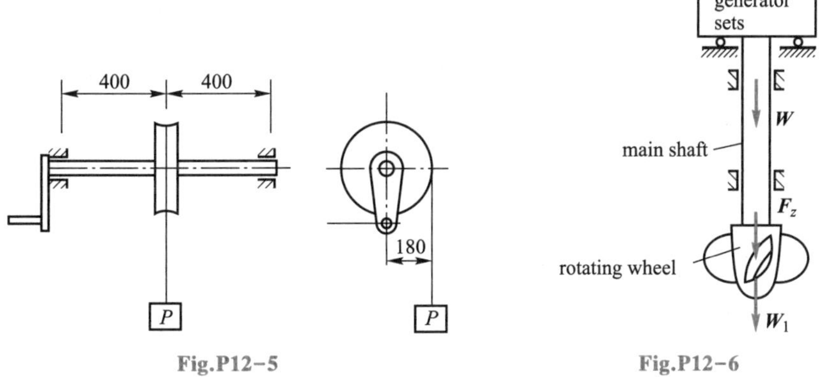

Fig.P12-5　　Fig.P12-6

12-7 Fig.P12-7 shows a diagram of a precision grinding machine with a grinding wheel shaft. The motor power is known as $P=3$ kW, rotor speed $n=1\ 400$ r/min. Rotor weight $W_1=101$ N, grinding wheel straight $D=250$ mm, grinding wheel weight $W_2=275$ N, grinding force $F_y : F_z=3 : 1$, grinding wheel shaft diameter $d=5$ cm. Material is bearing steel. Allowable stress is $M=60$ MPa.

(1) Try to represent the stress state at the point of danger by the unit body and find the principal stress and the maximum shear stress;

(2) Try to check the strength of the shaft by the third theories of strength.

Answer: (1) $\sigma_1=3.11$ MPa, $\sigma_2=0$, $\sigma_3=-0.22$ MPa, $\tau_{max}=1.67$ MPa;

(2) $\sigma_{xd3}=3.33$ MPa$<[\sigma]$, safe

12-8 Short columns of rectangular section beams subjected to eccentric pressures $F_1=25$ kN and transverse forces $F_2=5$ kN. The geometric scale of the short column is shown in the Fig.P12-8. Try to find the normal stresses at the four corner points A, B, C and D on the fixed end section and determine the position of the neutral axis of the section.

Answer: $\sigma_A=8.83$ MPa, $\sigma_B=3.83$ MPa, $\sigma_C=-12.17$ MPa, $\sigma_D=-7.17$ MPa, $a_y=15.7$ mm, $a_z=33.4$ mm

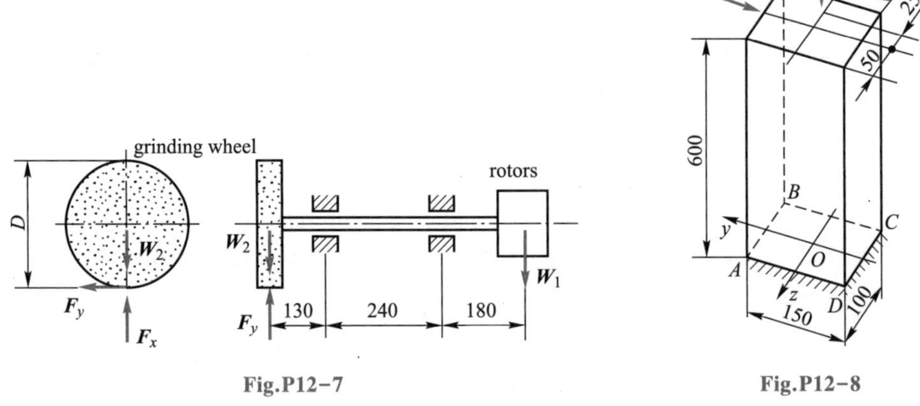

Fig.P12-7 **Fig.P12-8**

12-9 A circular shaft with a diameter of $d=30$ mm is subjected to the combined effect of torsional moment M_1 and couple moment M_2 in the horizontal plane Fig.P12-9. In order to measure M_1 and M_2, resistance strain gauges are applied to the shaft surface in the direction of the axis shown and the direction of 45° to the axis. If the measured strain values are $\varepsilon_0=500\times10^{-6}$, $\varepsilon_{45°}=426\times10^{-6}$, and we have known that $E=210$ GPa, $\mu=0.28$, try to find M_1 and M_2.

Answer: $M_1=214$ N · m, $M_2=278.3$ N · m

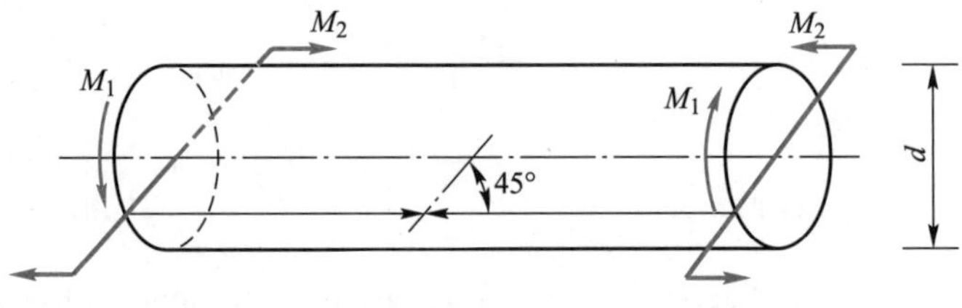

Fig.P12-9

Chapter 13 Energy Method

Teaching Scheme of Chapter 13

When a deformable solid deforms under the action of an external force, both the external and internal forces will do work. For linear elastic bodies, due to the reversible nature of the deformation, the work done by the external force at the corresponding displacement is numerically equal to the strain energy stored in the deformed body. When the external force is removed, all of this strain energy will be converted to other forms of energy, such as restoring the deformable body to its original state. This was discussed earlier in the previous basic deformation. The methods of solving for the displacement, deformation and internal forces of a deformable body using the concepts of work and energy are collectively known as the energy method.

The methods of solving problems using theorems and principles related to the concept of strain energy are known as the strain energy method. The energy method can be used to effectively study the deformation or displacement of bar structures such as rigid frames and curved rods, and also to solve for super-stationary structures. The energy method has a wide range of applications and is an important basis for computational solid mechanics.

This chapter first analyses the main properties of strain energy in terms of basic deformation of a bar, and then gives a general expression. We discuss several effective methods to solve structural deformation or displacement by strain energy, and finally give the reciprocity theorem of work and reciprocity theorem of displacement. Due to space limitations, the discussion in this chapter is confined to the linear elastic range.

13.1 Strain energy of a bar

This book studies the internal effects of forces on members and views them as a deformable body. The external force does work on the corresponding displacement. At the same time, the relative positions of the points within the deformable body also change and a new state of equilibrium is reached. The potential energy inside a deformable body as it reaches and maintains a new equilibrium state is called the strain energy.

As the external force on the deformed body slowly increases from zero to its final value, the deformation of the object also slowly increases from zero to its final value. During the whole deformation process, all forms of energy except the strain energy are very small and negligible. According to the law of conservation of energy, all the work W done by the external force is transformed into the strain energy U inside the object

$$U=W \tag{13-1}$$

This is the principle of strain energy.

It can be expressed as follows. Throughout the loading process, the strain energy of the object is numerically equal to the work done by the external force.The deformation of the object disappears with the release of the external force, and the strain energy is converted into work done by the external force and released. This is the reversible nature of elastic strain energy. For example, in the case of the clockwork of a mechanical clock, the deformation is produced by the external force when it is tightened and the strain energy is stored internally. When the external force is removed, the clockwork gradually returns to its original state by tightening and the strain energy is converted into work and released. After the elastic range has been exceeded, some of the non-recoverable energy is retained within the deformation body and only part of the strain energy is released as work.

13.1.1 Strain energy for basic deformation

The calculation of the strain energy under several basic deformations is now examined.

1. Axial tension or compression

For equally straight axial tension or compression, the external force is linearly related to the axial deformation of the bar within the elastic range. When the external force gradually increases from zero to the final value F, the axial deformation of the bar also gradually increases from zero to the final value Δl, as shown in Fig. 13-1. In the whole loading process, the work done by the external force is the area of the triangle OAB, i.e.

$$W=\frac{1}{2}F\Delta l \tag{13-2}$$

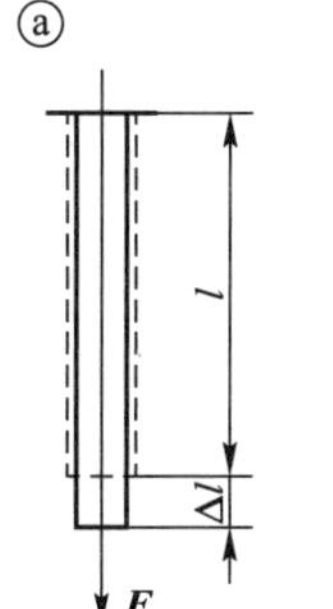

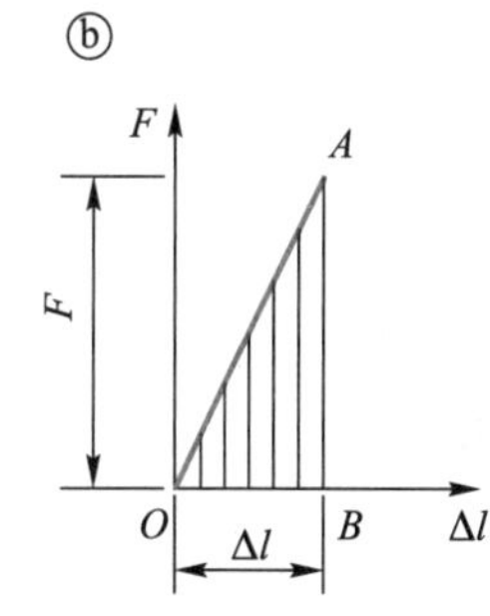

Fig.13-1

As can be seen from equation (13-1), this work is equal to the strain energy stored within the bar. When it is subjected to tension or pressure only at both ends, we have $F_N=F$, $\Delta l=\frac{F_N l}{EA}$. The strain energy of the bar can thus be written as

$$U=W=\frac{1}{2}F\Delta l=\frac{F_N^2 l}{2EA}=\frac{EA}{2l}(\Delta l)^2 \tag{13-3}$$

If the internal force varies continuously along the axis of the bar, i.e. $F_N=F_N(x)$, the strain energy dU within a micro-segment with axis length dx can be calculated first, and then dU can be integrated along the length of the bar. The strain energy of the bar can be

$$U=\int_l dU=\int_l \frac{F_N^2(x)\,dx}{2EA} \tag{13-4}$$

If the internal forces vary in steps, the strain energy of each segment can be calculated first, and then the strain energy of the whole structure can be obtained by summing

$$U=\sum_{i=1}^{m}U_i=\sum_{i=1}^{m}\frac{F_{Ni}^2 l_i}{2EA_i} \tag{13-5}$$

where m is the number of tension and compression bars.

The strain energy density (specific energy) per unit volume of a tension (compression) bar is

$$u=\frac{\mathrm{d}U}{\mathrm{d}V}=\frac{\sigma^2}{2E}=\frac{1}{2}\sigma\varepsilon=\frac{E}{2}\varepsilon^2 \tag{13-6}$$

2. Torsion of Circular shafts

For torsion of a circular shaft, as the external couple moment gradually increases from zero to the final value of M_e, the angle of twist also gradually increases from zero to the final value. The relationship between M and φ is also a straight line in the range of line elasticity, as shown in Fig.13–2. In the process of deformation, the work done by the torsional coupling can be expressed by the area of the triangle OAB, i.e.

$$W=\frac{1}{2}M_e\varphi \tag{13-7}$$

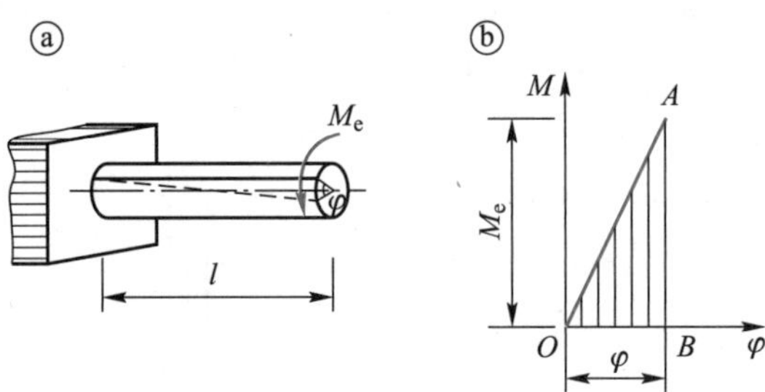

Fig.13–2

According to equation (13–1), this work is equal to the torsional strain energy stored in the circular shaft. When the circular shaft is subjected to external couples at both ends only, we have $M_n=M_e$ and $\varphi=\frac{M_n l}{GI_p}$. Thus, the torsional strain energy of the circular shaft can be written as

$$U=W=\frac{1}{2}M_e\varphi=\frac{M_n^2 l}{2GI_p}=\frac{GI_p}{2l}\varphi^2 \tag{13-8}$$

If the internal couple moment varies continuously along the axis of the circular axis, i.e. $M_n=M_n(x)$, the strain energy $\mathrm{d}U$ in a micro-segment of axis length $\mathrm{d}x$ can be calculated first, and then $\mathrm{d}U$ can be integrated along the axis to obtain the strain energy for the whole circular axis

$$U=\int_l \mathrm{d}U=\int_l \frac{M_n^2(x)\,\mathrm{d}x}{2GI_p} \tag{13-9}$$

If the internal couple moment varies in steps along the axis, the strain energy of each segment can be found first, and then the strain energy of the whole circular axis can be obtained by summing up

$$U=\sum_{i=1}^{m}U_i=\sum_{i=1}^{m}\frac{M_{ni}^2 l_i}{2GI_{pi}}u \tag{13-10}$$

The strain energy per unit volume of the circular shaft, i.e. the strain energy density in the pure shear state, is

$$u=\frac{\mathrm{d}U}{\mathrm{d}V}=\frac{1}{2}\tau\gamma=\frac{\tau^2}{2G}=\frac{G}{2}\gamma^2 \tag{13-11}$$

3. Bending

For the bending of a straight beam in plane, it subjectes to a concentrated couple moment at its free end. As the concentrated couple moment gradually increases from zero to its final value M_e, the angle of rotation at the free end of the cantilever beam also gradually increases from zero to its final value θ (Fig.13−3a). In the line elastic range, the bending deformation method can be applied to find $\theta=\frac{M_e l}{EI}$. Therefore, θ is linearly related to M_e. The work done by M_e can be expressed in terms of the area of the triangle OAB (Fig.13−3b), i.e.

$$W=\frac{1}{2}M_e\theta \tag{13-12}$$

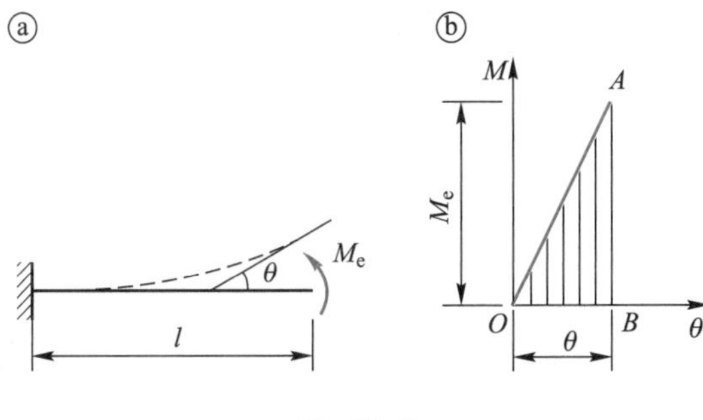

Fig.13−3

Since the cross-sectional moment of the beam in the purely bending state, the strain energy of the purely bending beam can be seen from equation (13−1) as

$$U=W=\frac{1}{2}M_e\theta=\frac{M_e^2 l}{2EI}=\frac{EI}{2l}\theta^2 \tag{13-13}$$

For the case of a straight beam in transverse-force bending, a micro-section with a long tail $\mathrm{d}x$ can be removed from the beam at x. Both sides of the micro-segment (Fig.13−4) act with bending moment and shear force, respectively. The bending moment and shear force should be a function of the section position coordinates x. Since $M(x)$ and $\theta(x)$ cause relative rotation and relative misalignment of the cross sections on both sides of the micro-segment, respectively, bending strain energy and shear strain energy are stored in the beam. However, in the case of slender beams, the shear strain energy is much smaller than the bending strain energy and can be omitted. Therefore, when calculating the strain energy of cross-force bending beams, the effect of shear on deformation is usually not considered. Thus, for the micro-segment shown in Fig.13−4b, using equation (13−13) we can find the strain energy

$$\mathrm{d}U=\frac{M^2(x)\,\mathrm{d}x}{2EI}$$

The integration of $\mathrm{d}U$ along the entire length of the beam gives the strain energy for the whole beam

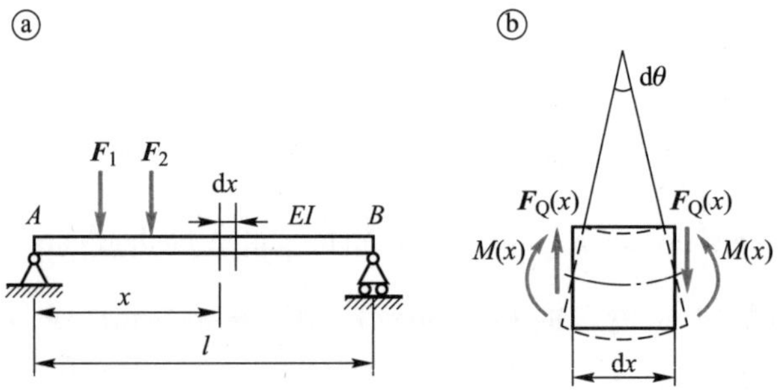

Fig.13-4

$$U = \int_l \mathrm{d}U = \int_l \frac{M^2(x)\,\mathrm{d}x}{2EI} \tag{13-14}$$

In summary, the strain energy of the bar is numerically equal to the work done by the external force during the bar. In the range of linear elasticity and under static load, the strain energy of a bar expressed by equations (13-2), (13-7) and (13-12) can be expressed uniformly as

$$U = W = \frac{1}{2}F\delta \tag{13-15}$$

Where F and δ denote the generalized force and its corresponding generalized displacement, respectively. When F denotes force, δ denotes the displacement; when F denotes force couple moment, δ denotes the angular displacement. In the case of linear elastic body, the relationship between the generalized force and the generalized displacement is linear.

13.1.2 Characteristics of elastic strain energy

From the above discussion, it can be seen that the elastic strain energy is a quadratic function of the generalized internal force or generalized displacement of the bar. M_1 and M_2 denote the bending moments caused by the two external forces acting alone in Fig13-4a, respectively. When they act together, the bending moments of the beam should be M_1+M_2 according to the superposition principle. The strain energy of the beam can then be found from equation (13-14) as

$$\begin{aligned} U &= \int_l \frac{M^2(x)\,\mathrm{d}x}{2EI} = \int_l \frac{M_1^2(x)\,\mathrm{d}x}{2EI} + \int_l \frac{M_2^2(x)\,\mathrm{d}x}{2EI} + \int_l \frac{M_1(x)M_2(x)\,\mathrm{d}x}{EI} \\ &= U_1 + U_2 + \int_l \frac{M_1(x)M_2(x)\,\mathrm{d}x}{EI} \end{aligned} \tag{13-16}$$

where U_1 and U_2 represent the strain energy caused by the action of F_1 and F_2 alone, respectively.

Obviously $U \neq U_1+U_2$, the deformation energies cannot be superimposed in general. However, in some cases, the strain energy can be superimposed. If a number of loads act on the bar, and any load does not do work on the deformation or displacement caused by other loads, that is, the strain energy caused by any load is independent. In this case, the total strain energy of the bar is equal to the sum of the deformation energies caused by each of the above-mentioned loads individually.

Another main characteristic of the elastic strain energy is that it is independent of the order of loading and depends entirely on the final value of the load and displacement. If the strain energy of a linear elastic body is related to the order of loading, then loading it in two different orders and unloading it in the same order will add or subtract some energy from the linear elastic body. This is contrary to the law of conservation of energy.

In addition, as the strain energy is the potential energy stored in the linear elastic body, when the external force is released, the linear elastic body releases the potential energy to do normal work externally. So, the elastic strain energy is always normal. This can be seen from the expressions for the strain energy for the various basic deformations above.

13.1.3 The Clapeyron's theorem

The calculation of the strain energy is now extended to the general case. Since the strain energy is independent of the order of loading and depends only on the final value of the external force and displacement, it can be assumed that each load increases gradually from zero to its final value in the same proportion. If the material is linearly elastic and the deformation is small, the relationship between the deformation of the structure and the external forces is also linearly elastic. They will also increase gradually from zero to the final value in the same proportion as the external force. As shown in Fig. 13−5, δ_i denotes the generalized displacement of the generalized force F_i at the point of action along its direction of action. δ_i can be written as

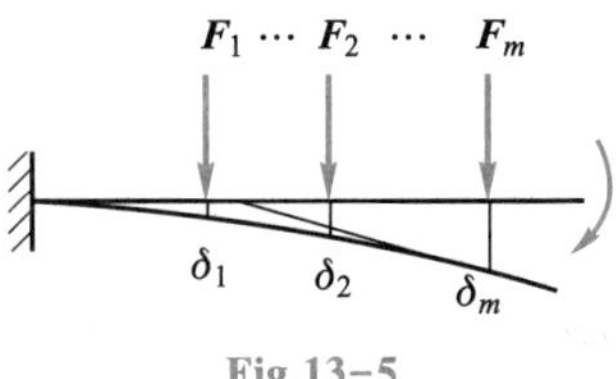

Fig.13−5

$$
\begin{aligned}
\delta_i &= \delta_{i1}+\delta_{i2}+\cdots+\delta_{ii}+\cdots+\delta_{im} \\
&= \beta_1 F_1+\beta_2 F_2+\cdots+\beta_i F_i+\cdots+\beta_m F_m \\
&= F_i\left(\beta_1 \frac{F_1}{F_i}+\beta_2 \frac{F_2}{F_i}+\cdots+\beta_i \frac{F_i}{F_i}+\cdots+\beta_m \frac{F_m}{F_i}\right)
\end{aligned}
\tag{13-17}
$$

where δ_{i1} represents the generalized displacement at the point of F_i along its action direction. It is caused by the generalized force F_1. The rest are similar. $\beta_1, \cdots, \beta_m$ is a constant related to the structure. In the scaled loading process, $\frac{F_1}{F_i}, \cdots, \frac{F_m}{F_i}$ are constant, so that the relationship δ_i with F_i is also linear.

Thus, the work done by F_i is $\frac{1}{2}\delta_i F_i$. The sum of the work done by each load is numerically equal to the strain energy of the structure,

$$
U = W = \sum_{i=1}^{m} \frac{1}{2}\delta_i F_i \tag{13-18}
$$

This conclusion is called Clapeyron's theorem. It is only applicable to calculate the strain energy and work in linear elastic structures.

13.1.4 Strain energy for combined deformation

Using the general expression for strain energy (13−18), the variable energy of a bar subjected to the combined action of bending, torsion and axial tension can be obtained. Now intercept a micro-segment of length dx in the bar (Fig.13−6), if the axial force, bending moment and torque in the cross section are $F_N(x)$, $M(x)$ and $M_n(x)$ (for the micro-section dx, $F_N(x)$, $M(x)$ and $M_n(x)$ should be regarded as external forces).

Fig.13−6

The relative axial displacement, rotation angle and torsion angle between the two end cross sections are $d(\Delta l)$, $d\theta$ and $d\varphi$, respectively. Since the deformations caused by each of $F_N(x)$, $M(x)$ and $M_n(x)$ are independent of each other, then according to equation (13−18), the variation within the micro-segment dx should be

$$\begin{aligned} dU &= \frac{1}{2}F_N(x)\,d(\Delta l) + \frac{1}{2}M(x)\,d\theta + \frac{1}{2}M_n(x)\,d\varphi \\ &= \frac{F_N^2(x)\,dx}{2EA} + \frac{M^2(x)\,d\theta}{2EI} + \frac{M_n^2\,dx}{2GI_n} \end{aligned}$$

The deformation of the entire combined deformed bar can then be integrated into the above equation,

$$U = \int_l dU = \int_l \frac{F_N^2(x)\,dx}{2EA} + \int_l \frac{M^2(x)\,d\theta}{2EI} + \int_l \frac{M_n^2\,dx}{2GI_n} \tag{13-19}$$

Here, GI_n it is the torsional rigidity of the bar. In the case of a bar of circular section, I_n should be replaced by I_p.

Example 13−1 Try to find the strain energy of the square truss structure shown in Fig.13−7 and find the relative displacements at points A and C. It is known that each bar has the same tensile and compressive rigidity EA.

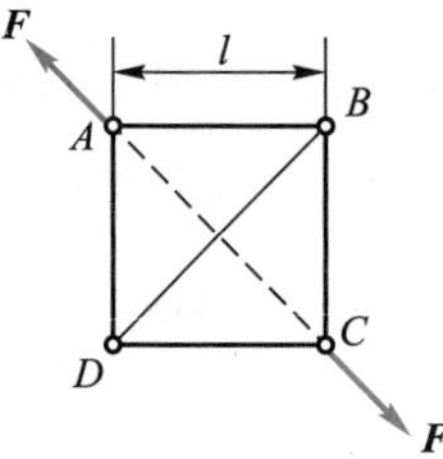

Fig.13−7

Solution First find the axial force of each bar of the Fig.13−7

$$F_{NAB} = F_{NBC} = F_{NCD} = F_{NAD} = \frac{\sqrt{2}}{2}F$$

$$F_{NBD} = -F$$

The strain energy of the whole structure can be calculated according to the following formula, namely

$$U = \sum_{i=1}^{5} \frac{F_{Ni}^2 l_i}{2EA} = 4\sum_{i=1}^{5} \frac{F_{NAB}^2 l_i}{2EA} + \sum_{i=1}^{5} \frac{F_{NBD}^2 \sqrt{2}\, l_i}{2EA} = \left(1 + \frac{\sqrt{2}}{2}\right)\frac{F^2 l}{EA}$$

It can be assumed that point C is fixed, then the relative displacement δ_{AC} of points A and C is the displacement of point A along $\boldsymbol{F}$. Thus, during the deformation of the structure, the work done by the external force is

$$W=\frac{1}{2}F\delta_{AC}$$

Since $U=W$, we have

$$F\delta_{AC}=(2+\sqrt{2})\frac{F^2l}{EA}$$

From this we can find

$$\delta_{AC}=(2+\sqrt{2})\frac{Fl}{EA}$$

Structures in the form of frames consisting of two or more bars are often encountered in engineering. In such structures, if no relative rotation occurs at the joint between the bars, or if the angle between the connecting bars remains constant before and after deformation, this structure is called a rigid frame. Rigid connection points are called rigid nodes and are sometimes indicated by small black corners. A simple steel frame is shown in Fig. 13-8, where point B is the steel joint. Because the two bars connected by the steel joint do not rotate relative to each other at the node, the internal forces at the joint are forces and moments. The axis of each bar is in the same plane, and the plane is the principal centroidal plane. This kind of rigid frame is called plane rigid frame. If the external forces are acting in the above plane, and the bending deformation occurred in the rigid frame must be plane bending.

Example 13-2 Fig. 13-8 shows a plane rigid frame with a fixed end C. A vertical downward concentrated force F is loaded at end A. The bending and tensile rigidities of the frame are known to be EI and EA, respectively. Try to find the vertical displacement δ_A of end A.

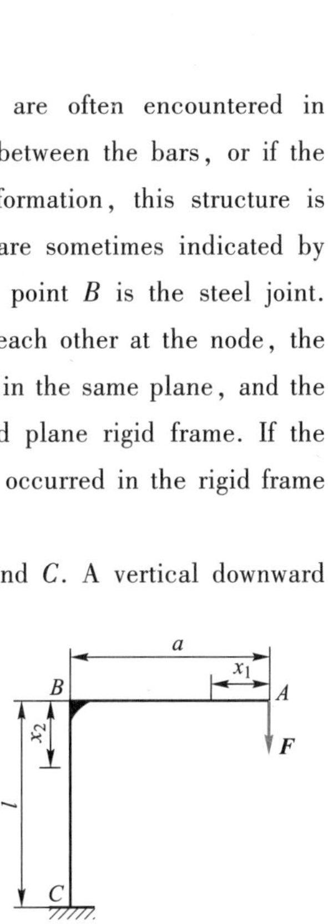

Fig.13-8

Solution Firstly, the internal force values of each segment of the steel frame are found by the section method. The axial force is positive in tension and negative in compression, while the bending moment is not specified as positive or negative. From Fig.13-8, we can find

Section AB: $M(x_1)=Fx_1$, $F_N(x_1)=0$

Section BC: $M(x_2)=Fa$, $F_N(x_2)=-F$

The strain energy of the entire steel frame can be calculated in subparagraphs according to equation (13-14) and then found,

$$\begin{aligned}U&=\int_0^a\frac{M^2(x_1)\,dx_1}{2EI}+\int_0^l\frac{M^2(x_2)\,dx_2}{2EI}+\int_0^l\frac{F_N^2(x_2)\,dx_2}{EA}\\&=\int_0^a\frac{(Fx_1)^2dx_1}{2EI}+\int_0^l\frac{(Fa)^2dx_2}{2EI}+\int_0^l\frac{F^2dx_2}{EA}\\&=\frac{(Fa)^2}{2EI}\left(1+\frac{a}{3}\right)+\frac{F^2l}{2EA}\end{aligned}$$

The work done by the concentrated force F at point A should be equal to the strain energy of the steel frame, i.e.

$$W=\frac{1}{2}F\delta_A=U=\frac{(Fa)^2}{2EI}\left(1+\frac{a}{3}\right)+\frac{F^2l}{2EA}$$

The vertical displacement of section A is thus found to be

$$\delta_A=\frac{Fa^2}{2EI}\left(1+\frac{a}{3}\right)+\frac{Fl}{2EA}$$

The first term in equation corresponds to the displacement caused by the bending deformation and the second term corresponds to the displacement caused by the axial tensile and compressive deformation. If $a=l$ and cross section diameter is $d(l=10d)$, then

$$\delta_A=\frac{4}{3}\frac{Fl^3}{EI}+\frac{Fl}{EA}=\frac{4}{3}\frac{Fl^3}{EI}\left(1+\frac{3I}{4Al^2}\right)=\frac{4}{3}\frac{Fl^3}{EI}\left(1+\frac{3}{6\ 400}\right)$$

The second term in brackets is less than 0.05%. So, the effect of axial forces can generally be neglected when solving for deformations or displacements in bending resistant bar structures.

Example 13 – 3 A plane curved bar with a semi-circular axis is shown in Fig. 13 – 9a. A concentrated force perpendicular to the plane in which the axis is located is acting at the free end A. Try to find the vertical displacement of section A.

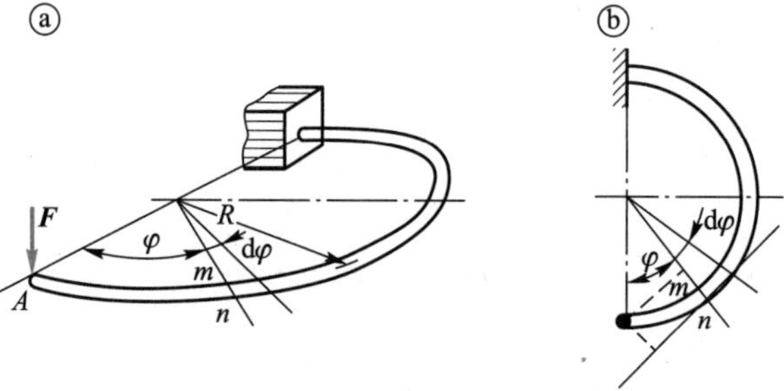

Fig.13–9

Solution The position of any cross section $m—n$ can be determined by the angle. From Fig.13–9b, it can be seen that the torsion and bending on the cross section $m—n$ are

$$M_n=FR(1-\cos\varphi)$$

$$M=FR\sin\varphi$$

And the strain energy in the micro-segment of mn length $R\mathrm{d}\varphi$ is

$$\mathrm{d}U=\frac{M_n^2R\mathrm{d}\varphi}{2GI_p}+\frac{M^2R\mathrm{d}\varphi}{2EI}=\frac{F^2R^3(1-\cos\varphi)^2\mathrm{d}\varphi}{2GI_p}+\frac{F^2R^3\sin\varphi\mathrm{d}\varphi}{2EI}$$

Integration of the above equation gives the strain energy of the whole curved bar

$$U=\int_l\mathrm{d}U=\int_l\frac{F^2R^3(1-\cos\varphi)^2\mathrm{d}\varphi}{2GI_p}+\int_l\frac{F^2R^3\sin\varphi\mathrm{d}\varphi}{2EI}$$

$$=\frac{3F^2R^3\pi}{4GI_p}+\frac{F^2R^3\pi}{4E}$$

Let the F force act on the vertical displacement position of point A. During the deformation, the work done by the external force is numerically equal to the strain energy of the curved bar, i.e.

$$W=\frac{1}{2}F\delta_A=U=\frac{3F^2R^3\pi}{4GI_p}+\frac{F^2R^3\pi}{4EI}$$

The result is

$$\delta_A=\frac{3FR^3\pi}{2GI_p}+\frac{FR^3\pi}{2EI}$$

13.2 Mohr theorem

Mohr theorem is an effective tool for determining displacement at any point and in any direction. It is also known as Mohr method, unit force method or unit load method. The concept and properties of strain energy are now used to derive Mohr theorem, using a beam as an example.

Suppose the beam is bent and deformed under the action of an external force $F_1, F_2, \cdots$, as shown in Fig.13-10a. We calculate the deflection δ at any point C on the beam under the action of the above external force.

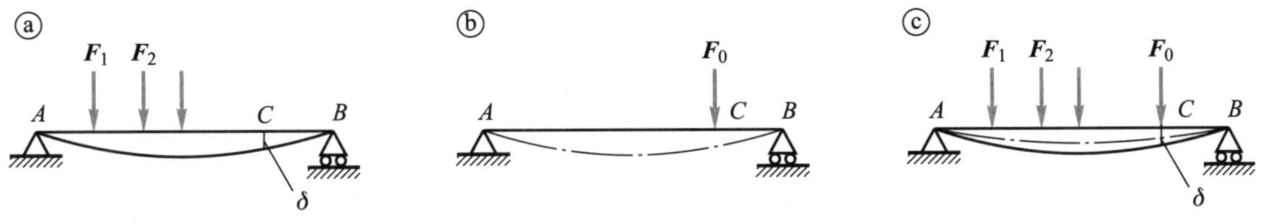

Fig.13-10

Firstly, the bending moment $M(x)$ of the beam can be found from the external force, and then the strain energy caused by $M(x)$ can be found according to equation (13-16)

$$U=\int_l\frac{M^2(x)\,dx}{2EI} \tag{13-20}$$

Then imagine that the load is removed from the beam and a unit force $F_0=1$ is applied at point C in the direction of deflection (Fig.13-10b). At this time, the bending moment of the beam is $M^0(x)$, and the deformation stored in the beam can be

$$U=\int_l\frac{[M^0(x)]^2\,dx}{2EI} \tag{13-21}$$

Finally, $F_1, F_2, \cdots$, is added back to the beam on top of Fig.13-10b (Fig.13-10c). At this point, the beam is further deformed from the dashed position to the solid position. If the material is within the range of linear elasticity and the deformation is small, the work done by $F_1, F_2, \cdots$, is the same as in the case of Fig.13-10a. That is, the strain energy increased by the action of the external force $F_1, F_2, \cdots$, and the deflection increased at point C are not changed by the action of the unit force F_0.

The strain energy increased by the action of these external forces on the beam is still equation (13-20). And the increased deflection at point C due to these external forces is still δ.Thus, during the reloading process of $F_1, F_2, \cdots$, the unit force F_0 completes the work with the value $F_0\delta$ again. In the case of Fig.13-10c, the strain energy of the beam can be

$$U_1 = U + U_0 + F_0\delta \tag{13-22}$$

Since the bending moment under the joint action of F_0 and $F_1, F_2, \cdots$, is $M(x) + M^0(x)$, the strain energy of the beam shown in Fig.13-10c can also be expressed as

$$U_1 = \int_l \frac{[M(x) + M^0(x)]^2}{2EI} dx \tag{13-23}$$

The deformation energies expressed in equations (13-22) and (13-23) are equal, i.e.

$$U + U_0 + F_0\delta = \frac{[M(x) + M^0(x)]^2}{2EI} dx$$

Substituting equations (13-20) and (13-21) into the left-hand side of the above equation and considering $F_0 = 1$, we get

$$F_0\delta = \delta = \int_l \frac{M(x)M^0(x)}{EI} dx \tag{13-24}$$

This is Mohr theorem also known as the Mohr integration. The sign of the bending moment $M^0(x)$ of the beam caused by the unit force is specified in the same way as $M(x)$.

The Mohr integration equation (13-24) is not restricted to finding the deflection of a section on a beam. If the angle of rotation at point C in Fig.13-10a is required, it is sufficient to replace the unit force at point C with a unit couple moment. By following the same procedure as above, the Mohr integration (13-24) can be derived. At this time, the equation $M^0(x)$ represents the bending moment of the beam due to the unit couple moment, and the resulting δ represents the angle of rotation of section C. The displacement δ in equation (13-24) is generalized. The method of solving structural displacements using Mohr theorem is also known as the unit load method, where the unit load is in a generalized sense.

Mohr theorem can also be used to solve the bending deformation of plane curved bar. For small curvature curved bar, if ignoring the effect of shear and axial forces on the deformation (the effect is very small), the Mohr integral formula for straight beam can be extended to obtain the Mohr integral for the bending deformation of the curved bar

$$\delta = \int_s \frac{M(s)M^0(s)}{EI} ds \tag{13-25}$$

where δ can be either the deflection or the angle of rotation of the curved bar; s is the arc length coordinate of the curved bar axis; $M(s)$ and $M^0(s)$ denote the bending moment on the cross section under the action of loads and generalized unit force, respectively.

It is also very convenient to calculate the nodal displacements of a truss structure using Mohr theorem. The truss structure is composed of many bars subject to axial tension or compression. The axial force of each bar is constant along the rod length and the strain energy of the whole truss structure can be calculated by equation (13-5). To calculate the displacement of a node in a certain direction, a unit force pointing to that direction can be applied to the node. The formula for calculating the displacement of the node can be derived according to the above method

$$\delta = \sum_{i=1}^{n} \frac{F_{Ni} F_{Ni}^{0} l_i}{EA_i} \tag{13-26}$$

where F_{Ni}^{0} denotes the axial force of the ith bar of the truss caused by the unit force; n is the number of bars in truss.

The following equation for calculating structural deformation (displacement) using Mohr theorem is extended to the general case.

Let the bars in a structure be subjected to the combined loads of tension (or compression), torsion and bending. If you want to calculate the displacement of a point in a certain direction, you can apply a unit load at that point along the same direction. Furthermore, the bending energy of each bar under the action of loads and unit force can be calculated according to equation (13-19). The Mohr formula for calculating the displacement of a combined deformed structure is

$$\delta = \sum_{i=1}^{m} \int_l \frac{M_i(x) M_i^0(x)}{EI_i} \mathrm{d}x + \sum_{i=1}^{m} \int_l \frac{M_{ni}(x) M_{ni}^0(x)}{GI_{ni}} \mathrm{d}x + \sum_{i=1}^{m} \int_l \frac{F_{Ni}(x) F_{Ni}^0(x)}{EA_i} \mathrm{d}x \tag{13-27}$$

Example 13-4 The cantilever beam subjected to uniform load is shown in Fig.13-11a. If EI is a constant, try to use Mohr theorem to calculate the deflection and deflection angle of section A at the free end.

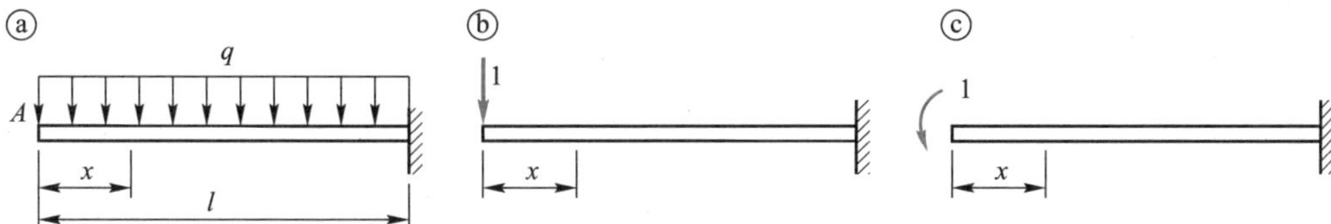

Fig.13-11

Solution Firstly, the moment equation of the cantilever beam under the action of the uniform laydown load is obtained according to the coordinates taken in Fig.13-11a.

$$M(x) = -\frac{qx^2}{2}$$

In calculating the deflection of section A, a unit force is applied to the section. This is shown in Fig.13-11b. The bending moment caused by the unit force is

$$M^0(x) = -x$$

By Mohr theorem the deflection of section A is

$$f_A = \int_0^l \frac{M(x) M^0(x)}{EI} \mathrm{d}x = \int_0^l \left(-\frac{qx^2}{2}\right)(-x)\frac{\mathrm{d}x}{EI} = \frac{ql^4}{8EI}$$

The positive result indicates that f_A is the same direction as the unit force. The direction of deflection of section A is downwards.

In calculating the deflection angle of section A, a unit couple should be applied to the section as shown in Fig.13-11c. The bending moment is $M^0(x) = -1$. According to Mohr theorem, the deflection angle of section A is

$$\theta_A = \int_0^l \frac{M(x)M^0(x)}{EI}\mathrm{d}x = \frac{1}{EI}\int_0^l \left(-\frac{qx^2}{2}\right)(-1)\mathrm{d}x = \frac{ql^3}{6EI}$$

θ_A is positive, indicating θ_A the same direction as the unit force couple. The angle of rotation of section A is counterclockwise.

Example 13-5 A simple truss structure is subjected to forces as shown in Fig.13-12a. Let the tensile (compressive) rigidity EA of each bar be the same. Try to find the relative displacement between points B and D.

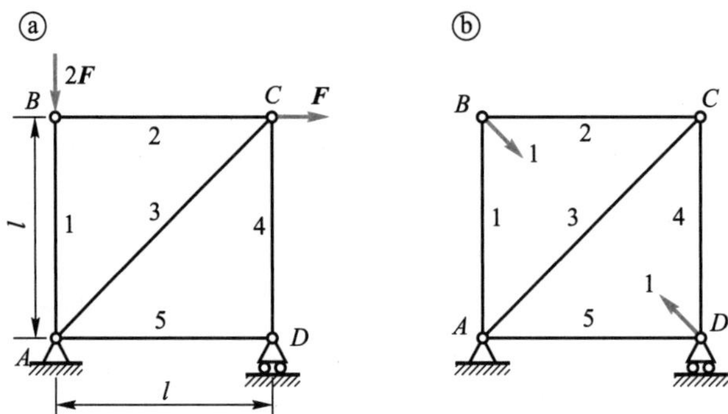

Fig.13-12

Solution First, according to Fig.13-12a, use the equilibrium equation of the nodes to find out the axial force F_{Ni} caused by the load of each bar, listed in table 13-1. To calculate the relative displacement δ_{BD} between nodes B and D, a unit force should be applied at each of points B and D along the BD line with opposite directions (Fig.13-12b). The axial forces of each bar caused by such a pair of unit forces F_{Ni}^0 are also listed in table 13-1. Substituting the data listed in table 13-1 into equation (13-5), the relative displacement between the two points is calculated as

$$\delta_{BD} = \sum_{i=1}^{5} \frac{F_{Ni}F_{Ni}^0 l_i}{EA} = \left(2 + \frac{3}{2}\sqrt{2}\right)\frac{Fl}{EA} \approx 4.12\frac{Fl}{EA}$$

The sign of δ_{BD} is positive, indicating that the relative displacement of the two points is in the same direction as the unit force.

Table 13-1

Rod number	F_{Ni}	F_{Ni}^0	l_i	$F_{Ni}F_{Ni}^0 l_i$
1	$-2F$	$-\frac{\sqrt{2}}{2}$	l	$\sqrt{2}Fl$
2	0	$-\frac{\sqrt{2}}{2}$	l	0

Table 13-1 (continued)

Rod number	F_{Ni}	F^0_{Ni}	l_i	$F_{Ni}F^0_{Ni}l_i$
3	$\sqrt{2}F$	1	$\sqrt{2}l$	$2Fl$
4	$-F$	$-\frac{\sqrt{2}}{2}$	l	$\frac{\sqrt{2}}{2}Fl$
5	0	$-\frac{\sqrt{2}}{2}$	l	0

Example 13-6 A steel frame of circular section is subjected to forces as shown in Fig.13-13a. The torsional rigidity of the whole frame is GI_p and EI, respectively. If the effect of shear on deformation is excluded, try to find the displacement δ_C of section C along the vertical direction.

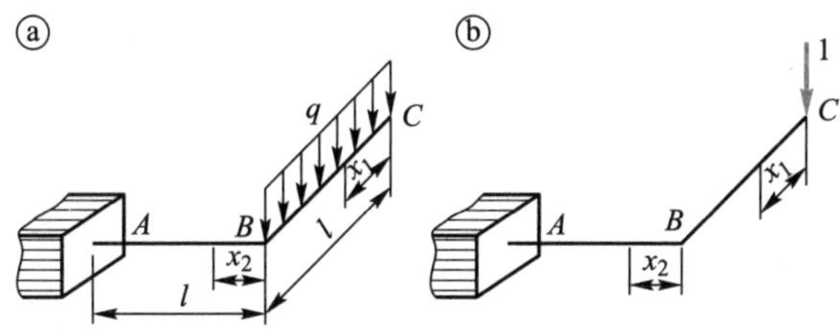

Fig.13-13

Solution The vertical displacement of section C is required by applying a unit force in section C along the vertical direction (Fig.13-13b). The internal forces in each segment of the steel frame can be calculated according to the coordinates taken in Fig. 13 - 13a and b. The positive and negative internal forces in each segment can still follow the sign regulations for the internal forces in the bar under various basic deformations. Thus there are

Section BC: $M(x_1)=-\frac{qx_1^2}{2}$, $\quad M^0(x_1)=-x_1$

Section AB: $M(x_2)=-qlx_2$, $\quad M_n(x_2)=-\frac{ql^2}{2}$

$M^0(x_2)=-x_2$, $\quad M^0_n(x_2)=-l$

Using equation (13-27) the numerical displacement of section$C$ can be found as

$$\delta_C=\int_0^l\frac{M(x_1)M^0(x_1)}{EI}\mathrm{d}x_1+\int_0^l\frac{M(x_2)M^0(x_2)}{EI}\mathrm{d}x_2+\int_0^l\frac{M_n(x_2)M^0_n(x_2)\mathrm{d}x_2}{GI_p}$$

$$=\int_0^l\frac{\left(-\frac{qx_1^2}{2}\right)(-x_1)}{EI}\mathrm{d}x_1+\int_0^l\frac{(-qlx_2)(-x_2)}{EI}\mathrm{d}x_2+\int_0^l\frac{\left(-\frac{ql^2}{2}\right)(-l)\mathrm{d}x_2}{GI_p}$$

$$=\frac{11ql^4}{24EI}+\frac{ql^4}{2GI_p}$$

The calculations show that the vertical displacement at point C is in the downward direction.

Example 13－7 The small curvature bar is shown in Fig. 13－14a. Try to find the vertical displacement and the angle of rotation of the free end A. The EI is a constant.

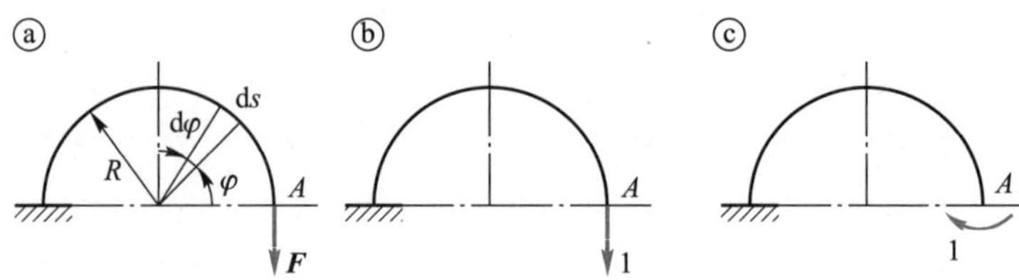

Fig.13－14

Solution The internal forces on a plane crank are generally axial force, shear force and bending moment. But the axial force and shear force on the deformation of the impact is very small, which are ignored. According to the coordinates taken in Fig.13－14a, the bending moment is

$$M(\varphi)=FR(1-\cos\varphi)$$

In finding the vertical displacement at point A, the corresponding bending moment of unit force at point A (Fig.13－14b) is

$$M^0(\varphi)=R(1-\cos\varphi)$$

Then using equation (13－25), find the vertical displacement of point A as

$$\delta_A=\int_l\frac{M(s)M^0(s)}{EI}\mathrm{d}s=\int_0^{\pi}\frac{M(\varphi)M^0(\varphi)R\mathrm{d}\varphi}{EI}\mathrm{d}s$$

$$=\frac{1}{EI}\int_0^{\pi}FR^2(1-\cos\varphi)^2R\mathrm{d}\varphi=\frac{3\pi PR^3}{2EI}$$

The calculations show that the vertical displacement at point A is in the downward direction.

In calculating the angle of rotation at point A, a unit couple moment (Fig. 13－14c) should be applied at point A, which can be found $M^0(\varphi)=1$.

Using equation (13－25), we get

$$\theta_A=\int_s\frac{M(s)M^0(s)}{EI}\mathrm{d}s=\int_0^{\pi}\frac{M(\varphi)M^0(\varphi)R\mathrm{d}\varphi}{EI}=\int_0^{\pi}\frac{FR(1-\cos\varphi)1R\mathrm{d}\varphi}{EI}=\frac{\pi FR^2}{EI}$$

The calculations show that the angle of rotation at point A is clockwise.

Based on the above discussion and analysis, steps of applying Mohr theorem to solve structural displacements are summarized as follows.

(1) Mohr theorem only applies to linear elastic structures where the material obeys Hooke law and deformations are small.

(2) When calculating the generalized displacement along a direction at a point, a generalized unit load corresponding to the requested displacement shall be applied at that point along that direction; if calculating the generalized relative displacement at two points in a structure, a generalized unit force corresponding to the requested displacement shall be applied at each of those points in the direction of the requested displacement. The pair of unit forces shall be in opposite directions (for the calculation of

relative angular displacements, see example 13-8 in section 13.3).

(3) In the calculation of the Mohr integration, if the internal force function is segmented continuous smooth, the integration should also be divided into segments. The coordinates of each segment must be the same.

13.3 Diagram multiplication method for Mohr integration

In calculating the Mohr integration, most members are straight bars of equal section. For the bending deformation of beams, EI is constant, so that equation (13-24) becomes

$$\delta = \frac{1}{EI}\int_l M(x)M^0(x)\,\mathrm{d}x \qquad (13\text{-}28)$$

$M^0(x)$ is the internal force caused by the unit load. It must consist of a straight line or a broken line. The $M(x)$ and $M^0(x)$ diagrams of Fig. 13-15 are the moments caused by the loads and unit force, respectively. where the graph of $M^0(x)$ is a section of oblique straight line. The equation of this straight line is

$$M^0(x) = kx+b$$

Substituting the above equation into equation (13-28), we get

$$\begin{aligned}\delta &= \frac{1}{EI}\int_l M(x)M^0(x)\,\mathrm{d}x \\ &= \frac{1}{EI}\left[k\int_l M(x)x\,\mathrm{d}x + b\int_l M(x)\,\mathrm{d}x\right] \qquad (13\text{-}29)\end{aligned}$$

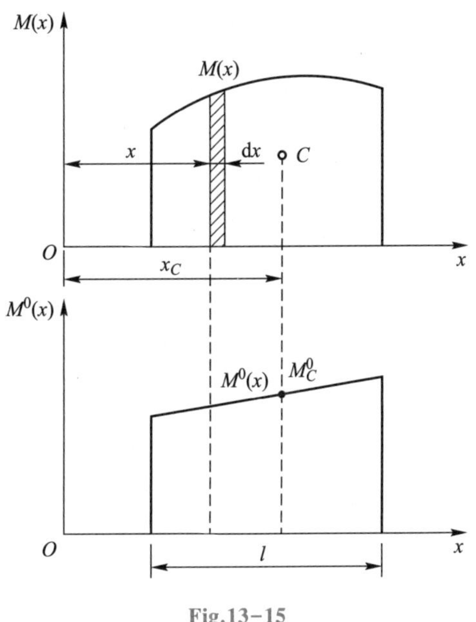

Fig.13-15

The first integral in the brackets of equation (13-29) is the entire $M(x)$ graph's static moment to the vertical axis, while the second integral represents the $M(x)$ area of the graph. If the area of the $M(x)$ graph is denoted by ω, the center coordinates of the $M(x)$ graph shape are denoted by x_C, then the equation (13-29) becomes

$$\delta = \frac{1}{EI}(k\omega x_C + b\omega) = \frac{\omega}{EI}(kx_C+b) \qquad (13\text{-}30)$$

The kx_C+b of equation (13-30) is vertical coordinate of the $M^0(x)$ diagram corresponding to the center C of the $M(x)$ diagram. If it is expressed in terms of M_C^0, Mohr integral equation (13-24) can be written as

$$\delta = \int_l \frac{M(x)M^0(x)}{EI}\mathrm{d}x = \frac{\omega M_C^0}{EI} \qquad (13\text{-}31)$$

This method of reducing the Mohr integration operation to an algebraic operation between graphs is known as the diagram multiplication method. When applying it to calculate the displacement of a structure, the following points should be noted.

(1) in equation (13-31), ω represents the area of graph of $M(x)$; M_C^0 is the coordinate value in the $M^0(x)$ diagram corresponding to the center of $M(x)$ diagram, not the center of $M^0(x)$ diagram itself.

(2) ω and M_C^0 are both generational quantities with the same positive and negative signs as $M(x)$ and $M^0(x)$.

(3) If $M(x)$ is a segmented smooth curve, or if $M^0(x)$ is a line, the graphical multiplication formula should be used for the segments up to the corresponding division points, and then find the algebraic sum. This method is not limited to the bending deformation, which can also be used to find the deformation or displacement of the straight bar with equal section.

In the practical application of the diagram multiplication method, it is often necessary to calculate the area and the position of the center of some figures. The area and the location of the centers of several common figures are given by Fig.13-16.

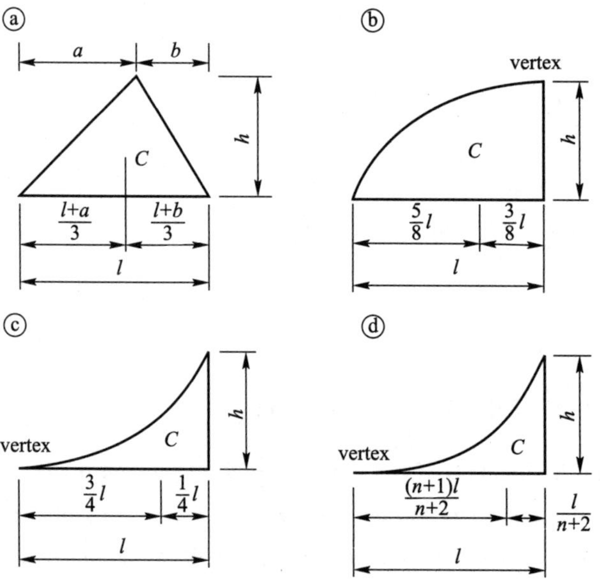

Fig.13-16

(a) Triangles $\omega=\frac{1}{2}lh$; (b) Quadratic parabola $\omega=\frac{2}{3}lh$; (c) Quadratic parabola $\omega=\frac{1}{3}lh$;

(d) Quadratic parabola $\omega=\frac{1}{n+1}lh$

Example 13-8 An externally overhanging beam is loaded as shown in Fig.13-17a. If EI is a constant, try to find the deflection at the free end C.

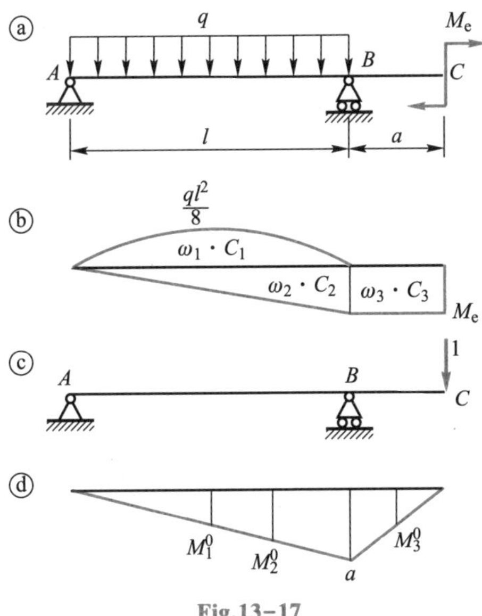

Fig.13-17

Solution In order to facilitate the calculation of the area and the center of the moment diagram, the superposition method can be used, as shown in Fig.13-17b. The parabolic part with area ω_1 is caused by the uniform load. The folded part with areas ω_2 and ω_3 is caused by the concentrated force couple. The diagram caused by the unit force is given $M^0(x)$ by Fig.13-17b. The value of M_C^0 corresponding to the shape centers of the three parts of the $M(x)$ diagram can be found using the proportional relationship between the line segments. Applying the graph multiplication equation (13-31) to the segment, and then finding the algebraic sum, the deflection of the C section can be found as

$$
\begin{aligned}
f_C &= \frac{1}{EI}(\omega_1 M_1^0 + \omega_2 M_2^0 + \omega_3 M_3^0) \\
&= \frac{1}{EI}\left[\frac{2}{3}\frac{ql^2}{8}l\left(-\frac{a}{2}\right) - \frac{1}{2}M_e l\left(-\frac{2a}{3}\right) - M_e a\left(-\frac{a}{2}\right)\right] \\
&= \frac{M_e a}{EI}\left(\frac{l}{3}+\frac{a}{2}\right) - \frac{qal^3}{24EI}
\end{aligned}
$$

Example 13-9 A steel frame of constant EI is shown in Fig.13-18a, with beam BC subjected to a uniform load of set q. If the effect of shear and axial forces on deformation are not considered, try to find the vertical displacement of section A.

Solution First draw the bending moment diagram of the steel frame under load as shown in Fig.13-18b. To calculate the vertical displacement of section A, a unit force in the vertical direction is applied on section A (Fig.13-18c) and then the corresponding $M^0(x)$ diagram is drawn as shown in 13-18d. According to Fig.13-18d, and the corresponding formula in Fig.13-16, the area of the moment diagram of the two bars AB and BC can be found as

$$
\omega_1 = \frac{1}{2}\cdot 2a \cdot 2qa^2 = 2qa^3, \quad \omega_2 = \frac{2}{3}\cdot 2qa^2 \cdot 2a = \frac{8}{3}qa^3
$$

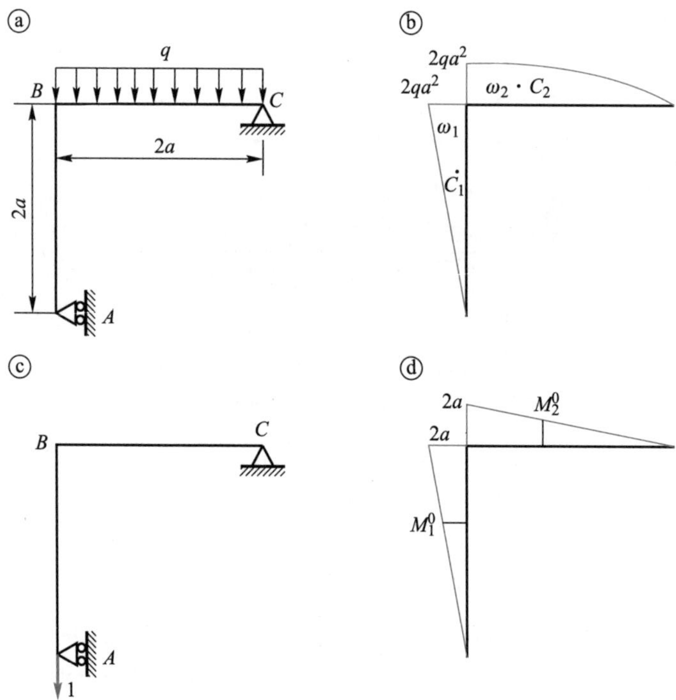

Fig.13-18

M_C^0 corresponding to the centroids of ω_1 and ω_2 in Fig.13-18d is

$$M_1^0=\frac{4}{3}a,\quad M_2^0=\frac{5}{4}a$$

Thus, from equation (13-31), the vertical displacement of section A can be found

$$f_A=\frac{\omega_1 M_1^0}{EI}+\frac{\omega_2 M_2^0}{EI}=\frac{1}{EI}\left(2qa^3\cdot\frac{4}{3}a+\frac{8}{3}qa^3\cdot\frac{5}{4}\right)=\frac{6qa^4}{EI}$$

Example 13-10 The structure of a static beam with intermediate pins is loaded as shown in Fig.13-19a. Knowing that EI is a constant, try to find the relative deflection angles of the sections on both sides of point C.

Solution Two unit couples are applied to each side of point C in opposite directions (Fig.13-19b). The bending moments of the child beam caused by the load are drawn in the form of Fig. 13-19c according to the superposition method, while the $M^0(x)$ diagram caused by the unit couple is given in Fig.13-19d. There is the graph multiplication equation (13-14) for calculating the Mohr integration to find the deflection angles of point C

$$\begin{aligned}\theta_C&=\frac{1}{EI}(\omega_1 M_1^0+\omega_2 M_2^0+\omega_3 M_3^0+\omega_4 M_4^0)\\&=\frac{1}{EI}\left[\frac{2}{3}\cdot\frac{qa^2}{8}\cdot a\cdot\frac{3}{4}-\frac{1}{2}\cdot\frac{Fa}{4}\cdot a-\frac{1}{2}\cdot\frac{Fa}{4}\cdot\frac{a}{2}\cdot\left(1+\frac{1}{2}+\frac{2}{3}\right)+\frac{1}{2}\cdot\frac{Fa}{4}\cdot a\cdot\frac{1}{2}\right]\\&=\frac{1}{EI}\left(\frac{qa^3}{16}-\frac{7Fa^2}{48}\right)\end{aligned}$$

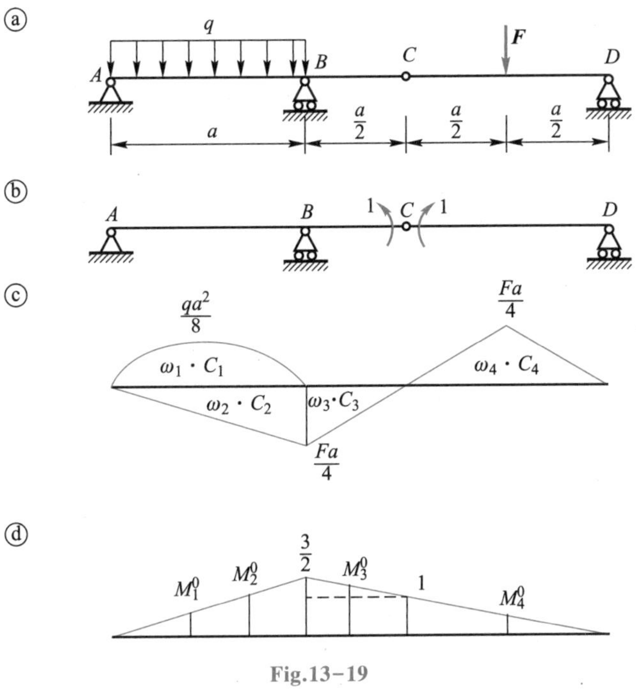

Fig.13-19

13.4 Castigliano's theorem

13.4.1 Castigliano's theorem

The Castigliano's theorem is also one of the effective tools for calculating the deformation of linear elastic structures. Let the free end A of a straight cantilever beam with EI be subjected to a concentrated force F(Fig.13-20). It is not difficult to find the strain energy stored in the cantilever beam

$$U=\int_0^l \frac{M^2(x)\,\mathrm{d}x}{2EI}=\int_0^l \frac{F^2x^2\,\mathrm{d}x}{2EI}=\frac{F^2l^3}{6EI}$$

The strain energy in the beam is numerically equal to the work of the external force W, i.e.

$$U=W=\frac{1}{2}Ff_A$$

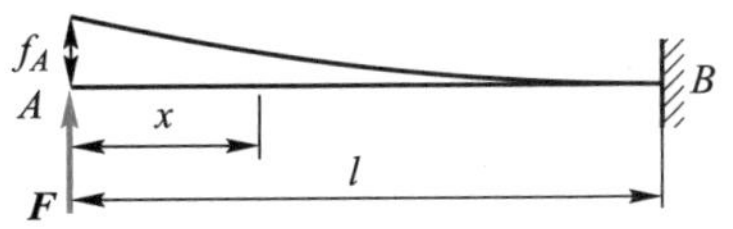

Fig.13-20

The deflection of the free end of the cantilever beam is

$$f_A=\frac{Fl^3}{3EI}$$

If we take the partial derivative of the strain energy U of the beam with respect to the concentrated force F at section A, we have

$$\frac{\partial U}{\partial F}=\frac{\partial}{\partial F}\left(\frac{F^2l^3}{6EI}\right)=\frac{Fl^3}{3EI}$$

This is exactly equal to the free end deflection. Therefore

$$\frac{\partial U}{\partial F}=f_A$$

The partial derivative of the strain energy with respect to F is equal to the displacement of the point of F along the force direction. The same conclusion can be drawn for the strain energy of the curved bar in example 13-2 with respect to the partial derivative of the load F. These conclusions reflect a general rule, which is known as Castigliano's theorem. It was proposed by the Italian engineer Castigliano in 1873 and is also known as Castigliano's second theorem.

$$\delta_n=\frac{\partial U}{\partial F_n} \tag{13-32}$$

The beam is now used to prove this theorem. Let a set of static loads $F_1, F_2, \cdots$, acting on a beam. The displacements in response to these loads are $\delta_1, \delta_2, \cdots$ (Fig.13-21).

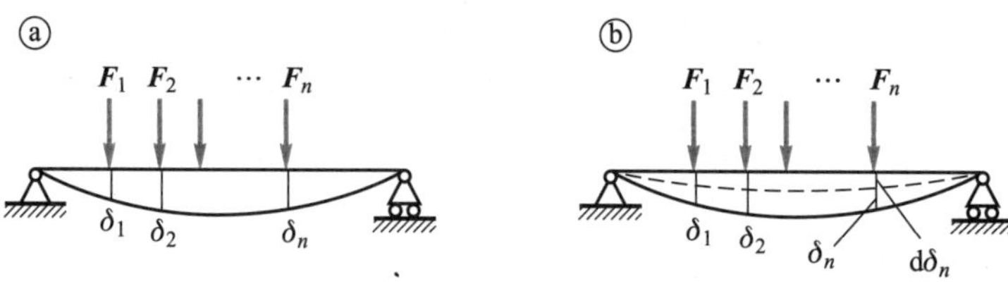

Fig.13-21

During the deformation process, the work done by the above load is equal to the strain energy stored in the beam. The strain energy U is a function of the load $F_1, F_2, \cdots$ and can be expressed as

$$U=U(F_1, F_2, \cdots) \tag{13-33}$$

If F_n is given an increment $\mathrm{d}F_n$, the strain energy U will also have an increment $\frac{\partial U}{\partial F_n}\mathrm{d}F_n$. The elastic strain energy of the beam can be written as

$$U_1=U+\frac{\partial U}{\partial F_n}\mathrm{d}F_n \tag{13-34}$$

Change the loading order by first adding $\mathrm{d}F_n$ to the beam and then acting $F_1, F_2, \cdots$. If the material obeys Hooke law and the deformation is small, the deformations caused by each external force are independent and do not affect each other. When first add $\mathrm{d}F_n$, it causes a displacement $\mathrm{d}\delta_n$ at its point along the same direction. The strain energy in the beam at this time should be $\frac{1}{2}\mathrm{d}F_n\mathrm{d}\delta_n$. In the $F_1, F_2, \cdots$ process, as F_n induces a deformation δ_n in the $\mathrm{d}F_n$ direction, $\mathrm{d}F_n$ continue to do the work $\mathrm{d}F_n \cdot \delta_n$. Because the strain energy caused by $F_1, F_2, \cdots$ is still the form of equation (13-33), the strain energy stored in the beam should be

$$U_2=\frac{1}{2}\mathrm{d}F_n\mathrm{d}\delta_n+\mathrm{d}F_n\delta_n+U \tag{13-35}$$

Since the strain energy within the linear elastic body is independent of the loading order, the strain energy caused by the two different loading orders expressed by equations (13-34) and (13-35) should be equal, i.e.

$$U_1 = U_2$$

$$U+\frac{\partial U}{\partial F_n}\mathrm{d}F_n = \frac{1}{2}\mathrm{d}F_n \cdot \mathrm{d}\delta_n + \mathrm{d}F_n \cdot \delta_n + U$$

Neglecting the second order micro-quantity, we get

$$\delta_n = \frac{\partial U}{\partial F_n}$$

This is precisely the expression of Castigliano's theorem of equation (13-32).

Although the above proof is given as an example of a beam, it does not involve the characteristics of bending deformation. So, the Castigliano's theorem is suitable for other deformations. In equation (13-32), the load F_n and displacement δ_n can also be expressed respectively couple moment and the corresponding angular displacement. It is a generalized. Since it is assumed that the material obeys Hooke law when calculating the strain energy, Castigliano's theorem is only applicable to linearly elastic structures.

13.4.2 Special forms of Castigliano's theorem

1. Truss

For a truss structure, the deformation of each member is uniaxial in tension or compression. If the whole truss consists of m bars, the strain energy of the whole structure can be calculated by equation (13-5), i.e.

$$U = \sum_{i=1}^{m} \frac{F_{Ni}^2 l_i}{2EA_i}$$

According to the Castigliano's theorem there are

$$\delta_n = \frac{\partial U}{\partial F_n} = \sum_{i=1}^{m} \frac{F_{Ni} l_i}{EA_i} \frac{\partial F_{Ni}}{\partial F_n} \tag{13-36}$$

2. Straight beams

For straight beams where plane bending occurs, the strain energy can be calculated using equation (13-14), i.e.

$$U = \int_l \frac{M^2(x)\,\mathrm{d}x}{2EI}$$

Applying Castigliano's theorem, we get

$$\delta_n = \frac{\partial U}{\partial F_n} = \frac{\partial}{\partial F_n}\left(\int_l \frac{M^2(x)\,\mathrm{d}x}{2EI}\right)$$

In the above equation, only the bending moment $M(x)$ is related to the load F_n. The integral variable x and F_n are not related.

$$\delta_n = \frac{\partial U}{\partial F_n} = \int_l \frac{M(x)}{EI}\frac{\partial M(x)}{\partial F_n}dx \tag{13-37}$$

3. Plane curved bars

The stress distribution of small curvature bar is similar to that of a straight beam. The bending strain energy can be written as

$$U = \int_s \frac{M^2(s)\,ds}{2EI}$$

According to the Castigliano's theorem, we get

$$\delta_n = \frac{\partial U}{\partial F_n} = \int_s \frac{M(s)}{EI}\frac{\partial M(s)}{\partial F_n}ds \tag{13-38}$$

4. Combination of deformed bars

For bars subjected to the combined action of tension (compression), bending and torsion, the strain energy can be written from equation (13-19), i.e.

$$U = \int_l \frac{F_N^2(x)\,dx}{2EA} + \int_l \frac{M^2(x)\,dx}{2EI} + \int_l \frac{M_n^2(x)\,dx}{2GI_n}$$

Applying Castigliano's theorem, we get

$$\delta_n = \frac{\partial U}{\partial F_n} = \int_l \frac{F_N(x)}{EA}\frac{\partial F_N(x)}{\partial F_n}dx + \int_l \frac{M(x)}{EI}\frac{\partial M(x)}{\partial F_n}dx + \int_l \frac{M_n(x)}{GI_n}\frac{\partial M_n(x)}{\partial F_n}dx \tag{13-39}$$

The above equation is also suitable for calculating the deformation of small curvature bars, simply by replacing the length coordinate x with the arc length coordinate s.

Example 13-11 A simply supported beam is loaded as shown in Fig.13-22. If EI is a constant, find the deflection angle of section A and the deflection at the mid-point C.

Solution Since there is a concentrated M_e moment at point A and a concentrated force F at the mid-point C, the deflection angle of section A and the deflection of section C can be solved by direct application of Castigliano's theorem. The moment equation $M(x)$ of the beam should first be established in sections. The partial derivatives of the moment with respect to the load M_e and F should be found. Finally, substituted them into the equation (13-37) for integration. According to the coordinates of Fig.13-23, we have

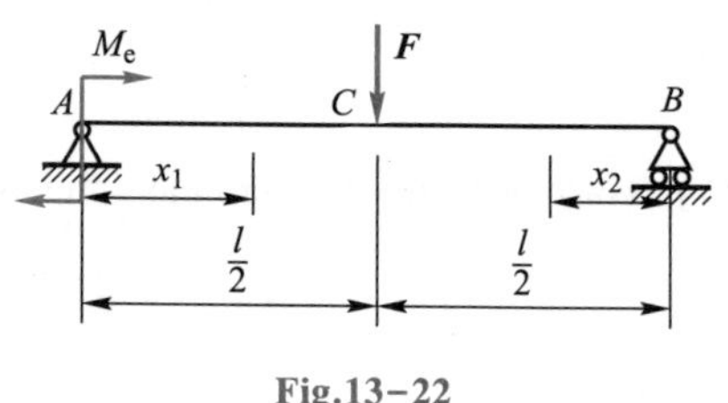

Fig.13-22

AC section: $M(x_1) = M_e + \left(\frac{F}{2} - \frac{M_e}{l}\right)x_1$

$$\frac{\partial M(x_1)}{\partial M_e} = 1 - \frac{x_1}{l} \quad \frac{\partial M(x_1)}{\partial F} = \frac{x_1}{2}$$

BC section: $M(x_2) = \left(\frac{F}{2} + \frac{M_e}{l}\right)x_2$

$$\frac{\partial M(x_2)}{\partial M_e}=\frac{x_2}{l}\frac{\partial M(x_2)}{\partial M_e}=\frac{x_2}{2}$$

According to equation (13-37), the deflection angle of section A and the deflection of section C can be found as

$$\begin{aligned}\theta_A&=\frac{\partial U}{\partial M_e}=\int_l\frac{M(x)}{EI}\frac{\partial M(x)}{\partial M_e}\mathrm{d}x\\&=\frac{1}{EI}\int_0^{\frac{l}{2}}\left[\left(\frac{F}{2}-\frac{M_e}{l}\right)x_1+M_e\right]\left(1-\frac{x_1}{l}\right)\mathrm{d}x_1+\\&\quad\frac{1}{EI}\int_0^{\frac{l}{2}}\left(\frac{F}{2}+\frac{M_e}{l}\right)x_2\frac{x_2}{l}\mathrm{d}x_2\\&=\frac{M_el}{3EI}+\frac{Fl^2}{16EI}\end{aligned}$$

$$\begin{aligned}f_C&=\frac{\partial U}{\partial F}=\int_l\frac{M(x)}{EI}\frac{\partial M(x)}{\partial F}\mathrm{d}x\\&=\frac{1}{EI}\int_0^{\frac{l}{2}}\left[\left(\frac{F}{2}-\frac{M_e}{l}\right)x_1+M_e\right]\frac{x_1}{2}\mathrm{d}x_1+\\&\quad\frac{1}{EI}\int_0^{\frac{l}{2}}\left(\frac{F}{2}+\frac{M_e}{l}\right)x_2\frac{x_2}{2}\mathrm{d}x_2\\&=\frac{M_el^2}{16EI}+\frac{Fl^3}{48EI}\end{aligned}$$

If they are all positive signs, indicating that they are oriented in the same direction as M_e and F, respectively.

13.4.3 Special treatment of Castigliano's theorem

If we use the Castigliano's theorem to calculate the generalized displacement, there must be the generalized external force corresponding to the form and direction of the requested generalized displacement. If there is no corresponding generalized force, the Castigliano's theorem cannot be directly applied. The method of additional forces is required. It is conceived to append a generalized force corresponding to the requested generalized displacement, and then apply Castigliano's theorem to solve it. This method is specified below by means of example problems.

Example 13-12 Try to find the horizontal displacement at point B and the deflection angle of point C of the rigid frame shown in Fig.13-23a.

Solution When calculating the horizontal displacement of section B, it is not possible to apply the Castigliano's theorem directly because there is no horizontal concentrated force acting at point B. For this reason, it is envisaged that a horizontal force F_f is attached to point B (Fig.13-23b). The bending moment of the rigid frame under the original external force and the F_f, and its partial derivatives of F_f are

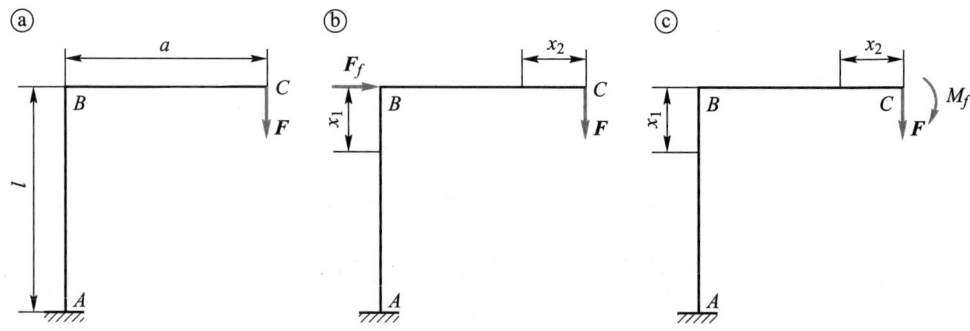

Fig.13-23

AB section:
$$M(x_1)=(Fa+F_f x_1)$$
$$\frac{\partial M(x_1)}{\partial F_f}=x_1$$

BC section:
$$M(x_2)=Fx_2$$
$$\frac{\partial M(x_2)}{\partial F_f}=0$$

Applying the Castigliano's theorem to calculate the horizontal displacement of the rigid frame in the case of Fig.13-23b, the horizontal displacement of section B is

$$\begin{aligned}\delta_B &= \int_l \frac{M(x)}{EI}\frac{\partial M(x)}{\partial F_f}\mathrm{d}s \\ &= \frac{1}{EI}\int_0^l (Fa+F_f x_1)(x_1)\mathrm{d}x_1 + \frac{1}{EI}\int_0^a Fx_2\cdot 0\mathrm{d}x_2 \\ &= \frac{1}{EI}\left(\frac{Fal^2}{2}+\frac{F_f l^3}{3}\right)\end{aligned} \tag{13-40}$$

The horizontal displacement of section B is the result of the combined action of the original load and F_f. The result is correct regardless of the value of F_f. Since F_f does not exist in the actual steel frame, the horizontal displacement of section B under the original load can be found by simply making F_f equal to 0 in the equation (13-40)

$$\delta_B=\frac{Fal^2}{2EI} \tag{13-41}$$

In calculating the deflection angle of section C, a couple moment M_f can be added near C (Fig.13-23c). The bending moment under the original load and the M_f combined action, and its partial derivative of M_f is

AB section: $M(x_1)=(Fa+M_f)$, $\quad \dfrac{\partial M(x_1)}{\partial M_f}=1$

BC section: $M(x_2)=(Fx_2+M_f)$, $\quad \dfrac{\partial M(x_2)}{\partial M_f}=1$

Applying Castigliano's theorem and making M_f equal to 0 before integration, the deflection angle of section C is found to be

$$\theta_C = \frac{1}{EI}\int_0^l (Fa)\cdot 1\mathrm{d}x_1 + \frac{1}{EI}\int_0^a (Fx_2)\cdot 1\mathrm{d}x_2$$
$$= \frac{Fa}{EI}\left(l + \frac{a}{2}\right)$$

The positive values of δ_B and θ_C indicate that their direction are the same as the direction of the additional force and additional couple moment.

Example 13–13 A plane curved bar with an axis of one-fourth circumference is shown in Fig13–24a. The end A of the bar is fixed and a vertical concentrated force F acts on the free end B. Find the vertical and horizontal displacement at point B. EI is known to be a constant.

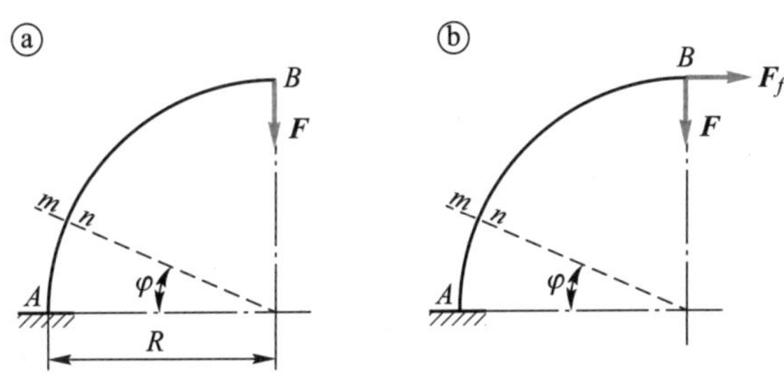

Fig.13–24

Solution First calculate the vertical displacement $(\delta_B)_{\text{vertical}}$ at point B. From the coordinates taken in Fig. 13–24a, the bending moment on any cross section $m—n$ of the curved bar can be obtained as

$$M = FR\cos\varphi$$

So
$$\frac{\partial M}{\partial P} = R\cos\varphi$$

Using the expression of the Castigliano's theorem (13–38) for calculating the deformation of the curved bar gives

$$(\delta_B)_{\text{vertical}} = \int_s \frac{M}{EI}\frac{\partial M}{\partial F}\mathrm{d}s = \frac{1}{EI}\int_0^{\frac{\pi}{2}} FR\cos\varphi\cdot R\cos\varphi\cdot R\mathrm{d}\varphi = \frac{FR^3\pi}{4EI}$$

Since there is no horizontal load at point B, a horizontal force F_f needs to be attached at point B, as shown in Fig.13–24b. At this time, the bending moment on the section mn and its partial derivative to F_f are

$$M = FR\cos\varphi + F_f R(1-\sin\varphi)$$
$$\frac{\partial M}{\partial F_f} = R(1-\sin\varphi)$$

Applying Castigliano's Theorem

$$(\delta_B)_{\text{leval}} = \left[\int_s \frac{M}{EI}\frac{\partial M}{\partial F_f}\mathrm{d}s\right]F_{f=0} = \frac{1}{EI}\int_0^{\frac{\pi}{2}} FR\cos\varphi\cdot R(1-\sin\varphi)\cdot R\mathrm{d}\varphi = \frac{FR^3}{2EI}$$

The following points should be noted with the application of Castigliano's theorem.

(1) The Castigliano's theorem is only applicable to linearly elastic structures with small deformations.

(2) When using the Castigliano's theorem to find the generalized displacement of a structure, the generalized force corresponding to the displacement needs to be available at that location. If there is no corresponding generalized force at that location, the additional force method is required.

(3) When using the additional force method to calculate the displacement of a structure, it is best to make the additional load in the corresponding expression zero after finding the partial derivative of the internal force against the additional load, which can simplify the calculation process.

To conclude this section, let us compare Mohr theorem with Castigliano's theorem. Both Mohr theorem and Castigliano's theorem are used to solve for the deformation or displacement of a linearly elastic bar structure. They are essentially the same. In the case of the bending deformation of a beam, the expression for solving for the displacement using Castigliano's theorem is

$$\delta_n = \int_l \frac{M(x)}{EI} \frac{\partial M}{\partial F_n} \mathrm{d}x$$

The Mohr integration expression is

$$\delta = \int_l \frac{M(x) M^0(x)}{EI} \mathrm{d}x$$

Since the partial derivative $\frac{\partial M(x)}{\partial F_n}$ actually represents the bending moment at $F_n = 1$, $\frac{\partial M(x)}{\partial F_n}$ and $M^0(x)$ are equal. Similar conclusions can be drawn for steel frames, trusses, curved bars or groups of deformed bars. The Mohr integration can be deduced from the Castigliano's theorem.

The Mohr theorem and the Castigliano's theorem have their own advantages and should be applied flexibly according to the specific questions. When there is a corresponding generalized force at the generalized displacement of the structure to be solved, it is more convenient to use Castigliano's theorem. Because it is not necessary to find the internal force caused by the unit load as in Mohr theorem. When there is no generalized force, it is simpler to use Mohr theorem. Because the calculation of the internal forces of the structure under the combination of the original load and the additional load is often more complex.

Problems

13-1 Two straight bars of circular section are made of the same material (Fig.P13-1), try to compare the strain energy of the two bars.

Answer: (a) $U=\frac{2F^2 l}{\pi E d^2}$; (b) $U=\frac{7F^2 l}{8\pi E d^2}$

13-2 The EA of each bar of the truss are same (Fig.P13-2). Try to find the strain energy of the truss under the F force.

Answer: $U=\frac{9+10\sqrt{3}}{12}\frac{F^2a}{EA}$

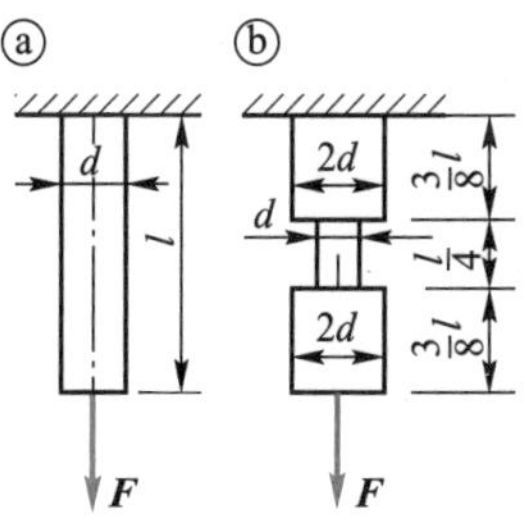

Fig.P13-1

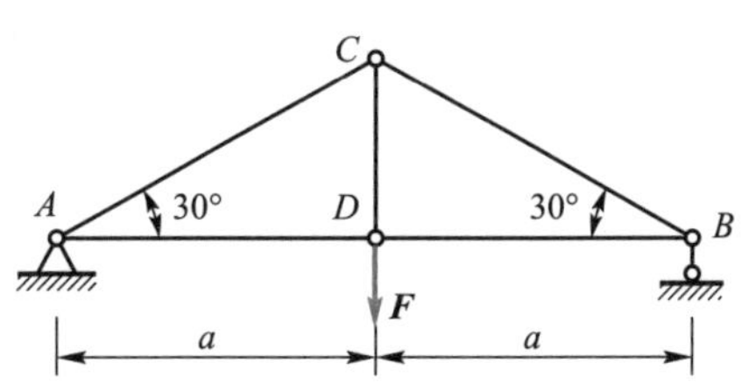

Fig.P13-2

13-3 Calculate the strain energy of each bar(Fig.P13-3).

Answer: (a) $\frac{5F^2l^2}{384EI}$; (b) $U=\frac{\pi F^2R^3}{8EI}$

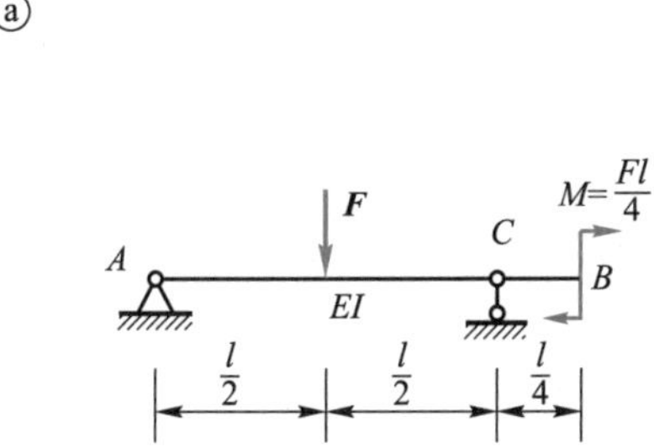

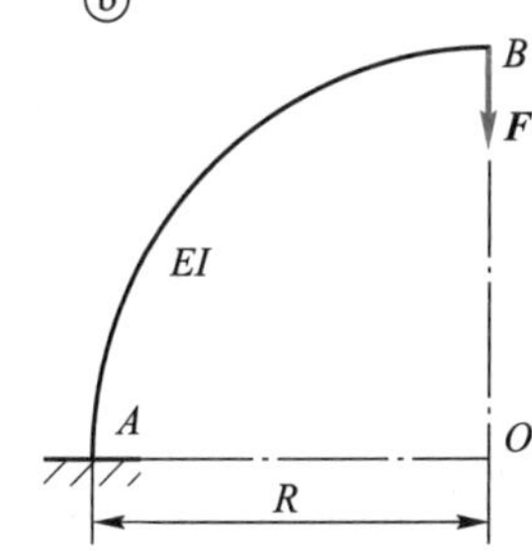

Fig.P13-3

13-4 Try to find the deflection angle at each point A of the beam shown in Fig.P13-4.

Answer: (a) $\delta_A=\frac{2qa^4}{3EI}(\downarrow)$, $\theta_A=\frac{5qa^3}{6EI}$(clockwise)

(b) $\delta_A=\frac{q_0l^4}{30EI}(\downarrow)$, $\theta_A=\frac{q_0l^3}{24EI}$(counterclockwise)

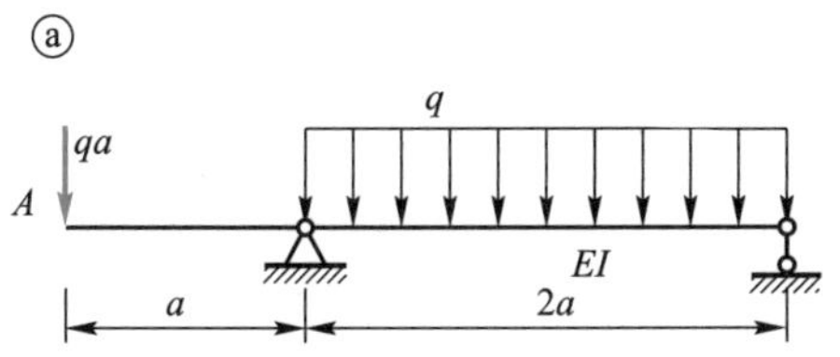

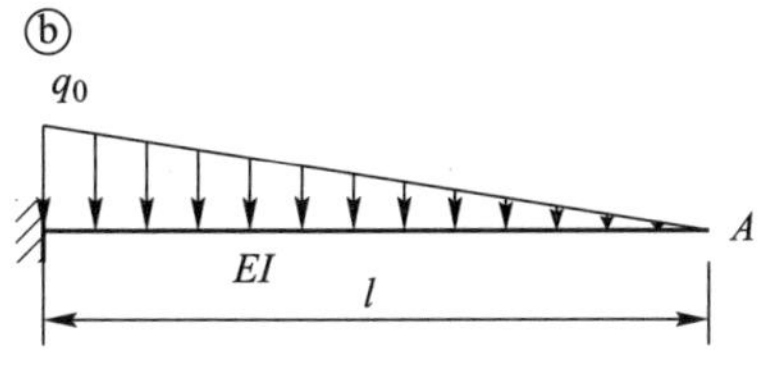

Fig.P13-4

13-5 Try to find the vertical displacement at each rigid frame point A, where the bars of the rigid frame are known to be EI(Fig.P13-5).

Answer: (a) $\delta_A=\frac{1}{EI}\left(\frac{Fa^3}{3}+Fa^2l+\frac{1}{6}ql^3a\right)$ (↓); (b) $\delta_A=\frac{11Fa^3}{6EI}$(↑)

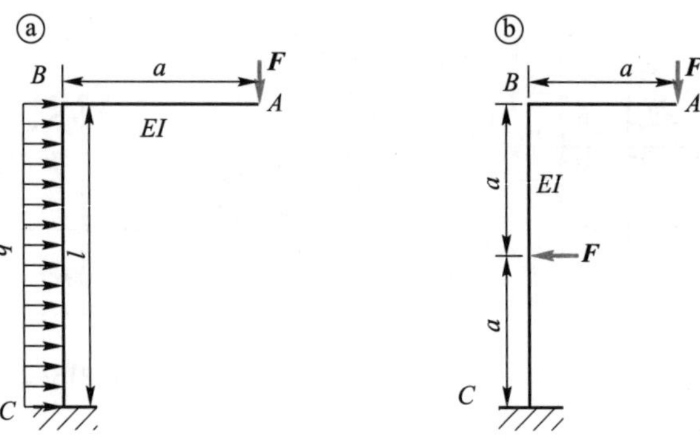

Fig.P13-5

13-6 The EA of each bars of the square truss are same(Fig.P13-6). Try to find the horizontal and vertical displacements at the nodes C.

Answer: $x_C=3.83\frac{Fl}{EA}$(←), $y_C=\frac{Fl}{EA}$(↑)

13-7 The same EA for each bar of the square truss shown in Fig.P13-7, try to find the relative displacement between the nodes B and D under load F.

Answer: $\delta_{BD}=2.71\frac{Fl}{EA}$(close)

13-8 A plane curved bar lies in the horizontal plane(Fig.P13-8). The end A is fixed and the free end B is subject to a vertical force F. Let EI and GI_p be constants and try to find the displacement of the section B in the vertical direction.

Answer: $\delta_B=FR^3\left(\frac{0.785}{EI}+\frac{0.356}{GI_p}\right)$ (↓)

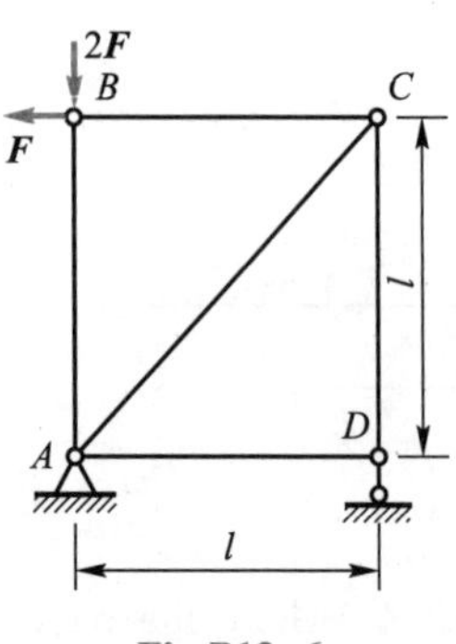

Fig.P13-6

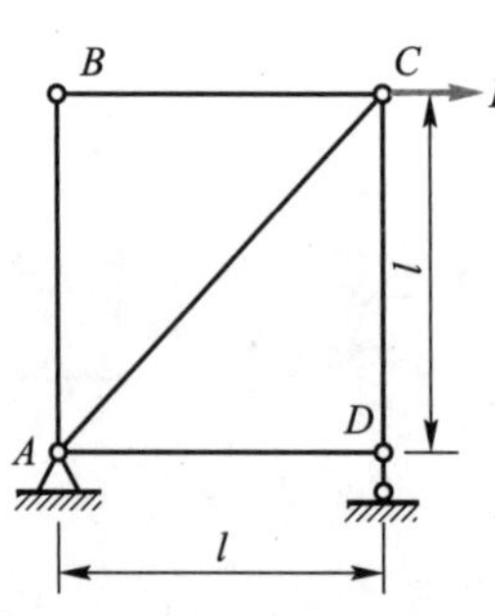

Fig.P13-7

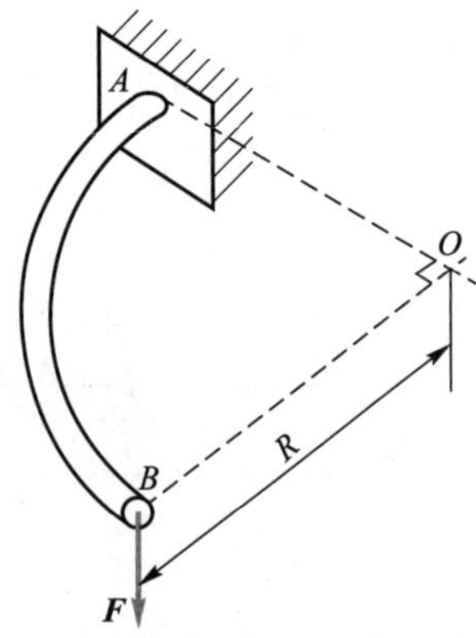

Fig.P13-8

13-9 Try to find the horizontal displacement at point B in the semicircular plane shown in Fig. P13-9. EI is a constant.

Answer: $\delta_B = \dfrac{FR^3}{2EI}(\rightarrow)$

13-10 As shown in Fig.P13-10, a rigid frame in the horizontal plane consists of circular section folding bars with right angles at all turns. Let EI and GI_p be constants. Find the displacement and known at point A along the direction of the vertical concentration force.

Answer: $\delta_A = \dfrac{F}{3EI}(8a^2+b^3)+\dfrac{Fab}{GI_p}(a+b)(\downarrow)$

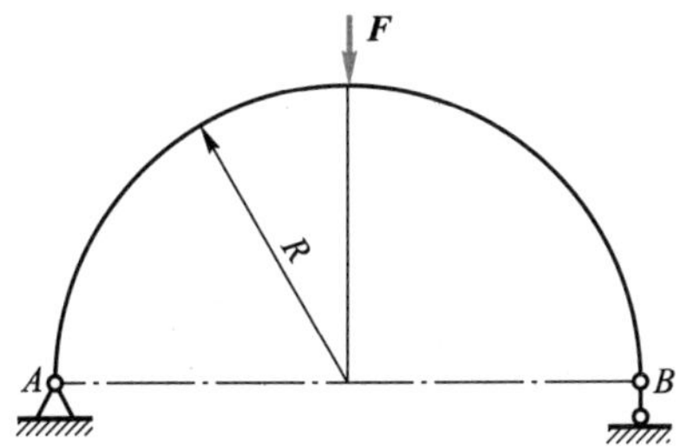

Fig.P13-9

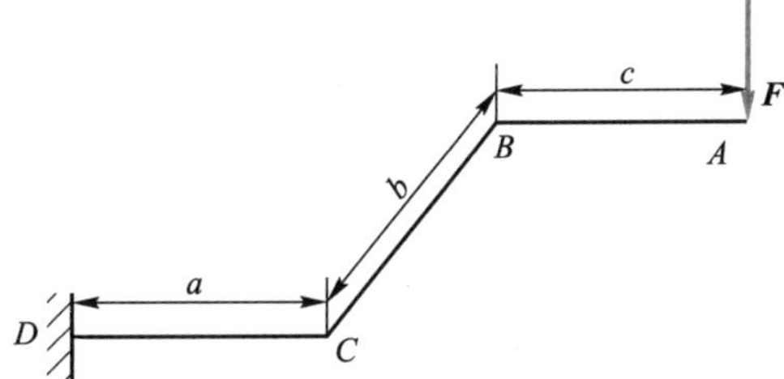

Fig.P13-10

Chapter 14 Stability of Columns

Teaching Scheme of Chapter 14

Stability is an aspect of the load-bearing capacity of a member. This chapter deals with the stability of straight bars subjected to axial compressive force, the so-called stability of column. Determining the critical force or critical stress of a column is the key to stability calculations. A clear understanding of the general diagram of critical stress is required in order to select the appropriate formulas according to the slenderness ratio of the column.

14.1 Introduction

In the past, when calculating the role of axial compressive force on a bar, it was always considered that the bar was in a straight line to maintain equilibrium. The damage to the bar is caused by insufficient strength. In engineering practice, some slender bars, such as the tappet in the engine distribution mechanism (Fig.14-1), are subjected to compressive force when it pushes the rocker arm to open the valve. This type of bar may be damaged before exceeding the strength. Because it can not maintain a straight shape under the balance and lose the ability to work. The problem of whether the column can maintain the original linear balance state is called the stability problem of the column. Fig.14-2 shows a slender column pin support at both ends, which is selected to illustrate the linear balance form of a bar under the axial compressive force. From the experiment, we can know that when the compressive force F gradually increased but less than F_{cr}, the column will always maintain a straight shape of equilibrium. Even if a transverse force is applied to temporarily bend it (Fig.14-2a), the column will return to the original straight state after the disturbance force is removed. If $F<F_{cr}$, the equilibrium of the linear column is stable. When the axial compressive force F increases to a critical value F_{cr}, the balance of the original straight column becomes unstable. In the case of $F=F_{cr}$, if a transverse disturbance force is applied to make a small bending deformation, after the lifting of the disturbance force, the column can not return to a straight shape but in a slightly bent state of equilibrium (Fig.14-2b). If the compressive force F continues to increase to exceed F_{cr}, the bending deformation will increase significantly until it breaks (Fig.14-2c).

Whether the linear balance form of the bar is stable or not is determined by the magnitude of the axial compressive force F. The balance is stable when F is less than the critical compressive force F_{cr}, and unstable when the axial compressive force F is equal to or greater than the critical compressive force F_{cr}. The process of the linear equilibrium of the column from stable to unstable is called buckling. Theoretical analysis and experimental results indicate that the magnitude of the critical compressive force of a column is not only related to the mechanical properties of material, but also to the shape and size of the cross section, the length and the constraint of the bar. The first step in the study of the stability of a column is to determine the value of F_{cr}. If the working compressive force is controlled within the range of F_{cr}, the bar will not become unstable.

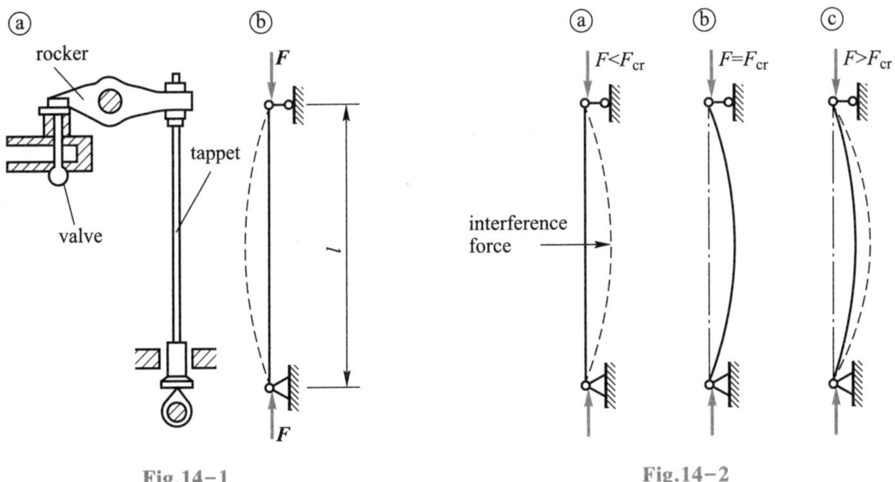

Fig.14-1　　　　　　Fig.14-2

In addition to columns, there are many other forms of members that have stability problems. For example, thin-walled cylinders subjected to uniform external compressive force become elliptical due to instability, beams of narrow rectangular section bend laterally due to instability (Fig.14-3), thin-walled twisted circular tubes will appear wrinkled on the wall due to instability. The reader can roll paper into a cylinder to test. These stability problems are beyond the scope of this book.

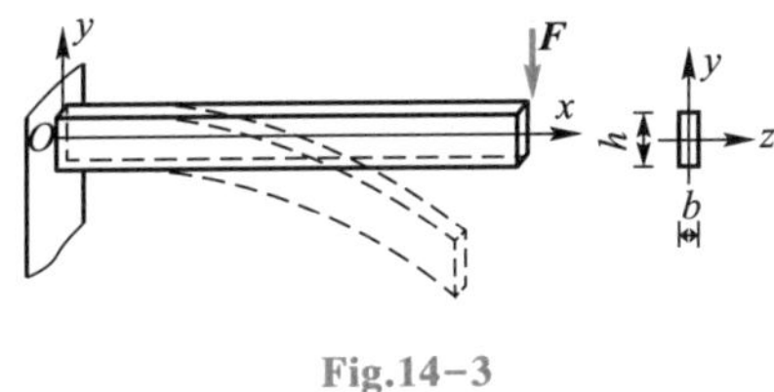

Fig.14-3

14.2 Critical compressive force of slender column with pin support

There is a slender column with length of l. Both ends of the column subjected to axial compressive force F are supported by the pin. It maintains equilibrium under a slightly bending deformation (Fig.14-4). The stress on the bar does not exceed the proportional limit of the material section. The original linear balance shape of the column will transform from stable to unstable under the action of critical compressive force. In other words, under critical compressive force, the column starts to remain in equilibrium in a slightly curved shape. The minimum compressive force F that keeps the bar in equilibrium in a slightly curved shape is therefore considered to be the critical compressive force F_{cr} of the column.

By taking the Cartesian coordinate system as shown in Fig. 14－4, the deflection of any cross section at a distance x from the origin is v and the bending moment is

$$M=-Fv \qquad (14-1)$$

Here F is an absolute value. With small deformations and stresses in the column not exceeding the material proportional limits, the approximate differential equation for the deflection curve of column is

$$\frac{d^2v}{dx^2}=-\frac{Fv}{EI} \qquad (14-2)$$

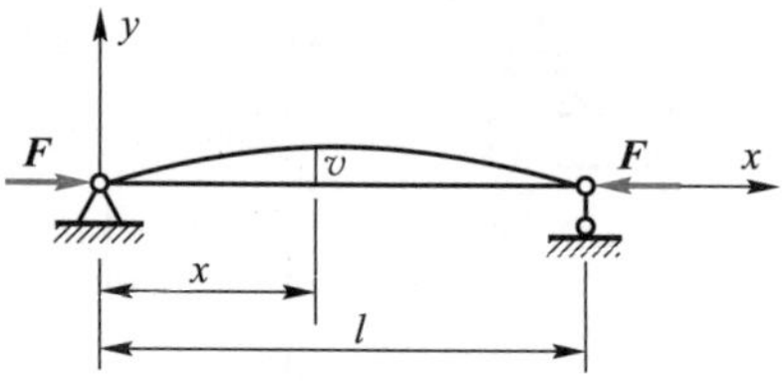

Fig.14-4

Let $k^2=\frac{F}{EI}$, then equation (14-2) can be written as

$$\frac{d^2v}{dx^2}+k^2v=0 \qquad (14-3)$$

The general solution of this differential equation is

$$v=a\sin kx+b\cos kx \qquad (14-4)$$

where a and b are two integration undetermined constants. Since the F_{cr} is unknown, k is also a value to be determined.

According to the boundary conditions at the column end, we have $x=0$ and $v=0$. By substituting it into equation (14-4), we get $b=0$, and then equation (14-4) can be rewritten as

$$y=a\sin kx \qquad (14-5)$$

Substituting the boundary condition at another column end ($x=l$, $v=0$) into equation (14-5), we get

$$a\sin kl=0 \qquad (14-6)$$

This requires a or $\sin kl$ is equal to zero. If $a=0$, it is known $v=0$ from equation (14-5). That means the deflection at all points on the axis of the column is equal to zero. This contradicts the fact that the column remains in equilibrium in a slightly bent state. So, we get $\sin kl=0$. The value of kl that satisfies this condition should be

$$kl=n\pi$$

where, $n=0,1,2,\cdots$, can be any integer, which gives

$$k=\sqrt{\frac{F}{EI}}=\frac{n\pi}{l}$$

or

$$F=\frac{n^2\pi^2EI}{l^2} \qquad (14-7)$$

Since n can be any integer, the above equation shows that the compressive force F that keeps the column in equilibrium in a curved shape is theoretically multivalent. However, the definition of F_{cr} shows that the minimum axial compressive force that keeps the column in equilibrium under the curve shape is the critical compressive force of the bar. When n and F is zero, it is meaningless. When $n=1$, the value of F is the minimum. So, F_{cr} of the slender column is

$$F_{cr}=\frac{\pi^2 EI}{l^2} \tag{14-8}$$

This is the formula for calculating the critical compressive force of a slender column with pin support at both ends, known as Euler equation.

The F_{cr} is proportional to the minimum bending rigidity EI of the column and inversely proportional to the square of the bar length l. This means that the slimmer the bar is, the smaller the critical compressive force is, and the more unstable the bar is.

Under the action of F_{cr}, we have $k=\frac{\pi}{l}$. Substitution it into equation (14-5), we get

$$v=a\sin\frac{\pi}{l}x \tag{14-9}$$

The above equation shows that the deflection curve of a slender column with pin support at both ends is a half-wave sine curve. If we make $x=\frac{l}{2}$, and substitute it into equation (14-9), we get

$$v_{x=\frac{l}{2}}=a\sin\left(\frac{\pi}{l}\cdot\frac{l}{2}\right)=a$$

Visible a is the deflection of the mid cross section of column, which can be any small value of displacement. There is no definite value of a. This is because in deriving the equation (14-5) for the deflection curve of the column, the approximate differential equation for the bending deflection curve of the column is basis. By exact differential equation for the deflection curve, we get

$$\pm\frac{v''}{[1+(v')^2]^{\frac{3}{2}}}=\frac{M(x)}{EI}$$

Solving for this gives the definite value of a.

It should be noted that an absolutely straight column does not exist and that the load cannot act along the axis of the bar as precisely as in the theoretical analysis. Nevertheless, the observed critical compressive force is still very close to the theoretical value F_{cr} in precisely conducted small specimen tests after eliminating the disturbing bending moment as much as possible.

14.3 Critical compressive force of slender column with other constraints

The critical compressive force can be found in the same way for slender columns with other constrained situations, and they can all be uniformly formed into

$$F_{cr}=\frac{\pi^2 EI}{(\mu l)^2} \tag{14-10}$$

This is the common form of Euler equation. μ is the length factor of the column under different constraints. μl is equivalent to the length of the column converted into pin support column, called the equivalent length. The length coefficients for several bar with ideal constraints are listed in table 14-1.

As seen from table 14-1, the length factor of a column with constraints at both ends is within the range from 0.5 to 1.0. In practice, it is difficult to completely fix the end of a column. As long as there is a possibility of slight rotation, the end can not be regarded as an ideal fixed end, but rather as an approximate pin. In this case, the length factor μ is often taken to be close to 1.0.

In addition, there may be other constraints for actual column, such as elastic constraints (bar ends fixed to other elastic members). The load acting on the column can also take various forms. For example, the compressive force may be distributed along the axis of the bar rather than concentrated at the ends. Such situations can also be reflected by different length factors, which are specified in general design manuals or codes.

It should also be noted that the constraint of column end may not act the same in different bending planes, and the resulting values of μ should also be different. Thus, a column may have two equivalent lengths. This must be taken into account when determining the critical compressive force.

Table14-1 Length factors for columns

Binding situation	Both ends hinged	One end free One end fixed	Both ends Fixed	One end hinged One end fixed
Deflection curve shape	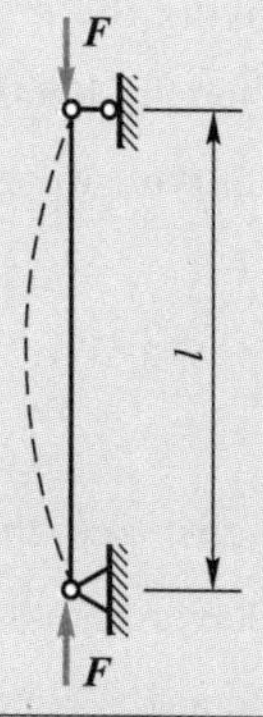 F, l, F	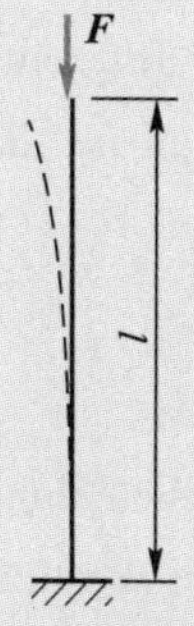 F, l	 F, l	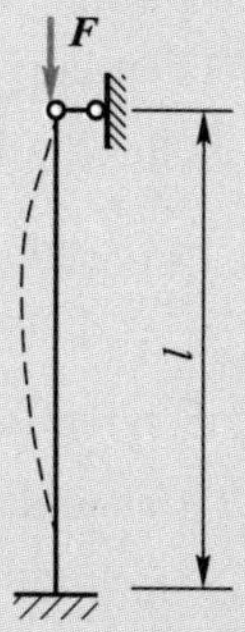 F, l
F_{cr}	$\dfrac{\pi^2 EI}{l^2}$	$\dfrac{\pi^2 EI}{(2l)^2}$	$\dfrac{\pi^2 EI}{(0.5l)^2}$	$\dfrac{\pi^2 EI}{(0.7l)^2}$
μ	1.0	2.0	0.5	0.7

Example 14-1 A cast iron column of circular cross section with one fixed end and one free end. It has a length $l=3$ m, a diameter $d=0.2$ m and a modulus of elasticity $E=120$ GPa. Calculate the critical compressive force of the column from equation (14-2).

Solution Check table 14-1 for the length factor $\mu=2$, and then the moment of inertia of the section is

$$I=\frac{\pi d^4}{64}=\frac{\pi\times(0.2\ \text{m})^4}{64}=7.85\times10^{-5}\ \text{m}^4$$

Therefore, the critical compressive force is

$$F_{cr}=\frac{\pi^2 EI}{(\mu l)^2}=\frac{3.14^2\times120\times10^9\times7.85\times10^{-5}}{(2\times3)^2}\text{ N}=2\ 580\text{ kN}$$

14.4 Diagram of critical stress

14.4.1 Critical stress and slenderness ratio

Divide the critical compressive force F_{cr} expressed in equation (14-10) by the cross section area A. Then we get the average stress on the cross section in the critical condition, called the critical stress σ_{cr} of the column. From the equation (14-10) we can get the critical stress of the slender column as

$$\sigma_{cr}=\frac{F_{cr}}{A}=\frac{\pi^2 EI}{(\mu l)^2 A} \tag{14-11}$$

Introducing the radius of gyration $i=\sqrt{I/A}$

$$\sigma_{cr}=\frac{\pi^2 E}{\left(\frac{\mu l}{i}\right)^2} \tag{14-12}$$

Introduction following symbol

$$\lambda=\frac{\mu l}{i} \tag{14-13}$$

The equation for the critical stress (14-12) can be written as

$$\sigma_{cr}=\frac{\pi^2 E}{\lambda^2} \tag{14-14}$$

where λ is a dimensionless quantity, called the slenderness ratio or flexibility of the column. It centrally reflects the influence of factors such as the length of the column, the constraint situation, the shape and size of the cross section to σ_{cr}. It is a very important quantity.

14.4.2 Applicability of Euler equation

Euler equation is derived from the approximate differential equation for the deflection curve of a bending bar, which is reasonable only if the deformation is small and the material obeys Hooke law. Therefore, Euler equation is only applicable when the critical stress σ_{cr} does not exceed σ_p. Thus, the conditions for the applicability of Euler equation are

$$\sigma_{cr}=\frac{\pi^2 E}{\lambda^2}\leqslant\sigma_p \tag{14-15}$$

Introduction following symbol

$$\lambda_p=\pi\sqrt{\frac{E}{\sigma_p}} \tag{14-16}$$

The conditions for the application of Euler equation (14-15) can be written as

$$\lambda \geqslant \lambda_p \tag{14-17}$$

That is, when the actual slenderness ratio λ of the bar is greater than λ_p, Euler equation (14-14) is applicable. This type of bar is called large flexibility or slender column. This type of stability problems is within the elastic range.

Take the commonly used material steel Q235 for example. It has modulus of elasticity $E = 200$ GPa and proportional limit $\sigma_p = 200$ MPa. Substitute these values into the equation (14-16) and obtain $\lambda_p \approx 100$. In other words, for a column made of steel Q235, the critical compressive force can be calculated by Euler equation only when its slenderness ratio is $\lambda > 100$.

14.4.3 Critical compressive force above the proportional limit

For the common column in engineering, such as internal combustion engine connecting rod, jack jacking rod, etc., the slenderness ratio λ is often less than λ_p. Their critical stress exceeds the proportional limit and can not be calculated by Euler equation. As the material of the column is in the elastic-plastic stage, the stability of such columns is also known as the elastic-plastic stability problem. At this point, due to the difficulties of theoretical analysis, we often calculation critical stress by the empirical formula established on experiments, such as the linear formula and parabolic formula. Among them, the linear formula is relatively simple and convenient. Its form is

$$\sigma_{cr} = a - b\lambda \tag{14-18}$$

where a and b are the constants related to the mechanical properties of the material.

The above empirical formula also has a range of application. For example, for columns made of plastic materials, it should also be required that the size of the critical stress shall not reach the yield stress σ_s of the material, i.e.

$$\sigma_{cr} = a - b\lambda \leqslant \sigma_s$$

or

$$\lambda \geqslant \frac{a - \sigma_s}{b} \tag{14-19}$$

Therefore, the minimum value of slenderness ratio λ_s in the above empirical formula is

$$\lambda_s = \frac{a - \sigma_s}{b} \tag{14-20}$$

Similar to above, for columns made of brittle materials, we have

$$\lambda_b = \frac{a - \sigma_b}{b} \tag{14-21}$$

where σ_b is the strength limit of the material. Therefore, the empirical equation (14-18) is applicable under $\lambda_s < \lambda \leqslant \lambda_p$ (or $\lambda_b < \lambda \leqslant \lambda_p$). That is, when the slenderness ratio of the column is between λ_p and λ_s (or λ_b), we can use empirical formulas to calculate the critical stress. This type of bar is called intermediate column or medium-length column.

When $\lambda < \lambda_s$ or $\leqslant \lambda_b$, the column is called small flexural column or short column. Damage to short columns can be considered to be caused by insufficient strength. For short columns made of plastic materials, if they are still formally treated as a stability problem, the yield stress σ_s should be used as the critical stress for such short columns, i.e. $\sigma_{cr}=\sigma_s$. In the case of brittle materials such as cast iron, the strength limit σ_b should be used as its critical stress, i.e. $\sigma_{cr}=\sigma_b$.

14.4.4 Diagram of critical stress

By plotting the relationship between the critical stress and the slenderness ratio of the column within the three slenderness ratio ranges in the Cartesian coordinate system, we can get the critical stress diagram of the column. It reflects the variation of the load bearing capacity of the bar against slenderness ratio. Fig. 14-5 shows the total critical stress diagram for plastic materials.

With a critical stresses diagram, we can multiply the critical stress by the cross-sectional area of the column to obtain the critical stress in each section, which is useful for stability calculation.

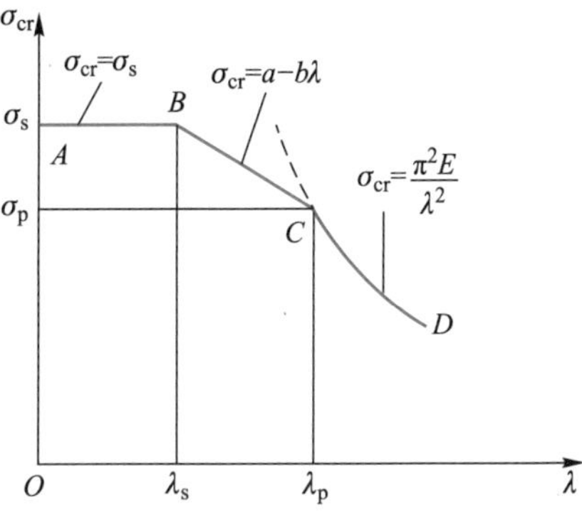

Fig.14-5

14.4.5 Parabolic formula and its stress general diagram

When the critical stress exceeds the proportional limit, we express the parabola relationship between the critical stress σ_{cr} and slenderness ratio as follows

$$\sigma_{cr}=a_1-b_1\lambda^2$$

where a_1 and b_1 are material-related constants. *Standard for design of steel structures* (GB 50017—2017) provides a parabolic formula established by our own experiments:

$$\sigma_{cr}=\sigma_s\left[1-a\left(\frac{\lambda}{\lambda_c}\right)^2\right]\quad \lambda\leqslant\lambda_c \tag{14-22}$$

where a is a factor, the value of which varies with different materials. For commonly used structural steels Q215 and Q235, steel 16Mn

$$a=0.43\lambda_c=\pi\sqrt{\frac{E}{0.57\sigma_s}} \tag{14-23}$$

σ_s is the yield stress of the material.

Example 14-2 A bar in compression, fixed at both ends, made of steel Q235, with a cross-sectional area of $32\times10^2\ \text{mm}^2$ (Fig. 14-6). Calculate the critical load when the cross section is rectangular and circular, respectively.

Solution (1) Rectangular section

by

$$A=b\times2b=32\times10^2\ \text{mm}^2$$

and get

$$b = 40 \text{ mm}$$

Minimum radius of gyration of the cross section

$$i = \sqrt{\frac{I_{\min}}{A}} = \sqrt{\frac{\dfrac{2b \times b^3}{12}}{2b \times b}} = \frac{b}{\sqrt{12}} = 11.55 \text{ mm}$$

Slenderness ratio of the column

$$\lambda = \frac{\mu l}{i} = \frac{0.5 \times 3 \times 10^3}{11.55} = 129.9 > \lambda_p$$

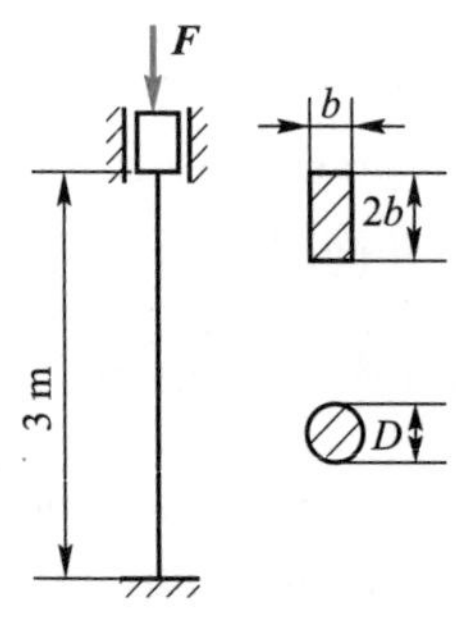

Fig.14-6

The column is a slender column and the critical compressive force is calculated using Euler equation

$$F_{cr} = \sigma_{cr} A = \frac{\pi^2 EA}{\lambda^2} = \frac{\pi^2 \times 210 \times 10^9}{129.9^2} \times 32 \times 10^2 \times 10^{-6} \text{ N} = 393 \times 10^3 \text{ N}$$

(2) Circular section

by

$$A = \frac{\pi d^2}{4} = 32 \times 10^2 \text{ mm}^2$$

Got

$$d = 63.8 \text{ mm}$$

Cross-sectional radius of gyration

$$i = \sqrt{\frac{I}{A}} = \sqrt{\frac{\dfrac{\pi d^4}{64}}{\dfrac{\pi d^2}{4}}} = \frac{d}{4} = 15.95 \text{ mm}$$

Slenderness ratio of the column

$$\lambda = \frac{\mu l}{i} = \frac{0.5 \times 3 \times 10^3}{15.95} = 94 < \lambda_p$$

Column is medium length

$$\begin{aligned} F_{cr} &= \sigma_{cr} A = (a - b\lambda) A \\ &= (304 - 1.12 \times 94) \times 10^6 \times 32 \times 10^2 \times 10^{-6} \text{ N} \\ &= 636 \times 10^3 \text{ N} \end{aligned}$$

(3) Discussion

Comparing these results shows that, when other conditions are same, the critical compressive force values differ due to the different shapes of the chosen cross sections. It is clear that the critical compressive force is greater for circular sections than for rectangular sections, i.e. circular sections are more resistant to instability than rectangular sections.

In addition, in order to calculate the critical compressive force, it is necessary to first calculate the slenderness ratio of the column, and then to select the appropriate formula. If the formula is applied incorrectly, regardless of the type of column, the wrong result will be obtained and it will be on the unsafe side. This can be explained by analyzing the critical stress diagram.

14.5 Calculation of stability of columns

In order to ensure that the column has sufficient stability, the working compressive force should be less than the critical compressive force of the column. In addition, considering a certain safety reserve, the stability conditions of the column are

$$F \leqslant \frac{F_{cr}}{n_w} \tag{14-24}$$

or expressed using safety factor for stability

$$n = \frac{F_{cr}}{F} \geqslant [n_w] \tag{14-25}$$

where n_w is the working safety factor for stability of the column and $[n_w]$ is the allowable safety factor for stability. Because the initial curvature of the column, the compressive force eccentricity, the non-uniformity of the material and other factors have a large influence on the critical compressive force, the allowable safety factor for stability should be taken as larger. Reference values of $[n_w]$ for several steel columns are listed below.

Columns in metal structures: $[n_w] = 1.8 \sim 3.0$;

Screws for machine tools: $[n_w] = 2.5 \sim 4.0$;

Tappet for low speed engines: $[n_w] = 4 \sim 6$;

Piston bar for grinding machine cylinders: $[n_w] = 4 \sim 6$;

Lifting spirals: $[n_w] = 3.5 \sim 5$.

When there is a partial section weakening of the bar, such as oil holes, screw holes, etc., as the critical compressive force of the bar is determined by the bending deformation, the local section weakening has little effect on the value of the critical compressive force and can be neglected in the stability calculation. All cross-sectional areas and minimum moments of inertia are calculated for the unweakened cross section.

Stability calculations for columns include stability checks, section design and determination of allowable loads. In general design, the section size of compressive bar is initially determined based on strength estimation, and then check its stability.

When applying the equation (14-10) for stability calculation, first calculate the slenderness ratio λ in each bending plane according to the actual size of the column and the constraint situation, then determine the formula for the critical compressive force according to the maximum slenderness ratio, and finally carry out stability calculation on.

Example 14-3 An I-beam is shown in Fig. 14-7a. We have known cross-sectional area $A = 720\ \text{mm}^2$, moment of inertia $I_z = 6.5 \times 10^4\ \text{mm}^4$ and $I_y = 3.8 \times 10^4\ \text{mm}^4$. The beam is made of silicon steel and is subject to compressive force $F = 85$ kN. Take the allowable safety factor for stability $n_w = 2.5$ and try to check the stability of the connecting bar.

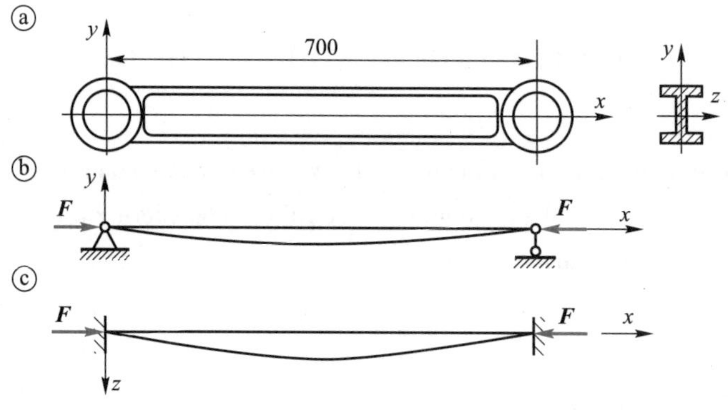

Fig.14-7

Solution (1) Calculate the slenderness ratio and determine the form of instability.

Whether the column is unstable in the $x-y$ plane or the $x-z$ plane can be determined by the slenderness ratio. If instability is in the $x-y$ plane (cross-sectional rotation about the z axis), the ends are pin support (Fig.14-7b) and $\mu=1$. The slenderness ratio of the connecting rod is

$$\lambda_z=\frac{\mu l}{i_z}=\frac{\mu l}{\sqrt{\dfrac{I_z}{A}}}=\frac{1\times700}{\sqrt{\dfrac{6.5\times10^4}{720}}}=73.7$$

If the connecting column is unstable in the $x-z$ plane (the cross section rotates around the y axis) and the ends are close to the fixed ends (Fig.14-7c) and $\mu=0.7$. The slenderness ratio of the connecting rod is

$$\lambda_y=\frac{\mu l}{i_y}=\frac{\mu l}{\sqrt{\dfrac{I_y}{A}}}=\frac{0.7\times700}{\sqrt{\dfrac{3.8\times10^4}{720}}}=67.4<\lambda_z$$

Due to the greater slenderness ratio in the $x-y$ plane, we just need to check the stability of the column in the $x-y$ plane.

(2) Stability check.

For silicon steel, we have $\lambda_p=100$, $\lambda_s=60$. The column is a intermediate column. The coefficient is $a=578$ MPa, $b=3.74$ MPa, and

$$F_{cr}=\sigma_{cr}A=(a-b\lambda_z)A=(578-3.74\times73.7)\times10^6\times720\times10^{-6}\ \text{N}=218\ \text{kN}$$

$$n=\frac{F_{cr}}{F}=\frac{218\times10^3}{85\times10^3}=2.56>n_w$$

So the stability requirements are met.

(3) In this example, if the stability of the connecting bar in the $x-y$ and $x-z$ planes is required to be the same, that is $\lambda_y=\lambda_z$

$$\frac{l}{\sqrt{\frac{I_z}{A}}}=\frac{0.7}{\sqrt{\frac{I_y}{A}}} \quad \text{or} \quad I_z=2.04I_y$$

This illustrates that in order to make the stability of the connecting bar close to the same in both directions, the relationship $I_z=2I_y$ should be roughly maintained when designing column cross section.

Problems

14-1 Fig.P14-1 shows a slender column with spherical hinged ends and a modulus of elasticity $E=200$ GPa. Calculate the critical compressive force using Euler equation for the following three sections. (1) circular section, $d=25$ mm, $l=1.0$ m; (2) rectangular section, $h=2b=40$ mm, $l=1.0$ m; (3) I-beam steel No.16, $l=2.0$ m.

Answer: (1) $F_{cr}=37.8$ kN; (2) $F_{cr}=52.6$ kN; (3) $F_{cr}=459$ kN

14-2 Fig. P14-2 shows an isotropic column with a fixed lower end and a free upper end subjected to axial forces at the free end. The length of the bar is l. It is possible to maintain equilibrium in the $x-y$ plane in a slightly bent state when the bar is destabilized under the action of F_{cr}. The moment of inertia of the cross-sectional area against the axis z is I. Try to derive Euler equation for its critical compressive force F_{cr} and find the equation for the deflection curve of the column.

Answer: $F_{cr}=\dfrac{\pi^2 EI}{(2l)^2}$, $v=\delta\left(1-\cos\dfrac{\pi x}{2l}\right)$

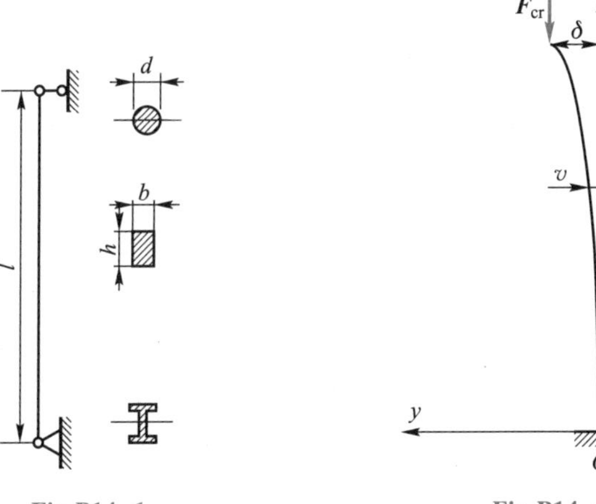

Fig.P14-1 Fig.P14-2

14-3 It is known a steel with $\sigma_p=230$ MPa, $\sigma_s=274$ MPa, $E=200$ GPa, $\sigma_{cr}=338-1.22\lambda$. Try to calculate λ_p and λ_s, and draw a general diagram of critical stress ($0\leqslant\lambda\leqslant150$).

Answer: $\lambda_s=92.6$, $\lambda_p=52.5$

14-4 Fig.P14-4 shows a column with a rectangular cross section ($h=80$ mm, $b=40$ mm, bar length $l=2$ m). It is made of high quality carbon steel ($E=210$ GPa). The constraint of ends: in the plane of the front view (a) is equivalent to a pin constraint; in the plane of the top view (b) is elastically fixed ($\mu=0.6$). Try to find F_{cr} of this bar.

Answer: $F_{cr}=613$ kN

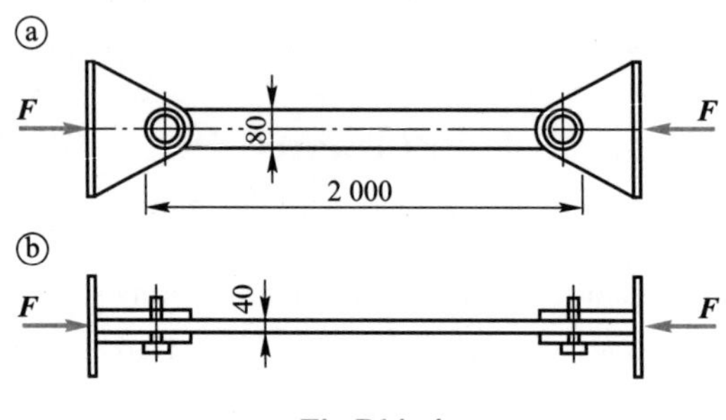

Fig.P14-4

14-5 A steel column consists of two same angle steel (Fig.P14-5). Its size are 56 mm×56 mm×8 mm, and the length of the bar is $l=1.5$ m. Two ends supported by spherical hinges subject to axial compressive force $F=150$ kN. Determine the critical stress and the working safety factor for stability of the column.

Answer: $\sigma_{cr}=182$ MPa, $n=2.03$

14-6 The column consists of two channel steel No.10 is shown in Fig.P14-6. We have known $l=6$ m. The column is fixed at the lower end and spherically hinged at the upper end. The modulus of elasticity of the material is $E=200$ GPa and the proportional limits is $\sigma_p=200$ MPa. When the critical compressive force of the column is the highest, what is its value?

Answer: $a=44$ mm, $F_{cr}=444$ kN

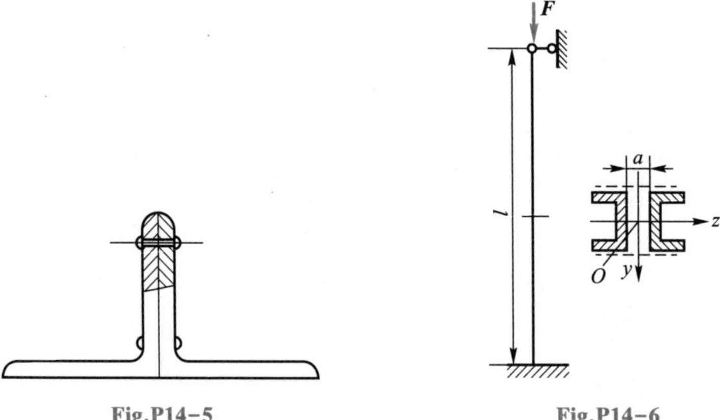

Fig.P14-5 Fig.P14-6

Reference

[1] 杨在林. 材料力学[M]. 2 版. 哈尔滨：哈尔滨工业大学出版社，2018.
[2] 李鸿. 理论力学[M]. 2 版. 哈尔滨：哈尔滨工程大学出版社，2021.
[3] 范钦珊. 材料力学[M]. 3 版. 北京：清华大学出版社，2014.
[4] 刘鸿文. 材料力学 I, II [M]. 6 版. 北京：高等教育出版社，2017.
[5] B. J. GOODNO, J. M. GERE. Mechanics of materials[M]. 9th ed. Boston：Cengage Learning，2018.